4th International Conference on Mechanical, Materials and Manufacturing (ICMMM 2017)

IOP Conference Series: Materials Science and Engineering
Volume 272

Atlanta, Georgia, USA
25 – 27 October 2017

ISBN: 978-1-5108-5404-8
ISSN: 1757-8981

Printed from e-media with permission by:

Curran Associates, Inc.
57 Morehouse Lane
Red Hook, NY 12571

Some format issues inherent in the e-media version may also appear in this print version.

Printed by Curran Associates, Inc. (2018)

For permission requests, please contact the Institute of Physics
at the address below.

Institute of Physics
Dirac House, Temple Back
Bristol BS1 6BE UK

Phone: 44 1 17 929 7481
Fax: 44 1 17 920 0979

techtracking@iop.org

Additional copies of this publication are available from:

Curran Associates, Inc.
57 Morehouse Lane
Red Hook, NY 12571 USA
Phone: 845-758-0400
Fax: 845-758-2633
Email: curran@proceedings.com
Web: www.proceedings.com

Table of contents

Volume 272

4th International Conference on Mechanical, Materials and Manufacturing (ICMMM 2017)

25–27 October 2017, Atlanta, USA

Accepted papers received: 15 November 2017
Published online: 21 December 2017

Preface

Papers

Mechanical and Mechatronics Engineering

Materials Science and Engineering

Building Materials and Structural Design

IOP Conf. Series: Materials Science and Engineering **272** (2017) 011001 doi:10.1088/1757-899X/272/1/011001

Preface

It is our great pleasure to welcome you to 2017 4th International Conference on Mechanical, Materials and Manufacturing (ICMMM 2017), which will be held in Atlanta, USA during October 25-27, 2017.

The objective of ICMMM 2017 has always been bringing together a large group of researchers, scientists, academics, and engineers in this related area from all over the world, who share much more than technical interests, among other things, culture and history.

This proceedings of ICMMM 2017 presents papers submitted to the conference from universities, research institutes and industries. The papers selected depended on their quality and their relevancy to the conference. The volume tends to present to the readers the recent advances in the field of composites, metal alloy materials and materials physics and chemistry, building materials and new energy materials, environmental friendly materials etc.

We would like to thank all the authors who have contributed to this volume and also to reviewers, speakers, chairpersons, sponsors and all the conference participants for their hard work and support to ICMMM 2017.

ICMMM 2017 Organizing Committee
October 25, 2017

IOP Conf. Series: Materials Science and Engineering **272** (2017) 011002 doi:10.1088/1757-899X/272/1/011002

Conference Committee

International Advisory Chair
Brian Vlcek, Georgia Southern University, USA

Conference Chair
Guanghsu Chang, Bakersfield college, USA

Conference Program Chairs
Sergei Alexandrov, Beihang University, China

Young Moon, Syracuse University, USA

Edward Gobina, Robert Gordon University, UK

Qing-Zhi Zhai, China Agricultural University, China

Conference Technical Committees
Guelailia Ahmed, Algerian Space Agency, Satellite Development Center, Algeria

Guimin Chen, Xidian University, China

Angelos Markopoulos, National Technical University of Athens, Greece

W Smitthipong, Kasetsart University, Thailand

Jesus Jaime Moreno Escobar, Instituto Politécnico Nacional, Mexico

Oswaldo Morales Matamoros, Instituto Politécnico Nacional, Mexico

Ricardo Tejeida Padilla, Instituto Politécnico Nacional, Mexico

Qiang Song, Qingdao Vocational and Technical College of Hotel Management, China

Zhongwei Yin, Shanghai Jiao Tong University, China

How Wei Benjamin Teo, Nanyang Technological University, Singapore

Arun Srinivasa, Texas A&M University, USA

Mohammad Tahersima, Oklahoma State University, USA

Trong-Phuoc Huynh, Can Tho University, Viet Nam

Jiraphon Srisertpol, Suranaree University of Technology, Thailand

Li Yang, University of Louisville, USA

Mark Jackson, Kansas State University, USA

Li Guoping, University of Jinan, China

Nitisak Numanoy, Suranaree University of Technology, Thailand

Yang Xi, Diodes, Inc., USA

Hariati Taib, Universiti tun hussein onn Malaysia, Malaysia

IOP Conf. Series: Materials Science and Engineering **272** (2017) 011002 doi:10.1088/1757-899X/272/1/011002

Wei (Will) Wang, University of Cincinnati, USA

Rampada Manna, Indian Institute of Technology Madras, India

N.C.Santhi Srinivas, Indian Institute of Technology (Banaras Hindu University) Varanasi, India

Cheng-Yen Wen, National Taiwan University, Taiwan

Mauricio Plaza, Universidad Militar Nueva Granada, Colombia

Mohammad Reza Forouzan, Isfahan University of Technology, Iran

Tei-Chen Chen, National Cheng Kung University, Taiwan

Somashekhar S. Hiremath, Indian Institute of Technology Madras, India

Koorosh Gharehbaghi, RMIT University, Australia

Mohd Hamdi Bin Abd Shukor, University of Malaya, Malaysia

Kamran Shavezipur, Southern Illinois University Edwardsville, USA

Soondo Kweon, Southern Illinois University Edwardsville, USA

Kamalesh Kumar Singh, Indian Institute of Technology (Banaras Hindu University) Varanasi, India

Viktor Rizov, University of Architecture, Civil Engineering and Geodesy, Bulgaria

Rahul Davis, Sam Higginbottom University of Agriculture, Technology & Sciences (SHUATS), India

Haijun Gong, Georgia Southern University, USA

Marina Rynkovskaya, RUDN University, Russian Federation

Jingwei Zhao, University of Wollongong, Australia

Tawfiq Al-Mughanam, King Faisal University, Saudi Arabia

Houriyeh Marashi, University of Malaya, Malaysia

Vibhash Jha, NXP Semiconductors, USA

Prabhudev M.S, Government Polytechnic, Kalgi, India

Ravi Kumar Puli, National Institute of Technology, India

Peer review statement

All papers published in this volume of *IOP Conference Series: Materials Science and Engineering* have been peer reviewed through processes administered by the proceedings Editors. Reviews were conducted by expert referees to the professional and scientific standards expected of a proceedings journal published by IOP Publishing.

IOP Conf. Series: Materials Science and Engineering **272** (2017) 012001 doi:10.1088/1757-899X/272/1/012001

Movement Intention Prediction to Find a New Exoskeleton Design with Light and Comfortable Materials

Ing. Mauricio Plaza Torres[1] ,Ing Fredy Bernal[2] and Ing. Andrés Cifuentes[3]

[1] Universidad Militar Nueva Granada
[2] Universidad Militar Nueva Granada
[3] Universidad Militar Nueva Granada
Abstract. This paper presents a discussion about the design a novel exoskeleton and the implementation of superficial magnetic sensor and pressure sensor for find patterns and parameters in order to predicting the intention of movement for people with musculoskeletal system problem in lower limbs. The information is obtained through reading position and movement by magnetic and pressure sensor installed in different parts of the body. The advantage of this sensor is the location of the position reference of the every part of the body. Using the sensor in some parts of the body is possible to calculate the muscle contraction for the prediction of intended movement and to find a new design with light and comfortable materials.

1. Introduction

The developments of exoskeleton for people with musculoskeletal system problem in lower members are increased in the last years. The most relevant design issue is the interaction between user and the machine. Today, it is possible to find speech, joysticks or electromyography controlled system. [1] Miniature accelerometers became available which measure acceleration forces. This sensor can be used to monitor inclination of body part, but the sensor can no detect the horizontal component of the orientation. [2] Other way, the magnetic sensor can measure orientation relative to the magnetic reference. The sensors are located in the way to detect the horizontal component and vertical component for complete orientation monitoring.[3] Apply intention of this technology is to allow monitor the body limbs orientation, in order to assess the intention of the movement. Orientation estimation in human motion tracking systems needs various types of sensor. There are commonly used: accelerometers, gyroscopes and magnetometers. [4] Used individual sensor have varying advantages and problems. Gyroscopes measure angular velocities by integrate the data over the time allowing to find the orientation, but the integration of gyroscope measurement produce errors on calculated orientation. Accelerometers measure the linear acceleration and gravity.

IOP Conf. Series: Materials Science and Engineering **272** (2017) 012001 doi:10.1088/1757-899X/272/1/012001

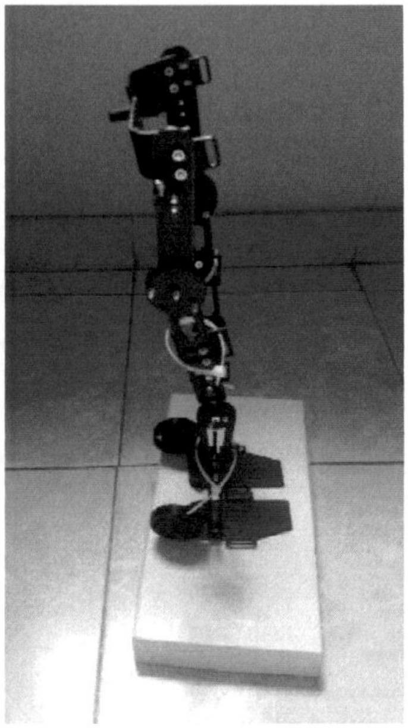

Figure 1. Exoskeleton controlled by movement intention.

Magnetometers are used to measure the local magnetic field vector allowing to the determinate of orientation relative to the vertical axis. The problem with magnetometers is the influence of magnetic interferences to the sensor frame or ferromagnetic materials around the sensor (Fig 1). To avoid all of this problem is necessary the use of filtering algorithm.[5]
Regarding to a filter technology, there are frequently filter design: deterministic, frequency approach and stochastic. The deterministic filter is used to find a rotation matrix between two coordinate systems. It can be used to find a best orientation from different sensor measurements. Orientation estimation filters for human motion tracking, the magnetic measurements are determinate by the gravity field or the magnetic reference into the exoskeleton. [6]. The most common motion capture systems for body motion application are based on either optical or inertial/magnetic sensing.[7] Optical tracking is based on the captured images processing of by multiple cameras. Identifying body parts to track, the system has made a following of the anatomical parts of interest and their detection. Infrared cameras produce streams of coordinates representing the using a reference system, with application in medical contexts. The reason is the reliable and accurate necessary to record the motions. In the optical tracking measures the orientation of the body parts of interest. It made by means of cameras operating in the visible spectrum and without marker. This system is less expensive and more portable than marker based on optical systems. [8] The system designed is wearable and reliable human motion capture system, based on the use of magnetic and pressure sensors units that are attached to the lower limbs in order to track their orientations. The motion tracking systems uses low cost pressure and magnetic sensors.
The use of inertial sensors was mostly investigated for biomechanical applications, like gait analysis, joint angle and body position. There is no inherent latency in this sensing technology or delays. Additional, the system does not use external source, whereas electromagnetic, acoustic, and optic devices require emissions from a source to track objects. It does not expose to limitations and interference problems of other sourced technologies.

2. Method
Position sensors measure the distance a body has moved from its reference and the resulting output is given as a feedback to the control system in the exoskeleton. Motion of a body can be rectilinear or curvilinear, for this reason the position sensors must detect linear position or angular position. The

IOP Conf. Series: Materials Science and Engineering **272** (2017) 012001 doi:10.1088/1757-899X/272/1/012001

sensors have accuracy limitations, the outputs signal is mixed with noise. Therefore, the quality needed to provide an accurate estimation is not necessary to be high. The sensor outputs can be filtering with algorithms that provide the good estimate of possible orientation. The final design take in account a good filters design for real-time applications that because accurate and computationally demanding algorithms required and it will may be too slow to work in real time.
Kalman filter produces estimates of the state variables and uncertainties to predict the position all time. The algorithm can run in real time using input measurement and the previously calculated state. The Kalman filter used, works with nonlinear systems associated with the process model. It is well used for orientation estimation problems in human motion tracking applications. Finally, offset angles are calculated representing orientations that have to be applied in order to overlap the sensor frame to the body frame. These angles are used throughout the motion tracking process to align the measured orientations to the body segments.

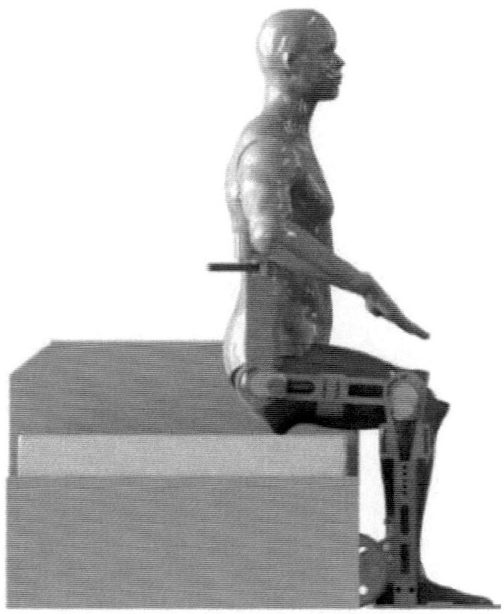

Figure 2. Sensor Position.

Additional aspect in the accuracy of the orientation estimation depends on the quality of result data provided by sensors (Fig 2). Because these reason high-end sensors are used that require a compensation of their systematic errors by sensor calibration procedure. The magnetic behavior can be influenced by electromagnetic interferences and magnetic materials in the surrounding that distort the measurement. To avoid this error measurement, compensating parameters must be applied to every magnetic sensor use in every environment. Fuzzy logic use a position degrade with previous acknowledge of the data to determine the final result. The data measure is continuing variable that is processed to control the exoskeleton properly.

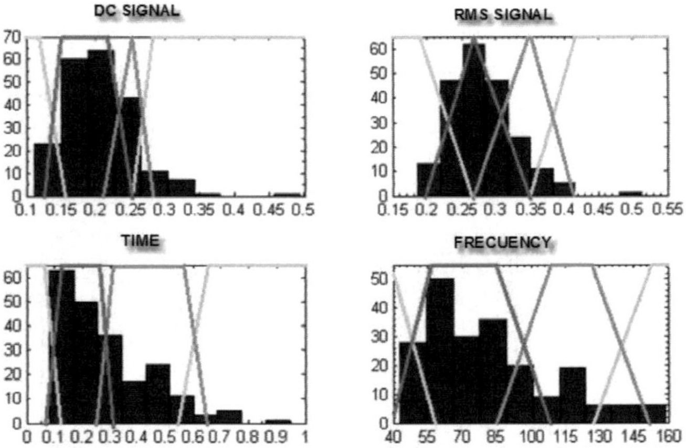

Figure 3. Histogram of the basic movement on stand up.

$$DC_{calculated} = \frac{\sum_{i=1}^{N} RMS(i)}{N} \quad (1)$$

$$RMS_{calculated} = \sqrt{\frac{\sum_{i=1}^{N} RMS(i)^2}{N}} \quad (2)$$

$$Time = t_{end} - t_{start} \quad (3)$$

To find the movement intention is necessary a data base to process (1),(2,3), compere and conclude based on de information, the movement. It is very similar to human decision. The human body need information missing a set of lose data to realize a movement. The human learn during the life specific movement, and this information is executing automatic. The same way, the fuzzy logic use a base data to decide how control an exoskeleton (Fig 3.). The fuzzy logic software use the information to transform in a diffuse group of data, related on some degrade with other group of data.

3. Conclusion

This method is accurate and applicable to detect intention of the movement. The present work aimed to use magnetic and pressure sensors to develop mechanics and electronics system for combined sensor, and use the fuzzy logic mathematic to control the exoskeleton with the movement intention of the user.

4. Discussion

The most common characteristics used because the low complexity and low computation time consumption, are the DC level RMS value, the signal time slot and the harmonic in frequency. The noise signal relations are obtained by means of filtering step.

This work represents an overview technologies used for human motion tracking with pressure and magnetic sensors and the fuzzy logic algorithm based on the result of the design of an exoskeleton for detect the motion intention for people with musculoskeletal tension. The result is used in the creation of an exoskeleton with intention motion control.[9] The methodology include motion intention estimation algorithm based on fuzzy logic for fast response on real time measurement and development of novel solutions for exoskeleton final design. The developed sensor and fuzzy logic mathematical processing do indeed monitor body position and prediction of the intention of the movement. [10] This seems sufficient for monitoring the intention of the movement for many applications. For other applications, control of the exoskeleton may have to be improved. The main Limitation is the sensor sensitive to movement accelerations and to fluctuations of the orientation of the environmental magnetic field. When taking into account these limitations, the intention of the movement can be easily monitored. [11]

5. Results

The result is measurement data of the orientation of the lower limb and simultaneously with pressure sensor, installed in the muscles of the lower limb, in order to allow the detection of the intention of the movement.

A brief comparison between fuzzy sets of data is done to determine the best system in generating reliable control orders for prototype robotic exoskeleton. The fuzzy system designed has nine rules in the inference engine, which had two stages of interaction with samples: one design and other validation. Analog and digital components for data acquisition processes are used to perform amplification, filtering, digitization and transmission of samples to fuzzy logic system.

6. Acknowledgements

Funding by research UMNG IMP ING 2135, Universidad Militar Nueva Granada

7. References

[1] Nene, A.V; Veltink, P.H. Accelerometer and rate gyroscope measurement of kinematics: An inexpensive alternative to optical motion analysis systems. J. Biomech. Vol 35. 2002 7

[2] B.-S. Liu, «Backpack load positioning and walking surface slope effects on physiological responses in infantry soldiers,» International Journal of Industries Ergonomics, pp. 2-6, 2007.

[3] Bergamini E.; Ligorio G.; Summa A.; Vannozzi G.; Cappozzo A.; Sabatini A.M.; Estimating Orientation Using Magnetic and Inertial Sensors and Different Sensor Fusion Approaches: Accuracy Assessment in Manual and Locomotion Tasks. Sensors Vol 14. 2014 28.

[4] Sabatini, A.M; Quaternion-Based Extended Kalman Filter for Determining Orientation by Inertial and Magnetic Sensing. IEEE Transactions on Biomedical Engineering, Vol. 53 No 78. July 2006 23.

[5] Šeketa G.; Dzaja, D; Zulj, S; Celic, L.; Lackovic I.; Magjarevic, R. Real-time evaluation of repetitive physical exercise using orientation estinmation from inertial and magnetic sensors. Work accepted for the 1 st European Biomedical Engineering Conference for Young Investigators to be held in May 2015 39.

[6] Seketa, G.; Ortiz, G; Wilches, C.; Perdomo O.; Celic, L.; Lackovic, I.; Zequera, M.; Magjarevic, R.. Simultaneous Measurement of Trunk Orientation and Centre of Pressure for Postural Stability Evaluation. 6 th European Conference of the International Federation for Medical and Biological Engineering, Dubrovnik, Croatia,. September 2014.

[7] Brigante M.N.C; Abbate N; Basile A. Toward Miniaturization of a MEMS-Based Wearable Motion Capture System. IEEE Transactions on Industrial Electronics, Vol 58, No 8. August 2011 2.

[8] S. D. Blacker, J. L. Fallowfield, J. L. Bilzon y M. E. Willems, «Neuromuscular function following prolonged load carriage on level and downhill gradients,» Aviat. Space Environ, vol. 81, pp. 745-753, 2010.

[9] Yadav N.; Bleakley C. Accurate Orientation Estimation Using AHRS under Conditions of Magnetic Distortion. Sensors, Vol 14, 2014 38.

[10] B. S. Liu y H. T. Chou, Compositores, Physiological responses during walking with combined dynamic and static loads. [Grabación de sonido]. Proceedings of the 33rd international conference on computers and industrial engineering. 2004.

[11] Yun X.; Bachmann E.R.; McGhee R.B. A Simplified QuaternionBased Algorithm for Orientation Estimation From Earth Gravity and Magnetic Field Measurements. IEEE Transactions on Instrumentation and Measurement Vol 57 No 3. March 2008. 26.

IOP Conf. Series: Materials Science and Engineering **272** (2017) 012002 doi:10.1088/1757-899X/272/1/012002

An artificial neural network controller based on MPSO-BFGS hybrid optimization for spherical flying robot

Xiaolin Liu[1], **Lanfei Li**[1] **and Hanxu Sun**[1]

[1]School of Automations, Beijing University of Posts and Telecommunications, No.10 Xitucheng Road, Beijing, China

Abstract. Spherical flying robot can perform various tasks in the complex and varied environment to reduce labor costs. However, it is difficult to guarantee the stability of the spherical flying robot in the case of strong coupling and time-varying disturbance. In this paper, an artificial neural network controller (ANNC) based on MPSO-BFGS hybrid optimization algorithm is proposed. The MPSO algorithm is used to optimize the initial weights of the controller to avoid the local optimal solution. The BFGS algorithm is introduced to improve the convergence ability of the network. We use Lyapunov method to analyze the stability of ANNC. The controller is simulated under the condition of nonlinear coupling disturbance. The experimental results show that the proposed controller can obtain the expected value in shoter time compared with the other considered methods.

1. Introduction

Because of simple structure, low manufacturing cost, strong adaptability to environment and flexible mobility [1], spherical aircraft are being paid more and more attention by organizations and researchers. With the development of MEMS manufacturing process and artificial intelligence technology [2] – [5], spherical flying robot has played an important role and shows great advantage in the field of agricultural photogrammetry, terrain exploration, mountain rescue and military scouting [6].

At present, there are some basic problems in spherical flying vehicles, including underactuated, nonlinear, time-varying, strong coupling and so on [7], [8]. As one of the key technologies to solve these control problems, autonomous control has been the focus of research in this field. Hatamleh et al. [9] described a parameter estimation method to improve the robustness of flight control under uncertain system parameters. Islam et al. [10] has studied a robust adaptive tracking algorithm to solve modeling error and uncertain disturbance. Aiming at solving coupling and underactuated problems in the control, Zhao et al. [11] proposed a nonlinear robust tracking control scheme based on RISE and I&I method. Diao et al. [12] study on a continuous time-varying adaptive controller based on underactuated vehicles. Peng et al. [13] designed a backstepping sliding mode controller based on RBF neural network to enhance the robustness of the flight height control. Xiong et al. [14] introduced a full drive subsystem controller based on TSMC algorithm to realize the high precision tracking control of position and attitude In order to the complexity and singularity problem caused by attitude representation, Lee [15] proposed a robust adaptive controller based on SO(3) to achieve a track of height and angular velocity in the case of unkown rigid body inertia matrix. Lin et al. [16] put forward an intelligent controller based on recursive wavelet neural network, which can realize the vehicle's motion control under the condition of control force attenuation and wind disturbance. WANG et al.

[17] studied a neural network adaptive inverse model control method to deal with the parameter variation and external disturbance.
In this paper, we propose an artificial neural network controller based on MPSO-BFGS hybrid optimization algorithm for the motion control of spherical flying robot. Firstly, in order to solve the learning object of the controller, the dynamics of robot is analyzed and its motion model is established. Secondly, an artificial neural network is constructed according to the motion model. Finally, in order to improve the learning efficiency and generalization ability of the network, a hybrid optimization algorithm based on MPSO and BFGS is used to select and update weights of the network. The simulation results show that the proposed controller can obtain the expected control in shorter time than the considerated controller in the presence of disturbance parameters and coupling.

2. Dynamic model of spherical flying robot

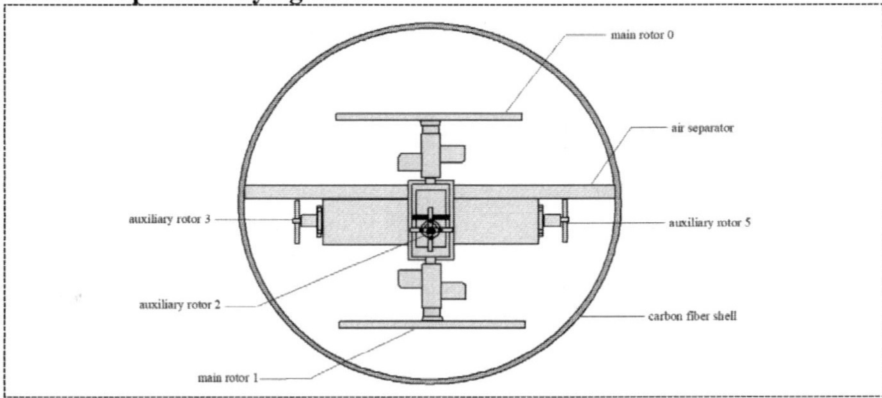

Figure 1. Structure of spherical flying robot.

The spherical flying robot is mainly composed of carbon fiber mesh shell, propeller, air separator and power device as shown in Figure 1. Servo motor, energy management system and control system are located in the power device. The two main propellers are used to realize the lift control of robot, and the four auxiliary propellers realize the attitude control by adjusting the rotational speed.
We analysis the external force of spherical flying robot by using Newton-Euler equation. The specific expression can be written as:

$$F = G_o + f + T_o \tag{1}$$

$$\Delta_b = \dot{\mu_b} + \delta_b \times \mu_b \tag{2}$$

The translational equation of spherical flying robot is obtained:

$$\dot{\mu_b} = -\delta_b \times \mu_b + \frac{G_o + f + T_o}{m} \tag{3}$$

The rotation of spherical flying robot satisfies the theorem of moment of momentum, as shown in equation (10). Considering the good symmetry of the sphere, the moment of inertia matrix can be written by equation (11).

$$J\dot{\delta_b} + \delta_b \times J\delta_b = M \tag{4}$$

The resultant moment of force relative to the center of robot's mass is

$$M = M_c + M_f \tag{5}$$

M_c indicates the outer moment of rotors acting on the spherical flying robot, M_f represents the aerodynamic drag torque of robot body.
In summary, the dynamic equation of spherical flying robot is as follows:

$$\begin{cases} \dot{u} = vr - wq - g\sin\varphi + \frac{f_x + T_{ox}}{m} \\ \dot{v} = wp - ur + g\cos\varphi\sin\gamma + \frac{f_y + T_{oy}}{m} \\ \dot{w} = uq - vp + g\cos\varphi\cos\gamma + \frac{f_z + T_{oz}}{m} \\ \dot{\alpha} = p + \tan\varphi\sin\gamma\, q + \tan\varphi\cos\gamma\, r \\ \dot{\beta} = q\cos\gamma - r\sin\gamma \\ \dot{\omega} = \frac{q\sin\gamma + r\cos\gamma}{\cos\varphi} \\ I_x\dot{p} = (I_y - I_z)qr + M_{cx} + M_{fx} \\ I_y\dot{q} = (I_z - I_x)rp + M_{cy} + M_{fy} \\ I_z\dot{r} = (I_x - I_y)pq + M_{cz} + M_{fz} \end{cases} \quad (6)$$

3. Artificial neural network controller

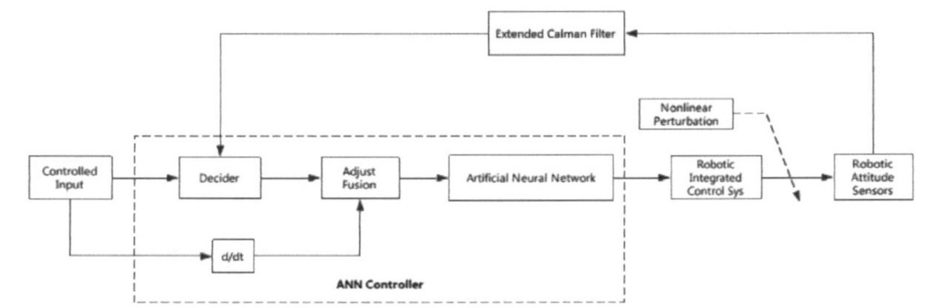

Figure 2. Structure of ANNC for spherical flying robot.

The proposed artificial neural network controller structure is shown in Figure 2. The controller includes feedback decision control, fusion tuning function and neural network to realize the flight control law. The control law is used to stabilize the output power of spherical robot by compensating the nonlinear dynamic disturbance of the system. Current attitude informations of robot are fed back to the decider after EKF filtering. The angular rate obtained by the differential and input datas are transformed into the training sets of neural network through the fusion function. Net learn data with its neuron compensation function to approximate th dynamic disturbances of spherical flying robot system. This neural network controller can be trained offline to construct the network structure and obtain the optimal control performance by automatically adjusting the parameters during the flight.

The controller consists of five feedforward layers. $E=[\psi;\theta]$ is the training data vector of controller, $\psi=[\psi_1, \psi_2, \ldots \psi_n]^T$ is the adjusting angle vector of control system, $\theta=[\theta_1, \theta_2, \ldots \theta_n]^T$ represents the feedback value of the current robot attitude data after noise reduction. The definition of ψ_i and θ_i is as follows. $\psi_i=[\alpha_i, \beta_i, \omega_i]$, $\theta_i=[\alpha_i', \beta_i', \omega_i']$, where α、β、θ respectively represents pitch angle , tilt angle and roll angle.

The 2n neurons of network input layer are used to input ψ and θ. Its hidden layer is divided into three layers. The first layer has 6n neurons which are used to store the angular vector members from the input layer. The neurons in this layer use linear activation function $f_i(x) = \mu^i x$, where μ^i is suppression parameter vector for correcting the overflow of peak data. The second layer has 9n neurons. There are three kinds of neuron activation functions in this layer, namely, the stable control activation function, the error compensation activation function, and the overshoot control activation function. And the third layer has 3n neurons.

The output layer has n neurons. The neuron activation function is a vector conversion function, which is used to the conversion of the hidden layer's data format.

4. Simulation

The simplified control model of spherical flying robot is defined as:

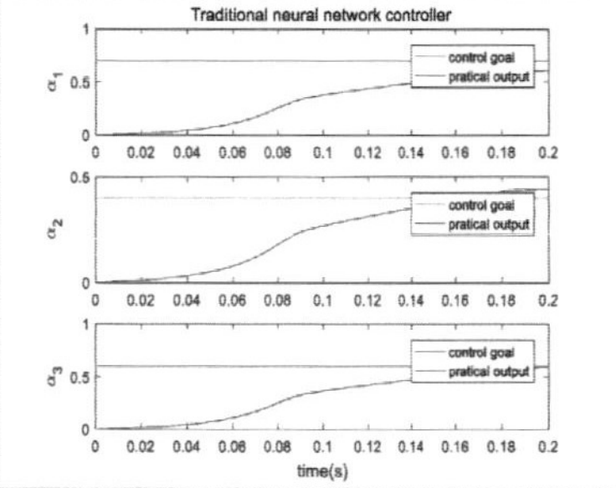

Figure 3. The output of traditional neural network controller.

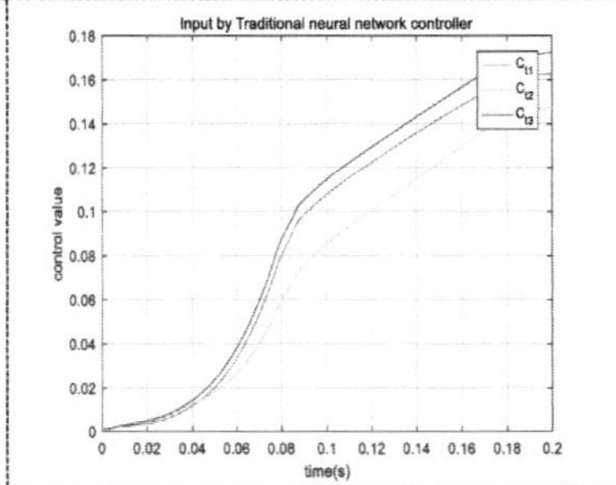

Figure 4. The system control variables of traditional neural network controller.

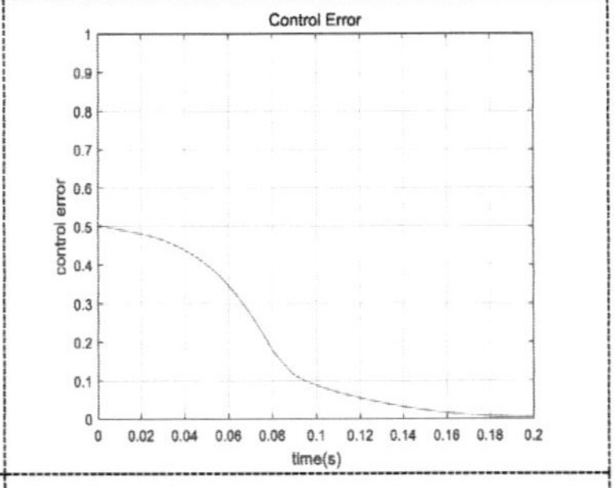

Figure 5. The control error of traditional neural network controller.

$$
\begin{cases}
\theta_i(k)^T = 0.6 \times \theta_i(k-1)^T + 0.5 \times \theta_{i+1}(k-1)^T + \frac{c_{ti}(k-1)}{c_{ti}(k-1)^2 + c_{ti}(k-1)+1} \times I_o + 0.3 \times c_{ti+1}(k-1)^3 \times I_o \\
\theta_{i+1}(k)^T = 0.5 \times \theta_{i+1}(k-1)^T + 0.4 \times \theta_{i+2}(k-1)^T + \frac{c_{ti+1}(k-1)}{c_{ti+1}(k-1)^2 + c_{ti+1}(k-1)+1} \times I_0 + 0.7 \times c_{ti}(k-1)^3 \times I_o \\
\theta_{i+2}(k)^T = 0.8 \times \theta_{i+2}(k-1)^T + 0.5 \times \theta_i(k-1)^T + \frac{c_{ti+2}(k-1)}{c_{ti+2}(k-1)^2 + c_{ti+2}(k-1)+1} \times I_o + 0.2 \times c_{ti+1}(k-1)^3 \times I_o
\end{cases}
$$

Where i = 1, 2 ... n, $I_0 = [1, 1, 1]$; $\theta_i(k) = [\alpha_i(k)', \ \beta_i(k)', \ \omega_i(k)']$, which represents the feedback angle signal of robot system in the k-th sampling period; c_{ti}(k) represents the control signal of robot system in the k-th sampling period. In order to facilitate the analysis and visualization, $\alpha(k)'$ is regarded as observation object. Set i=1, the simulation parameters of the model are as follows: (1) The initial value of the control is [0,0,0];(2) The output value of the system is $\alpha' = [0.7,0.4,0.6]$, randomly initialize residual variables;(3) The learning rate of network is set to 0.001;(4) The sampling period is set to 0.0001s~0.0005s.

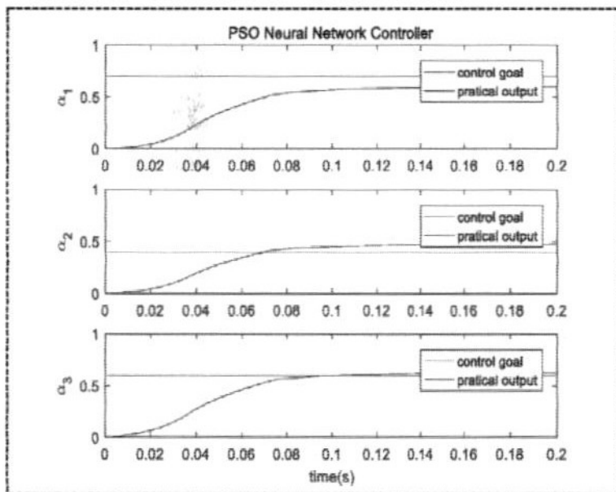

Figure 6. The output of PSO neural network controller.

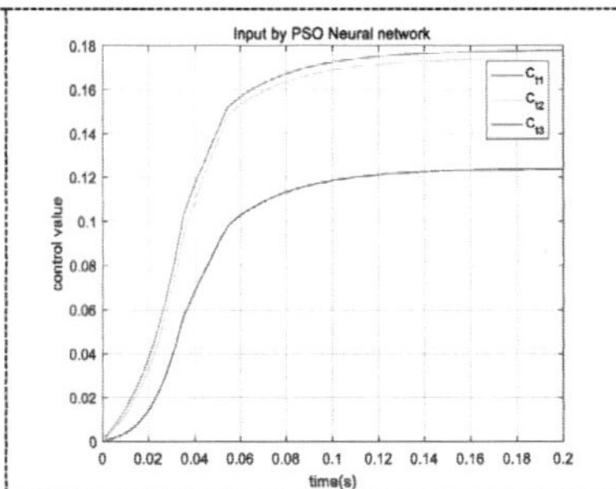

Figure 7. The system control variables of PSO neural network controller.

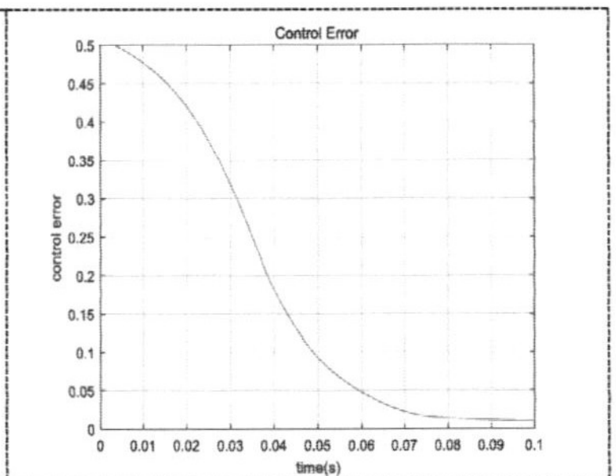

Figure 8. The control error of PSO neural network controller.

To illustrate the advantages of the controller, three different methods are used to control the motion of system. The first is the simulation results of control system through the traditional neural network. As shown in Figure 3 to 5, the simulation results show that the actual output is close to the target output, the control quantity of the system is stable and the error tends to 0. In a word, the controller can have some effect on the control of the system.

Secondly, the simulation results of PSO optimization neural network controller are presented. As shown in Figure 6 to 8, the simulation results show that the output of this method is close to the target output and the learning speed is better than the first two methods.

Finally, in the Figure 9 to 11, the simulation results of MPSO-BFGS neural network controller are presented.

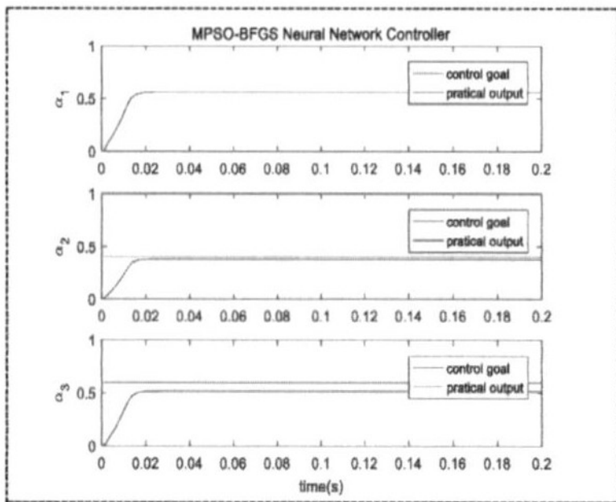

Figure 9. The output of MPSO-BFGS neural network controller.

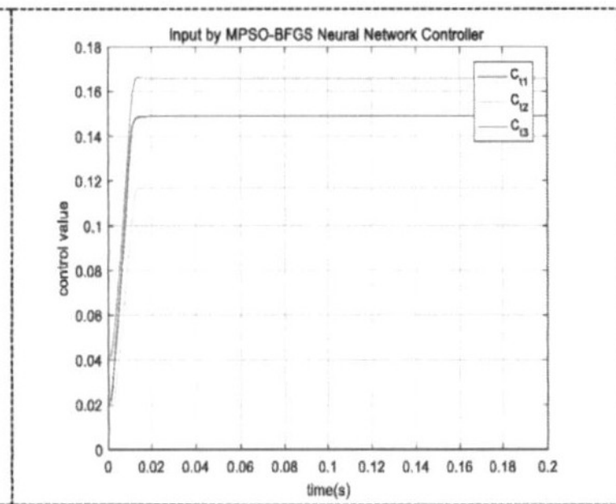

Figure 10. The system control variables of MPSO-BFGS neural network controller.

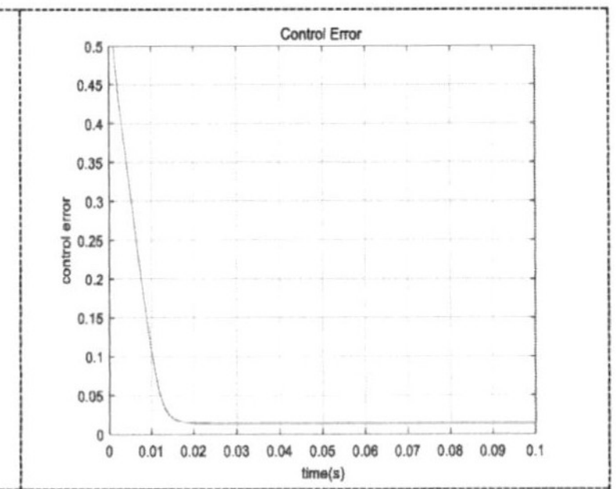

Figure 11. The control error of MPSO-BFGS neural network controller.

5. Conclusion

In this paper, an artificial neural network controller based on MPSO-BFGS hybrid optimization algorithm is proposed for the motion control of a spherical flying robot. The experimental results show that the ANNC is superior to other considered methods. The controller can obtain the expected output in shorter control period. However, due to the combination of the MPSO and BFGS algorithm, it results in occupying so many computing resources. One of the considered methods is to simplify the algorithm and optimize the corresponding hardware layer, which is the focus of future research work.

6. Reference

[1] Qiang, Z., Zengbo, L., & Yao, C. (2008). A back-stepping based trajectory tracking controller for a non-chained nonholonomic spherical robot. *Chinese Journal of Aeronautics*, *21*(5), 472-480.

[2] Ananda, C. M., Akula, P., & Prasad, S. (2014, November). MEMS sensor suites for Micro Air Vehicle (MAV) autopilot. In *Circuits, Communication, Control and Computing (I4C), 2014 International Conference on* (pp. 291-294). IEEE.

[3] Alpaydin, E. (2014). *Introduction to machine learning*. MIT press.

[4] Jordan, M. I., & Mitchell, T. M. (2015). Machine learning: Trends, perspectives, and prospects. *Science*, *349*(6245), 255-260.

[5] Schmidhuber, J. (2015). Deep learning in neural networks: An overview. *Neural networks*, *61*, 85-117.

[6] Dudley, C. J., Woods, A. C., & Leang, K. K. (2015, September). A micro spherical rolling and flying robot. In *Intelligent Robots and Systems (IROS), 2015 IEEE/RSJ International Conference on* (pp. 5863-5869). IEEE.

[7] Yue, M., & Liu, B. (2014). Adaptive control of an underactuated spherical robot with a dynamic stable equilibrium point using hierarchical sliding mode approach. *International journal of adaptive control and signal processing*, *28*(6), 523-535.

[8] Yu, Y., & Ding, X. (2015). On hybrid modeling and control of a multi-propeller multifunction aerial robot with flying-walking locomotion. *Autonomous Robots*, *38*(3), 225-242.

[9] Hatamleh, K. S., Al-Shabi, M., Al-Ghasem, A., & Asad, A. A. (2015). Unmanned aerial vehicles parameter estimation using artificial neural networks and iterative bi-section shooting method. *Applied Soft Computing*, *36*, 457-467.

[10] Islam, S., Liu, P. X., & El Saddik, A. (2015). Robust control of four-rotor unmanned aerial vehicle with disturbance uncertainty. *IEEE Transactions on Industrial Electronics*, *62*(3), 1563-1571.

[11] Zhao, B., Xian, B., Zhang, Y., & Zhang, X. (2015). Nonlinear robust adaptive tracking control of a quadrotor UAV via immersion and invariance methodology. *IEEE Transactions on Industrial Electronics*, *62*(5), 2891-2902.

[12] Diao, C., Xian, B., Yin, Q., Zeng, W., Li, H., & Yang, Y. (2011, May). A nonlinear adaptive control approach for quadrotor UAVs. In *Control Conference (ASCC), 2011 8th Asian* (pp. 223-228). IEEE.

[13] Peng, C., Bai, Y., Gong, X., Gao, Q., Zhao, C., & Tian, Y. (2015). Modeling and robust backstepping sliding mode control with Adaptive RBFNN for a novel coaxial eight-rotor UAV. *IEEE/CAA Journal of Automatica Sinica*, *2*(1), 56-64.

[14] Xiong, J. J., & Zheng, E. H. (2014). Position and attitude tracking control for a quadrotor UAV. *ISA transactions*, *53*(3), 725-731.

[15] Lee, T. (2013). Robust Adaptive Attitude Tracking on ${\ rm SO}(3) $ With an Application to a Quadrotor UAV. *IEEE Transactions on Control Systems Technology*, *21*(5), 1924-1930.

[16] Lin, C. M., Tai, C. F., & Chung, C. C. (2014). Intelligent control system design for UAV using a recurrent wavelet neural network. *Neural Computing and Applications*, *24*(2), 487-496.

[17] Wang, Y., Zhang, H., & Han, D. (2016, July). Neural network adaptive inverse model control method for quadrotor UAV. In *Control Conference (CCC), 2016 35th Chinese* (pp. 3653-3658). IEEE.

IOP Conf. Series: Materials Science and Engineering **272** (2017) 012003 doi:10.1088/1757-899X/272/1/012003

Path Planning Method in Multi-obstacle Marine Environment

Jinpeng Zhang and Hanxv Sun

Automation school of Beijing University of Posts and Telecommunications, China

Abstract. In this paper, an improved algorithm for particle swarm optimization is proposed for the application of underwater robot in the complex marine environment. Not only did consider to avoid obstacles when path planning, but also considered the current direction and the size effect on the performance of the robot dynamics. The algorithm uses the trunk binary tree structure to construct the path search space and A * heuristic search method is used in the search space to find a evaluation standard path. Then the particle swarm algorithm to optimize the path by adjusting evaluation function, which makes the underwater robot in the current navigation easier to control, and consume less energy.

1. Introduction

Path planning is one of the most important tasks in underwater robot navigation, and it represents a degree of intelligence on underwater robots. The overall path planning consists of three parts: 1. Model the environmental information of the robot in relation to the environment.2.Get a search space that contains environmental information.3.Search in the search space using the corresponding search algorithm [1]. Before the underwater robot path planning, it must first model environment, described the underwater robot activity space by the external environment of the original form through a series of processing into a suitable planning model of the algorithm. A reasonable environment is conducive to reducing the amount of search in the plan and the overhead of space and time [2]. Path search algorithm search the feasible space of the path from the environment model, while the path generated from search to the path of the feasible to generate a feasible path in the space. Path optimization is considering basis dynamic characteristics of the intelligent underwater robot, in order to let more conducive to the implementation of underwater robot path for a smooth path [3].

2. Environmental Expression

Obstacles such as islands and reefs in the Marine environment are more irregular. Obstacles are constructed as a relatively neat polygon to facilitate the mathematical representation of obstacles in the planning algorithm. Because of the convex polygons are more concise and convenient in the point of tangency, intersection and judgment of whether the obstacles are in the polygons or not and so on environment planning algorithm expression into a convex polygon obstacle. The principle of constructing obstacles into convex polygon in planning algorithm is that giving priority to fill a vacancy if it does not pose a threat to the precision of path planning for the case [4, 5]. If obstacles in the depth of the underwater robot work area are large and filling a vacancy of convex polygon will have an impact on the precision of path planning then using split method. The expansion is based on the size, navigation error and safety factor of the underwater robot. The path planning of a large voyage will not affect the accuracy of the plan. In the planning system, the convex polygons have been reformulated or separated and expanded such as Figure 1.

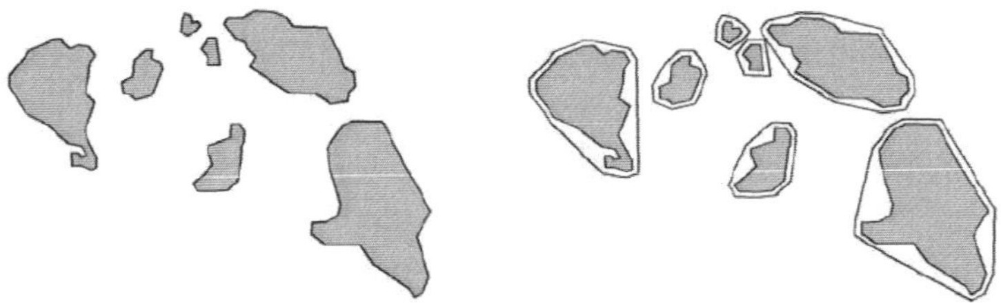

Figure 1. Obstacle extension diagram.

In order to verify the planning ability of the underwater robot's path planning algorithm in the ocean current, the numerical simulation simulates the current field. A series of uniform flows and a point vortex overlay are used to simulate the current of ocean. According to the theory of potential flow, the velocity potential of the flow field is obtained by solving the Laplace equation, and the velocity distribution of the flow field is obtained by the velocity potential.
According to the characteristics of velocity potential can be stacked, the complex flow velocity potential φ under polar coordinates can be calculated as follows:

$$\varphi = \sum_{p=1}^{p=m} v_p r \cos\theta - \sum_{q=1}^{q=n} \frac{i\Gamma_q}{2\pi}\theta \quad (1)$$

Where m, n are the number of uniform flow and vortices at a certain point, r is the distance from current position to the centre of the vortex,θ is the angle between compound velocity and uniform flow velocity.v_p is the pth speed of uniform flow and Γ_q is the velocity of the vortex at that point. The velocity distribution is obtained through the relationship of velocity and velocity distribution:

$$v_r = \frac{\delta\varphi}{\delta r} = \sum_{p=1}^{p=m} v_p \cos\theta \quad (2)$$

$$v_\theta = \frac{1}{r}\frac{\delta\varphi}{\delta\theta} = -\sum_{p=1}^{p=m} v_0 \sin\theta - \sum_{q=1}^{q=n} \frac{\Gamma_q}{2\pi r} \quad (3)$$

The upper part of the formula is calculated to get the current vector at the center of the grid, and the grid only records the information about the flow of the sea without any information about the obstacles. Each grid's current information is based on the centre of the grid. The size of the grid is only related to the complexity of the current instead of the dimensions of obstacles and underwater robot [6].
Current flow around obstacles can produce change. Due to the planning used obstruction information after expansion, the current changes around the obstacles do not take into account in the path planning.

3. Path searching

Use the A * heuristic search method to find a path that meets the criteria in the search space. Because of the presence of the current, it is not appropriate to use the shortest path as a path planning criterion. The algorithm uses the minimum energy consumption as the criterion of the path planning. The evaluation function is defined as

$$\mathrm{f(n) = g(start, n) + h(n, goal)} \quad (4)$$

g(start, n) denotes the path cost from initial node s to node n; h(n, goal) denotes estimate cost from node n to goal node G; h(n, goal) rely on information that is relevant to the problem, which is called heuristic function. g(start, n) consist of path cost L, current cost C and diversion cost T [7,8].

$$\begin{aligned}\mathrm{g(start, n)} &= \mathrm{L(start, n) + C(start, n) + T(n)} \\ &= \sum_{i=start;j=i+1}^{n} L_{ij} + \sum_{i=start;j=i+1}^{n} C_{ij} + \sum_{i=start+1}^{n-1} T_i \end{aligned} \quad (5)$$

$$\begin{aligned}\mathrm{h(n, goal)} &= \mathrm{L(n, goal) + C(n, goal) + O(n, goal)} \\ &= L_{n,goal} + C_{n,goal} + \sum_{i=1}^{ob} O_i \end{aligned} \quad (6)$$

The different steps of the A* algorithm can be summarized as follows:
Establish the backbone structure between the start node and the end node. Get the boy and girl point from the first obstacle among starting node and terminating node.

Calculate boy's and girl's evaluation function and set the lesser one as expanding node.
Create a backbone structure between the current extension node and the end node and get the boy and girl points of the next layer of tree structure.
Keep repeating the above steps until there is no barrier between the extension node and the terminated node.

4. Path optimization by Particle Swarm Optimization

The problem of path planning for underwater robots is the problem of constraint optimization, which can be described as follow:

$$\mathrm{Min}(F) \tag{7}$$

$$\mathrm{s.t.}\ \ P_i^w \notin \mathrm{obstacle} \tag{8}$$

Where F denotes the fitness function, which is also called target function. P_i^w denotes location of the ith dimension of the wth particle. The constraint is that P_i^w can not be located within obstacles.
There are two types of particles that do not meet the constraints of an obstacle: 1. The position of particle is outside of the obstacle, but the path of particles passes through the barrier.2. The position of a particle is in the obstacle [9].
Both situations need to be adjusted. Move along its own velocity direction to the adjusted position until it is visible to the previous and the next dimension. After calibration, the particle satisfies the constraint, but this processing changes the search process of the particle group. Due to the adjusted position is not clear the relationship between the global best position and their own best position, the adjustment speed is set to zero, waiting for its own best location and the global best position to guide.
The PSO algorithm path optimization implementation steps are as follows [10,11]:
Initialize the velocity and position of the particle in the search space. The optimal location of the individual, pBest, is its current location. The optimum position of all particles is determined by the fitness function, which is called gBest [12].
Update the flight speed of each particle. When the particles fly into an obstacle, the two criteria that are generated by random particle encoding are used to adjust the position of the particle.
Calculate the fitness of each particle. If the current fitness of particle is less than the fitness value of the individual's optimal location, update pBest by current position. If the fitness of individual location is less than the fitness value of gBest, update gBest by the exact particle's position.
Keep repeating the above steps until the maximum iteration number is reached.

5. Implementation

Simulation experiments were conducted in two simulated multi-obstacle Marine environments. Set the area of the complex current to be size (3500*2500).The number of particles is 20. Particle groups use the von neumann structure. The nonlinear adjustment inertia coefficient is set to 1.45. Search width is set to 200. In order to ensure the adjustment of the sea current, the minimum dimension is set to 5. And the maximum value of the dimension is set to 10 for the convergence of the particle swarm optimization algorithm.

IOP Conf. Series: Materials Science and Engineering **272** (2017) 012003 doi:10.1088/1757-899X/272/1/012003

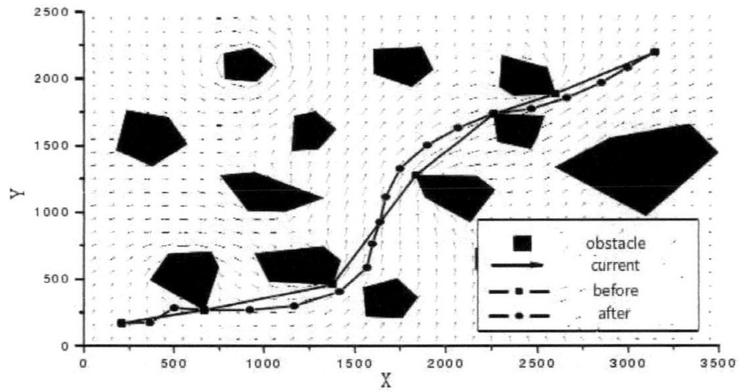

Figure 2. Path optimization in the first environment.

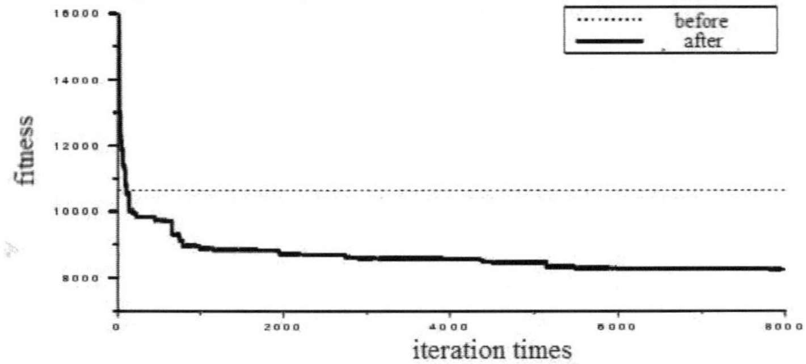

Figure 3. Optimization convergence in the first environment.

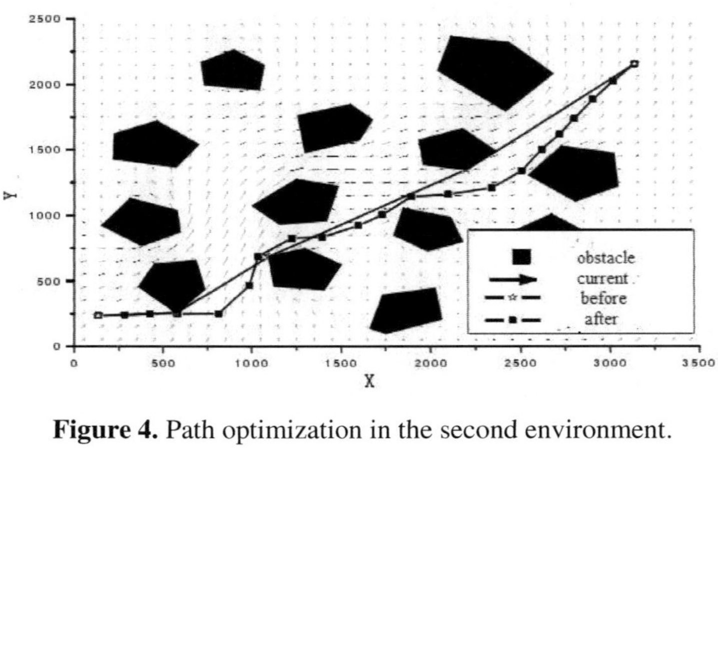

Figure 4. Path optimization in the second environment.

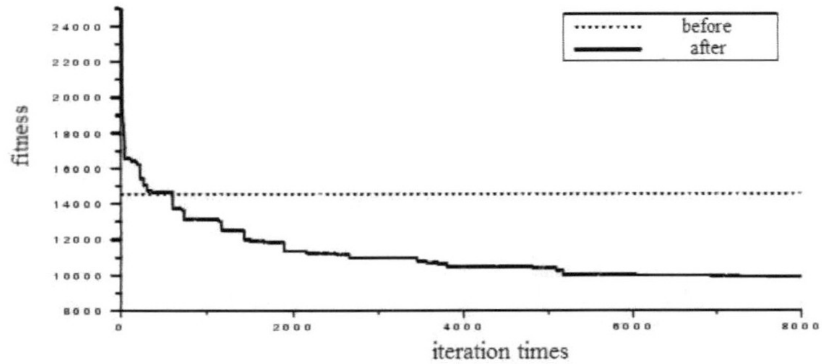

Figure 5. Optimization convergence in the second environment.

In Figure 2 and Figure 4, the black convex polygon represents the obstacle, and the arrow indicates the direction of the current. The cables represent the path before and after the optimization. In Figure 3 and Figure 5, the curve shows the change trend of fitness function as the number of iterations increases.

6. Conclusions & Future work

From the optimized path diagram (Figure 2, Figure 4)can be seen that the optimized path reduce the influence of lateral flow as far as possible, which can greatly reduce the energy consumption of underwater robot in the current. As you can see from the fitness convergence diagram (Figure3, Figure5) that the fitness has been plummeted. So the search speed is getting faster and the energy consumption is getting smaller. Because of this path segmentation optimization method considering the current environment without the obstacles, the dimension of the particle swarm will be reduced. Particle swarm optimization algorithm's convergence speed is faster than the particle swarm path planning algorithm and easier to converge.

The simulation results show that the environmental modeling using binary tree structure is simple, and A * algorithm search path is effective and feasible. Particle swarm optimization is more suitable for sailing in the current environment. The influence coefficient of marine current on underwater robot are obtained by experience and simulation experiment. As future work, the influence of different number needs further experiment.

7. References

[1] Xunqiang Hu. Application of PSO algorithm and surface with minimum threat in path planning of aircraft. 2012 *ICCEE* **1**.

[2] Wei Kan. Path Planning and Decision-making Control for AUV with Complex Environment[A]. IACSIT.proceedings of 2010 3rd International Conference on Computer and Electrical Engineering (ICCEE 2010 no.2)[C].IACSIT:,2012:**7**.

[3] Li Di. Robot path planning based on an improved multi-objective PSO method. 2016 *ICCIA*. Wuhan Zhicheng Times Cultural Development Co.:,2016:**6**.

[4] Jihee Han,Yoonho Seo. Mobile robot path planning with surrounding point set and path improvement[J]. Applied Soft Computing,2017,**57**:.

[5] Xuemin Hu,Long Chen,Bo Tang,Dongpu Cao,Haibo He. Dynamic Path Planning for Autonomous Driving on Various Roads with Avoidance of Static and Moving Obstacles[J]. Mechanical Systems and Signal Processing,2017,:.

[6] Yufeng Mao, Yongjie Pang, Zhaoli Wang .Underwater vehicle's long voyage path planning in complex sea condition. IEEE Beijing Section, China. State Key Laboratory of Autonomous Underwater Vehicle, Harbin Engineering University SKLAUV,HEU Harbin, China. 2009 *ICIS* Vol.**2**,2009:5.

[7] Thanapong Phanthong, Toshihiro Maki, Tamaki Ura, Takashi Sakamaki, Pattara Aiyarak. Application of A* algorithm for real-time path re-planning of an unmanned surface vehicle avoiding underwater obstacles. 2014 *JMSA* ,**01**:105-116.

[8] Lin Liu. Research on dynamic energy saving path planning algorithm based on real-time traffic information[A].Proceedings of 2016 4th International Conference on Advanced Materials and Information Technology Processing(AMITP 2016)[C].2016:**6**.

[9] Jian ZHANG. Research of Mobile Robot Path Planning Based on Genetic Algorithm[A]. Proceedings of 2nd Information Technology and Mechatronics Engineering Conference(ITOEC 2016)[C].,2016:**7**.

[10] Emad A.Elsheikh. Practical path planning and path following for a non-holonomic mobile robot based on visual servoing. IEEE Beijing Section、Global Union Academy of Science and Technology、Chongqing Global Union Academy of Science and Technology. 2016 *ITNEC* ,2016:**6**.

[11] Paul J.Werbos, Beyond Regression: New tools for prediction and analysis in the behavioral sciences,phD thesis,Harvard University,1974.

[12] Chao Shen. Model Predictive Control for an AUV with Dynamic Path Planning[A]. The Society of Instrument and Control Engineers（SICE）.Proceedings of the Society of Instrument and Control Engineers Annual Conference 2015（2）[C]. The Society of Instrument and Control Engineers (SICE): 2015:**6**.

Analysis and experiment of capacitor charging characteristics

Ma Ning, Liu Weijian, Qin Zhenjie, Lai Xiangdong, Dong Feng

Wuhan Electronic Information Institute, Wuhan, 430019, China

Abstract. This paper analyses capacitor charging characteristics which affects the delay time of DC output of Switch Mode Power Supply(SMPS). By charging experiments of three different capacitors, quantitative analysis of different situations influence and algorithm of the delay time due to ohmic leakage is made. Through theoretical calculation, software simulation and physical circuit experiment, finds a reasonable solution under different charging conditions.

Keyword. Switch Mode Power Supply, capacitor, charging characteristics, ohmic leakage

1. Charge characteristic analysis of capacitor

Common switching power supply after input AC, DC output is not immediately available, usually there is a certain delay time. As shown in figure 1, a is an AC voltage input for a switching power supply, b is the DC voltage output. From the data read-out of the graph, there is a 2.86s delay between the inputs and outputs. Figure 2 is the actual circuit diagram of this switching power supply. The input is sinusoidal AC voltage, through the diode full bridge rectifier and filter, the chip IC1 5 feet connected capacitor C4 is charged. When the charging voltage reached Uset, Chip IC1 starts external direct current power supply. Figure 3 is a switching power supply of another type, In the case of pulsating DC current after half wave rectification, the capacitor C10 is charged to start the chip. The red arrow is the direction of the voltage signal[1].

Aluminum electrolytic capacitors are usually used to charge for starting a chip. If capacitance leakage resistance is different, the charging time will be affected, thus affecting the delay time of the DC output of the Switching Power Supply. In this paper, three experiments are conducted to analyse quantitatively the influence of leakage resistance in different cases on delay time.

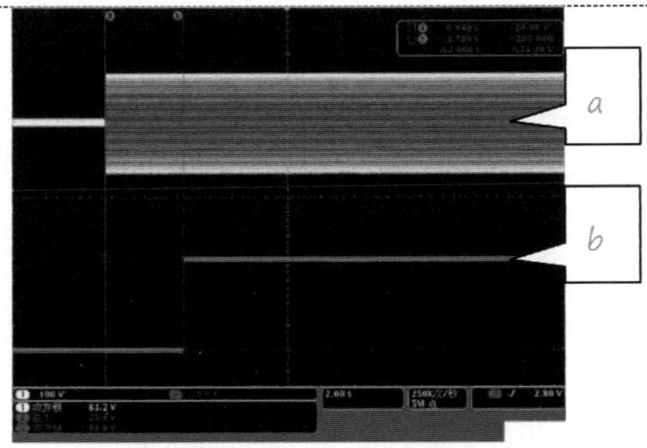

Figure 1. switching power input.

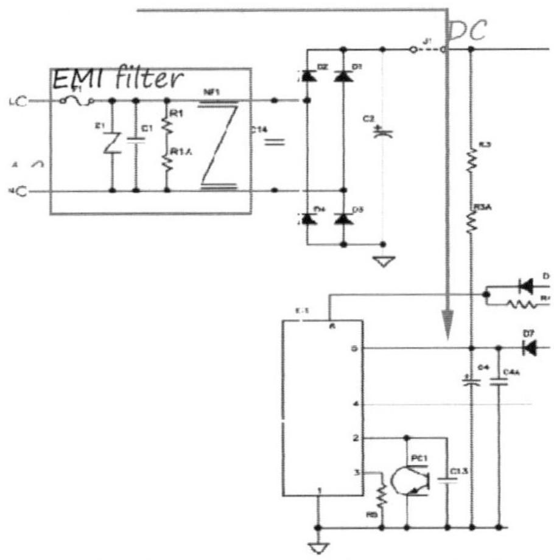

Figure 2. actual circuit diagram.

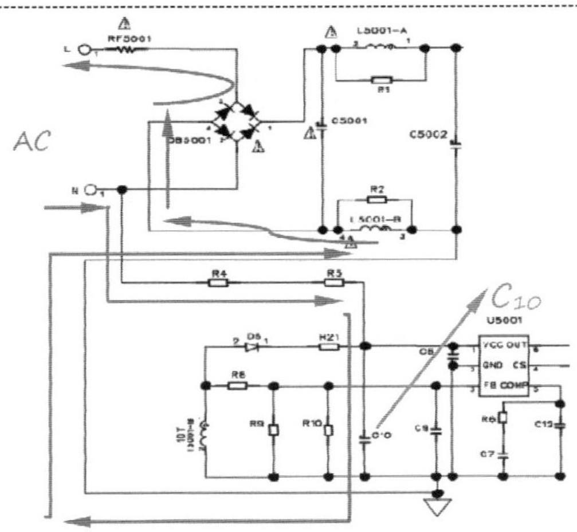

Figure 3. another switching power charging circuit.

2. Experimental verification

The first experiment is the DC charging of the ideal capacitor without leakage. Experiment two is the direct current charging of the capacitor with leakage resistance. Experiment three is the charging of a resistive capacitor with half wave sinusoidal voltage input. By theoretical calculation, software simulation and physical circuit experiment, algorithms and solutions for different charging scenarios is found.

2.1. Experiment 1: Measurement of the leakage of the capacitor under DC voltage conditions

Experimental objective: Theoretical calculation method, simulation method and actual measurement method are adopted to measure the relationship between the DC voltage and the time on the capacitor C1.

Experimental illustrations: The experimental circuit is shown in figure 4. The actual power supply V1 uses 50Hz amplitude 100V sinusoidal AC voltage source. After full bridge rectification, filtered by 3.3uF capacitor, the voltage waveform is shown in figure 5. The measured capacity of film capacitor C1 is 0.95uF, the measured leakage current is 0. That is to say, the leakage resistance tends to infinity. Measured impedance of oscillograph used in actual measurement is R3=10MΩ, parallel with C1. When the voltage on the capacitor C1 is Uset=10V, measures the time took to recharge.

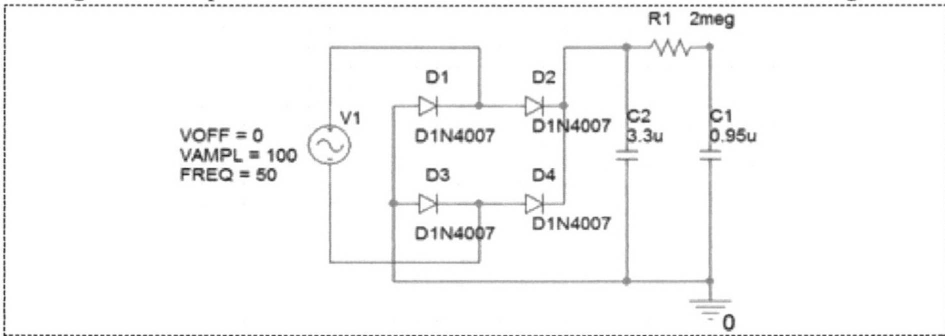

Figure 4. no leakage Capacitor DC charging circuit.

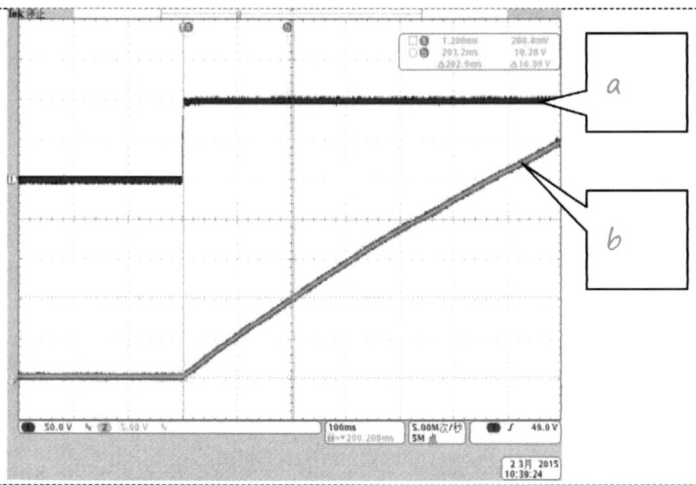

Figure 5. oscillograph waveform.

IOP Conf. Series: Materials Science and Engineering **272** (2017) 012004 doi:10.1088/1757-899X/272/1/012004

2.1.1. Theoretical arithmetic. The theoretical calculation circuit is shown in figure 6, DC Voltage Source Vdc=100V, ignoring the influence of capacitor leakage resistance and oscilloscope probe on the circuit.

According to the three element method: $u_c(t) = U_{c(\infty)} \times (1 - e^{-\frac{t}{R_1C_1}}) = U_{set}$,

that is
$$t = -R_1C_1\, ln\left[1 - \frac{U_{set}}{U_{c(\infty)}}\right]$$

(1)

thereinto $U_{c(\infty)} = 100V, R_1 = 2M\Omega, C_1 = 0.95uF, U_{set} = 10V$. Substitution formula $t = 200.2ms$.

2.1.2. Software emulation. The simulation circuit is shown in figure 7. In order to reduce the influence of the rising edge of the voltage source and the leakage resistance, the DC voltage source V1 uses ideal voltage source, parameter for initial value is 0, step high level is 100V, time-delay is 0s, rising edge is 0s, trailing edge is 0s, high level time is 1000s, pulse period is 1001s, R1=2MΩ, C1=0.95uF. Using Orcad9.2 version of simulation, the result is shown in figure 8 of the A coordinate. When charging 202.6ms, voltage on C1 is 10.006V.

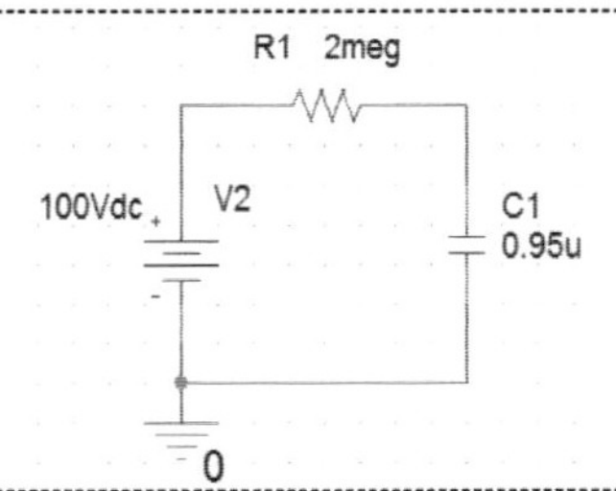

Figure 6. Experiment 1.

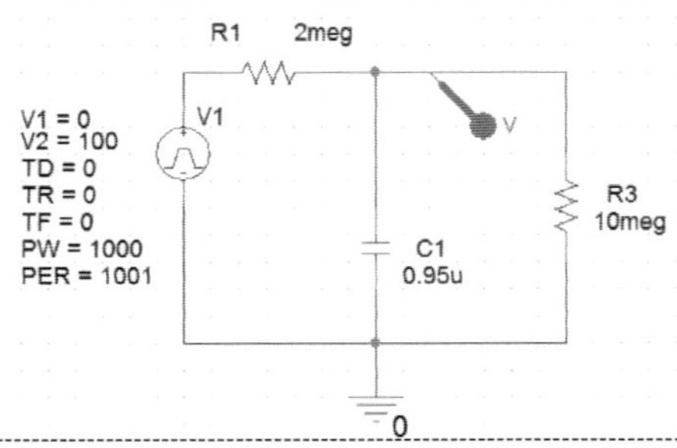

Figure 7. no leakage simulation circuit.

IOP Conf. Series: Materials Science and Engineering **272** (2017) 012004 doi:10.1088/1757-899X/272/1/012004

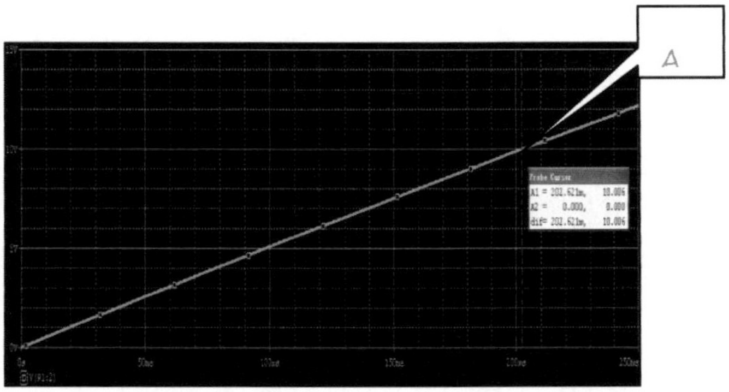

Figure 8. simulation charging.

2.1.3. Physical circuit experiment. Oscilloscope settings trigger, observed waveform is shown in figure 5. Where a is the DC supply voltage, b is the capacitor(capacitor cannot leak electricity; there is a residual charge) charging voltage. Starts clocking from 200mV, read from the chart: when $\Delta V = 10V$, $\Delta t = 202ms$.

Experiment 1 Conclusion: when capacitive DC voltage is U_{set}=10V, formula calculator t=200.2ms, simulation result t=202.6ms, experimental measuring t=202ms, the time difference between the three is very small. Formula (1) $t = -R_1C_1 \, ln\left[1 - \frac{U_{set}}{U_{c(\infty)}}\right]$ can be used in actual circuit calculations, moreover, the simulation results can accurately predict the measured values.

Error analysis: In the calculation method, the influence of 10M ohm resistance leakage on oscilloscope probe is not considered, caused about 1% error. In order to be close to the actual conditions, the experimental 2 is improved.

2.2. Experiment 2

Considering the influence of capacitance leakage resistance and oscillograph probe, measures the charging of capacitor under DC voltage.

Experimental objective: Theoretical calculation method, simulation method and actual measurement method are adopted to measure the relationship between the DC voltage and the time on the capacitor C1 in case of leakage of electricity.

Experimental illustrations: The experimental circuit is shown in Figure 9. To reduce the impact of other factors, use figure 10 equivalent. R2=1MΩ equivalls leakage resistance of capacitance C1, measured impedance of oscilloscope probe is R3=10MΩ, the other experimental conditions are the same as those of Experiment 1.When the voltage on the capacitor C1 is Uset=10V, measures the time took to recharge.

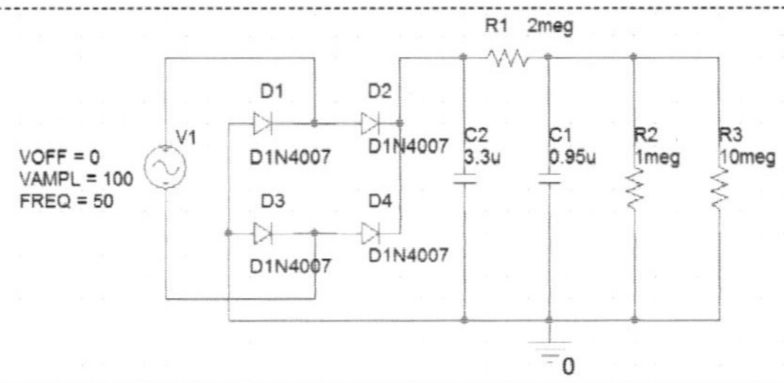

Figure 9. Experiment 2 circuit.

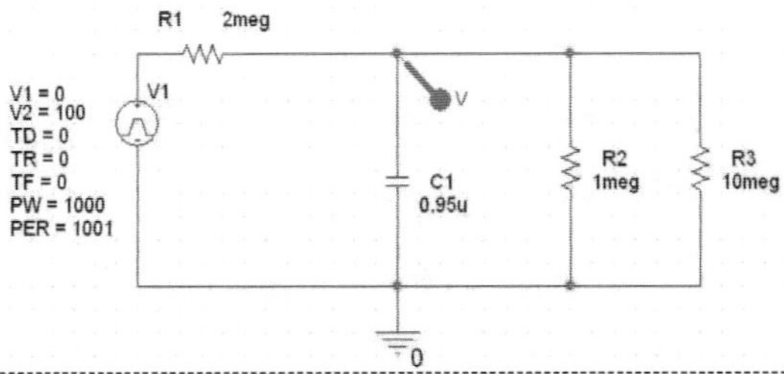

Figure 10. equivalent simulation circuit.

2.2.1. Theoretical arithmetic. The circuit is shown in Figure 11, considers capacitor C1 as an external circuit, the rest is the internal circuit. By Thevenin's theorem, the equivalent circuit is shown in Figure 12. Output resistance $R_O = R_1 // R_2 // R_3$, thereinto $R_1 = 2M\Omega,\ R_2 = 1M\Omega,\ R_3 = 10M\Omega$. So $R_O = 0.625M\Omega$, orders $R' = R_2 // R_3 = 0.909M\Omega$, open circuit voltage $U_{OC} = U_{C(\infty)} = V_1 \dfrac{R'}{R' + R_O}$.

Using three factor method, brings the above conditions into the formula（1）

gets:

$$t = -\frac{R_1 \times R'}{R_1 + R'} \times C_1 \times \ln\left[1 - \frac{U_{set} \times (R_1 + R')}{R' \times V_1}\right] \qquad (2)$$

brings data into formula (2): t =229ms.

2.2.2. Software emulation. The simulation circuit is shown in figure 11, among them, V1, R1 and C1 are the same as the experiment 1.

Drain resistance and resistance of film capacitor R2=1M Ω, and parallel with oscilloscope probe R3=10M Ω. The simulation results are shown in figure 11 in the A point coordinates, when charging 229.2ms, voltage on C1 is 10.0V.

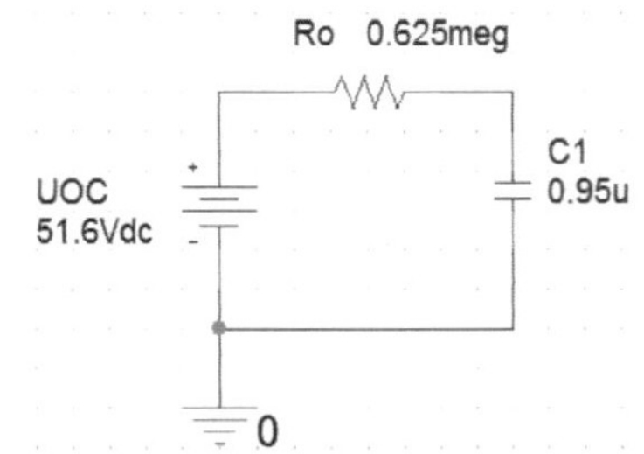

Figure 11. Thevenin equivalent circuit.

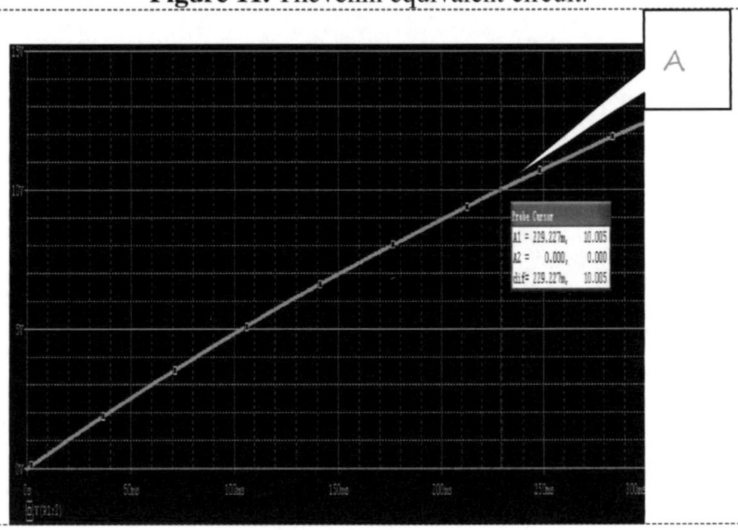

Figure 12. simulation result.

2.2.3. Physical circuit experiment. Oscilloscope settings trigger, observed waveform is shown in figure 12. Where a is the DC supply voltage, b is the capacitor(capacitor can leak electricity; there is no residual charge) charging voltage. Starts clocking from 0V, read from the chart: when $\Delta V = 10.00V$, $\Delta t = 229.0ms$.

Experiment 2 Conclusion: when capacitive DC voltage is U_{set}=10V, formula calculator t=229ms, simulation result t=229.2ms, experimental measuring t=229ms, the time difference between the three is very small. Formula (2) $t = -\frac{R_1 \times R'}{R_1 + R'} \times C_1 \times \ln\left[1 - \frac{U_{set} \times (R_1 + R')}{R' \times V_{CC}}\right]$ can be used in actual circuit calculations, moreover, the simulation results can accurately predict the measured values[2], thereinto $R' = R_2 // R_3$.

2.3. Experiment 3

Considering the influence of capacitance leakage resistance and oscillograph probe, measures the charging of capacitor under the condition of half wave sinusoidal voltage.

IOP Conf. Series: Materials Science and Engineering **272** (2017) 012004 doi:10.1088/1757-899X/272/1/012004

Experimental objective: Simulation method and actual measurement method are adopted to measure the relationship between the voltage and the time on the capacitor C1.

Experimental illustrations: The experimental circuit is the same as the simulation circuit, shown in figure 13. Actual voltage source is 50Hz, V_m=100V AC Voltage Source. After half wave rectification, the capacitor C1 is charged by the resistance R1.

2.3.1. Software emulation. The simulation circuit is shown in figure 14 among them, thereinto V1, C1, R1, R2 and R3 are the same as those of Experiment 2. D1 is diode 1N4007, takes half wave rectification of sinusoidal alternating current. The simulation results are shown in figure 15 in the A point coordinates, when charging 1326.5ms, voltage on C1 is 10.03V.

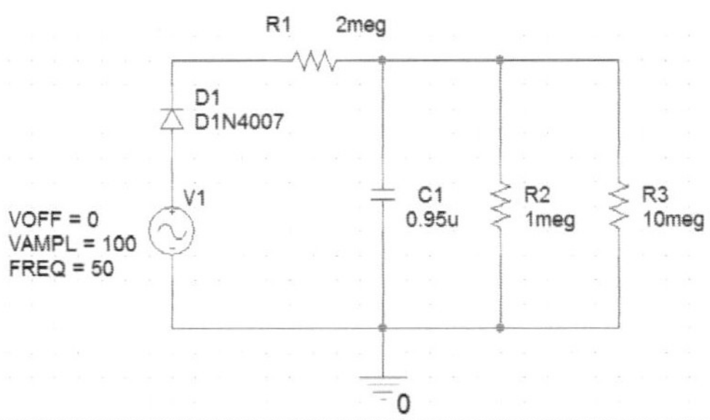

Figure 13. Experiment 3 circuit.

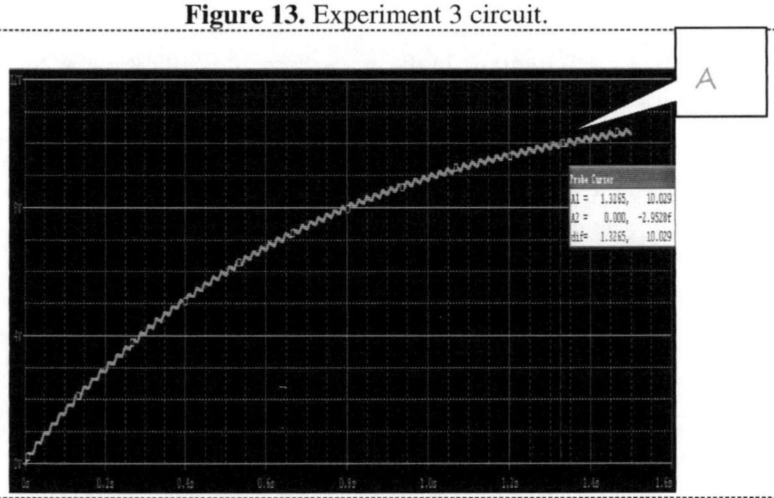

Figure 14. simulation result.

2.3.2. Physical circuit experiment. Oscilloscope settings trigger, observed waveform is shown in figure 15. Where a is the AC voltage source voltage, b is the capacitor(capacitor can leak electricity; there is no residual charge) charging voltage. Starts clocking from 0V, read from the chart: when $\Delta V = 10V$, $\Delta t = 1324ms$.

Experiment 3 Conclusion: In half wave sinusoidal charging, when capacitive DC voltage Uset=10V, simulation result t=1326.5ms, experimental measuring t=1324ms, the time difference between the three is very small (less than 0.2%). The simulation results can accurately predict the measured values[3].

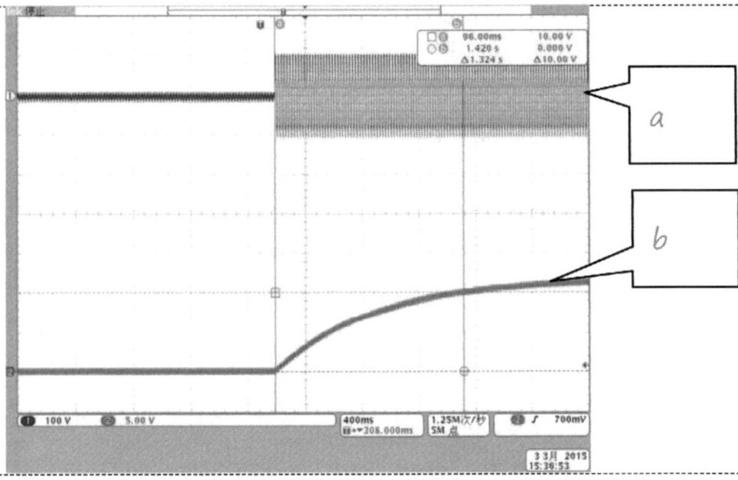

Figure 15. oscillograph waveform.

3. Interpretation of result

Comprehensive conclusions and Solutions: In the case of the above three capacitors charging, you can use formulas (1) and (2) or simulations to get results quickly in experiment 1 and 2; simulation results can be used as an important basis and conclusion for similar experiments in experiment 3. Different algorithms and solutions should be chosen under different charging conditions.

4. References

[1] Portet C, Taberna P L, Simon P, et a1.High power density electrodes for carbon supercapacitor applications[J]. Electrochimica Aeta, 2005, 50:4174-4184.

[2] Conway B E, Pe11 W G. Power limitations of supercapacitor operation associated with resistance and capacitance distribution in porous electrode devices[J].J Power Sources, 2002, l05(1):169-181.

[3] Niu J J, Pell W G, Conway B E. Requirements for performance characterization of C double-layer supercapacitors: applications to a high specific-area C-cloth material[J].Journal of Power Sources, 2006, 156:725-740.

[4] ShenYuliang, Xuweixin, Zhaowei. Study of Control Rules of Single Input Single Phrase SPWM Modulation Photovoltaic Synchronization Power Generation system [J]. Solar Energy Transaction. 2004.25(6):794-797.

[5] Songqiang, Liuwenhua, Yangangui. The Design of The LC Filter from High-Capacity PWM Voltage Source Inverter. [J] Qinghua University Transaction(science) 2003.43(3):345-348.

[6] WANG Xuehua, RUAN Xinbo,LIU Shangwei(College of Electric and Electronic Engineering,Huazhong University of Science & Technology, Wuhan 430074,Hubei Province,China);Control Strategy for Grid-connected

Inverter to Suppress Current Distortion Effected by Background Harmonics in Grid Voltage[J];Proceedings of the CSEE;2011-06.

[7] ZHANG Guoju, TANG Xisheng, ZHOU Long, QI Zhiping(Institute of Electrical Engineering,Chinese Academy of Sciences,Haidian District,Beijing 100190,China);Research on Complementary PWM Controlled Buck/Boost Bi-directional Converter in Supercapacitor Energy Storage[J];Proceedings of the CSEE;2011-06.

[8] Torabzad S, Babaei E, Kalantari M (2010) "Z-Source Inverter based Dynamic Voltage Restorer" 1st Power Electronic & Drive Systems & Technologies Conference.

IOP Conf. Series: Materials Science and Engineering **272** (2017) 012005 doi:10.1088/1757-899X/272/1/012005

Research Progresses and Suggestions of Manufacturing Technologies of Engine Bearing Bushes

J Cao, Z W Yin*, H L, Li and G Y Gao

School of Mechanical Engineering, Shanghai JiaoTong University, Shanghai , China

*Corresponding author: yinzw1972@163.com (Z.W. Yin)

Abstract: Bearing bush is a key part of diesel engine, and its performance directly influences the life of whole machine. Several manufacturing technologies of bearing bush such as centrifugal casting, sintering, electroplating and magnetron sputtering have been overviewed. Their bond strength, porosity, production efficient, layer thickness, frictional coefficient and corresponding materials analyzed and compared. Results show that the porosity and oxidation of sintering and centrifugal casting are higher than that of other two methods. However, the production efficiency and coating thickness are better than that of electroplating and magnetron sputtering. Based on above comparisons and discussions, the improvements of all manufacturing technologies are suggested and supersonic cold spraying is suggested. It is proved that cold spraying technology is the best choice in the future with the developing of low frictional materials.

1. Introduction

Bearing bush is a kind of journal bearing which widely used in automotive engineering, aerospace and many industrial applications. Roller bearings replaced by plain bearings are becoming popular because its life, manufacture cost, structure, vibration and wear are higher than that of plain bearings [1]. The yaw roller bearings are replaced by plain bearings due to the smaller structure and lower cost in the wind power system. In Pratt & Whitney Group, plain bearings had been used in aero-engine gear box [2]. With the development of advanced materials and manufacturing technologies, the low starting torque of plain bearings will be obtained, and roller bearings can be widely replaced in the future.

Bearing bushes must meet several requirements such as compliance, structure strength, low CoF, high bond strength and long service life [3]. In this paper, four common manufacturing technologies like centrifugal casting, sintering, electroplating and magnetron sputtering of plain bearing and their bond strength, porosity, production efficient, layer thickness, frictional coefficient and corresponding materials are introduced and compared. Based on the above summaries, the improvement suggestions of centrifugal casting, sintering, electroplating and magnetron sputtering are given. In addition, new manufacturing technology of the cold spraying is introduced. Through analyzing the mechanical properties and manufacturing efficiency, it proves that cold spraying can be applied for producing bearing bushes.

2. Manufacturing Technologies of Bearing Bush

The review ranges are narrowed to the engine bearing bushes because they are the most widely used plain bearings. The structures of engine bearing bushes usually are double layers and three layers which structure shown as figure 1 and figure 2. The double layers consist of a layer of steel back and a layer of alloy. The bearing of three layers consists of steel back, a transition layer and a antifriction layer. The common material of transition layer is nickel which can enhance the bond strength of back steel and antifriction layer [4]. The materials of top layer are usually the copper or aluminium alloys which have low CoF (Coefficient of Frictional) and the capacity of bearing heavy load.

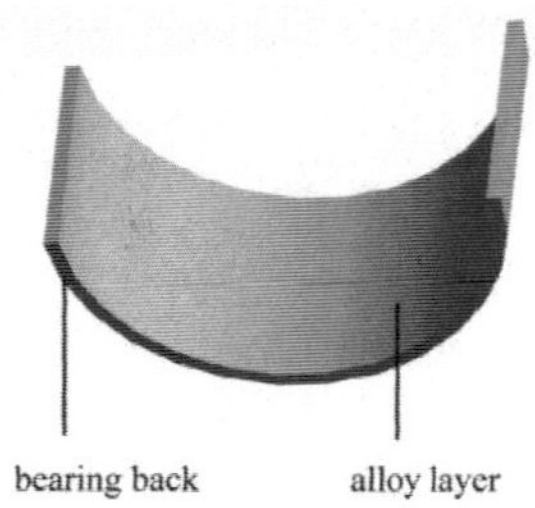

Figure 1. Bimetallic bearing bush

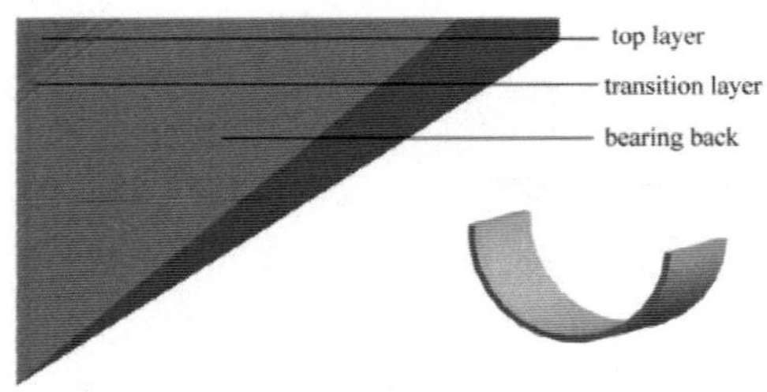

Figure 2. Triple metal bearing bush

2.1. Properties centrifugal casting

The centrifugal casting technology is shown as figure 3. With the centrifugal force, the molten liquids of alloy metals are bonded on the surface of bearing back steel. The properties of bearing bush had been studied and many excellent results obtained. The porosity was about 2.4%, and the CoF was more than 1 of aluminium bronze bearing bush made by centrifugal casting which indentified by the study of Alam [5]. The traditional material such as Pb has the problem of environment pollution. Some green materials had been studied to improve the frictional properties of bearing bush. The material of $Cu_6Sb_{11}Sn_{83}$ and its proper coating thickness was from 0.05 to 0.15 mm which CoF was from 0.03 to 0.04 [6]. To improve mechanical properties of bearing, the additive materials had been studied. It was proved that with the additive of WC, anti-wear performance of bearing bush made by centrifugal casting was improved [7]. Samadi had studied the temperature and thickness which influenced Al_2Cu layer hardness and density. It is shown that the properties of 10 mm coating thickness were better than that of 16 mm [8]. The bond strength of babbitt alloy and bearing back was from 40 MPa to 60 MPa which shown by Diouf [9].

2.2. Properties of sintering

Sintering is the most popular technology to produce bearing bushes. The principle of sintering of bearing bush is shown in figure 4. Bearing alloys are melted in the furnace, and the bearing back is preheated. The bearing back and liquid alloy are ground together by a water cooling roller. Cao has studied the mechanical properties of Ti_6Al_4V which made by sintering, and its porosity is 3.5% [10]. Mironov had shown the mechanical properties of bearing bush which was made of low alloyed materials. The CoF was 0.26 and its porosity was about 15% [11].

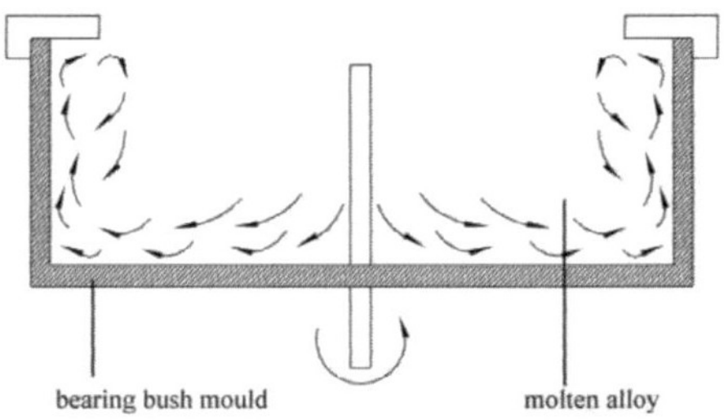

Figure 3. The principle of centrifugal casting

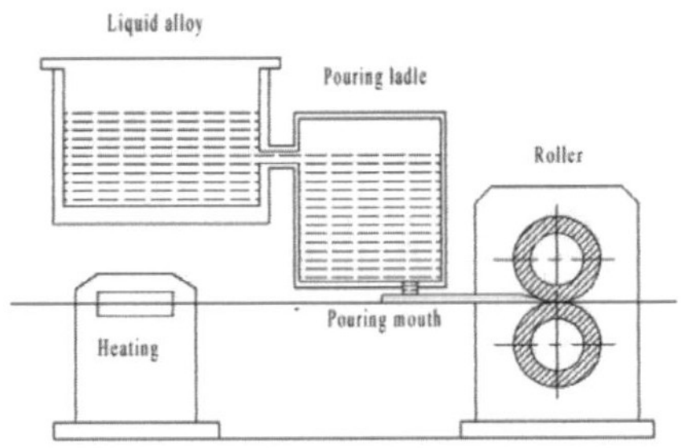

Figure 4. The principle of sintering

Dourandish had studied 3Y-TZP/Cr composite bilayers produced by sintering. The results shown that the bond strength was 41 MPa, and its porosity was from 10 % to 16 % [12]. In order to improve the mechanical properties of Cu alloys, some materials like W, Ta, Mo, Ag had been added [13, 14]. The composite alloys were studied by researchers also. $Sn_{4.0}Ag_{0.5}Cu_{1.0}Ni$ alloy was produced and its microstructure properties were studied [15]. The layer thickness of bearing bush can be produced more than 1 mm, and its produce efficiency was high. However, the oxidation and high porosity can not be avoided [16].

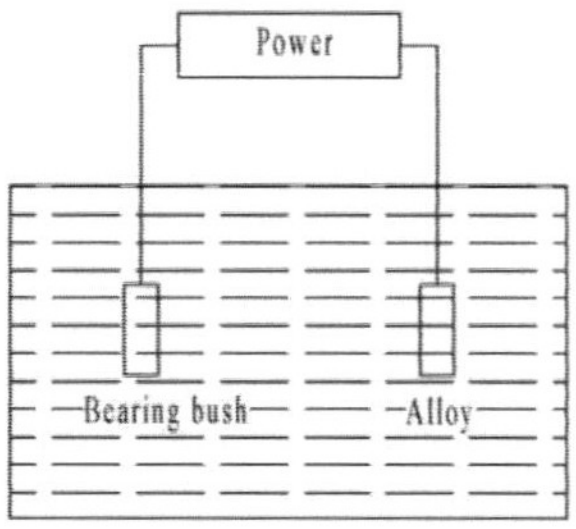

Figure 5. The principle of electroplating

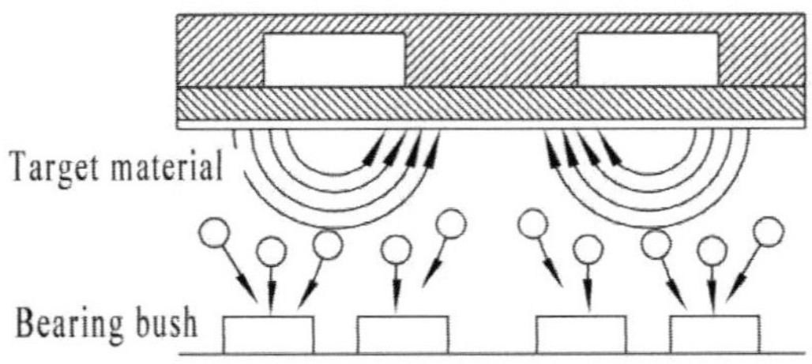

Figure 6. The principle of magnetron sputtering

2.3. Properties of electroplating

In order to improve the tribological properties and decrease the oxidation of bearing bushes, three elements, four elements and even five elements chemical plating are applied [17-19]. Electroplating of bearing bush is shown in figure 5. The bearing bush was set as cathode, and the material of coating was set as anode. The more thick the coating thickness, the more porosity would be obtained. The coating thickness should not beyond 5 μm was suggested by Yu [20]. The bond strength of coating produced by electroplating was from 26.8 MPa to 33.6 MPa which studied by Elsaka [21]. The CoF of Ag coating made by electroplating was 0.18 which shown by Arash. Its bond strength was from 18 MPa to 24 MPa [22]. The porosity of Ni-Al based coatings via electroplating was from 4.29 % to 10.55 %. The bond strength is lower than that of centrifugal casting and sintering. In order to improve performances of electroplating, some rare earth materials such as Ln, La and Ta were studied and added to improved performances of materials [23, 24]. However, the plating solution is harmful to the environment, and it will be restricted in the future.

2.4. Properties of magnetron sputtering

Different from electroplating, magnetron sputtering does not pollute environment. The bond strength of Ti/ZrSiN with different Si content was from 25.2 MPa to 57.3 MPa which shown by Zhang [25]. The shortages of magnetron sputtering are the low efficiency and thin coating thickness. The proper thickness of $CuCr_{0.93}Mg_{0.07}O_2$ thin film deposited by reactive magnetron was from 70 nm to 280 nm [26]. Magnetron sputtering is an effective way that many materials can be deposited on bearing bushes. For example, Ti/Cu/N coatings, Cu_xSn_y, CrN/Ag, TiN/Cu coatings can be deposited, and their CoFs were lower than 0.15 [27, 28]. The porosity of Cr single coating and CrN coating was 3.72 % and 5.83 % which studied by Wu. However, with the filling action, the porosity of two composite coatings was 0.39 % [29]. The fatigue strength will be improved 10 %-30 % if the porosity is decreased 1%-2% [30]. Thus, the low porosity of bearing bush should be obtained.

IOP Conf. Series: Materials Science and Engineering **272** (2017) 012005 doi:10.1088/1757-899X/272/1/012005

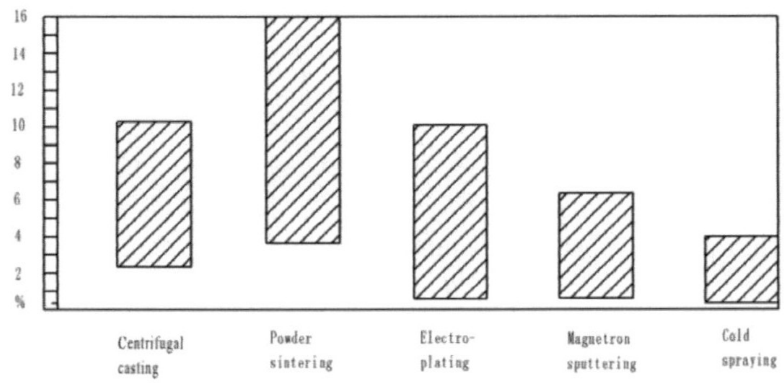

Figure 7. Porosity of different manufacturing technologies

3. Comparisons of different manufacturing technologies

The performances of bond strength, porosity, layer thickness, frictional coefficient and corresponding materials of centrifugal casting, sintering, electroplating and magnetron sputtering are briefly introduced respectively. In order to describe properties of different manufacturing technologies clearly, and find the best method, the high pressure cold spraying is discussed. Based on the reviews of references, the porosity of different manufacturing technologies is shown in figure 7 [31-33].

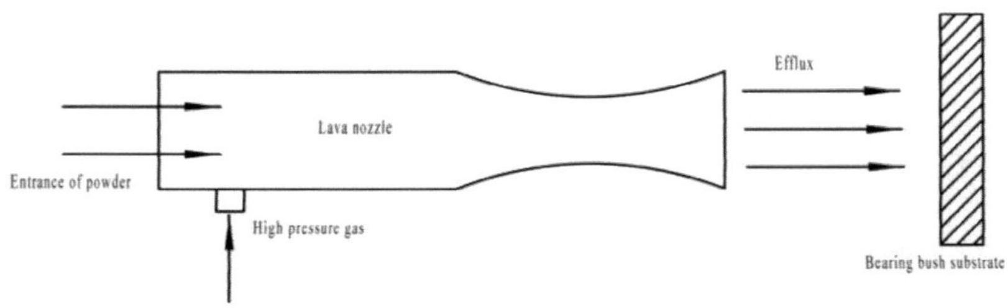

Figure 8. The principle of cold spraying

It is found that the porosity of cold spraying is the smallest. The principle of cold spraying is shown in figure 8. With the supersonic speed of carried gas, the coating is consolidated with the next flaying powders. Thus, its porosity is condensed and low. The bond strengths of all manufacturing technologies are shown in figure 9 [34]. The temperature of centrifugal casting and sintering are higher than that of electroplating, magnetron sputtering and cold spraying. Without inert gas protection or vacuum environment, the oxidation of casting and sintering are higher than that of other three manufacturing technologies. However, coating thickness and production efficiency of electroplating and magnetron sputtering are lower than that of casting, sintering and cold spraying. Take the cold spraying as an example. As a deposition efficiency of 90 %, and powder carried speed of 30 g/min the

30300 mm^2 per minutes can be obtained if the coating thickness is 1mm. The comparisons of different manufacturing technologies are shown in table 1 [35, 36].

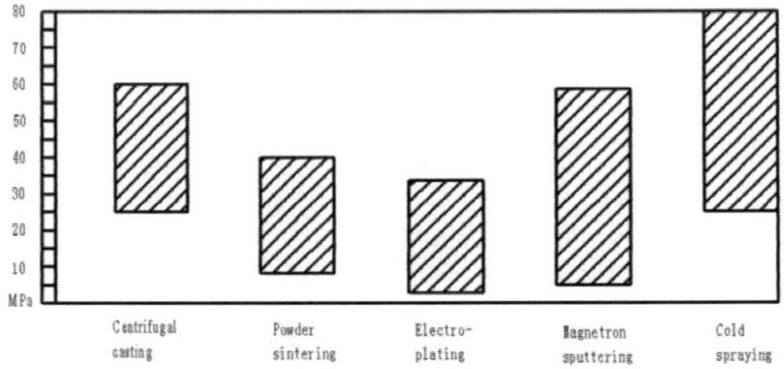

Figure 9. Bond strength of different manufacturing technologies

Table.1 Performances of different coatings

Manufacturing technologies	Production efficiencies	Oxidation	Layer thickness
Centrifugal casting	Middle	High	>1mm
Sintering	Middle	High	>1mm
Electroplating	Low	Middle	<20μm
Magnetron sputtering	Low	Low	<15μm
Cold spraying	High	Low	>1mm

The microstructures of copper and tin alloy of different manufacturing are shown in figure 10 [37-40]. It is found coating made by casting and sintering has some micro cracks. The compactness of coating made by cold spraying, magnetron sputtering are higher than that of electroplating, sintering and casting.

4. Conclusion and suggestion

From above discussions, it is found that the porosity and oxidation of centrifugal casting and sintering are higher than that of other three manufacturing technologies. However, the vacuum environment can reduce the porosity. The other method is special module design, with the full chamber of molten alloy liquid, the air was expelled and the porosity was down [41]. The oxidation and porosity of electroplating and magnetron sputtering are lower than that of casting and sintering. However, their production efficiencies are lower too. The green plating should be studied because the common solutions are environment pollutions. Cold spraying is a potential way to produce bearing bush. The main shortages of cold spraying are the high CoFs of coating. However, with the development of new materials, the common bearing bush manufacturing technologies may be replaced by cold spraying.

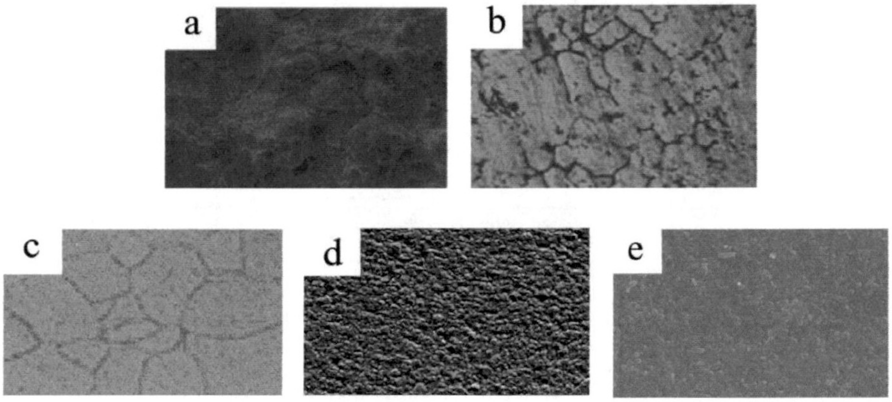

Figure 10. Microstructure of Cu-Sn alloy of different manufacturing technologies: (a) Cold spraying; (b) centrifugal casting; (c) Sintering; (d) Electroplating; (e) Magnetron sputtering

5. Reference

[1] Cao J., Yin Z. W., Cui Y. Q., Li H. L and Wang X. B. 2017 *Ind Lubr Tribol* **69** 95.
[2] Zeng Z. Q., Wang Y. K. and Wang Z. H. 2012 *Appl. Mech. Mater.* **121-126** 3087.
[3] Yang H. L. 2011 *Adv. Mater. R*es.**189-193** 1443.
[4] Lee J. W., Lee Z. H .and Lee H. M.2005 *Mater. Trans.* **46** 2344.
[5] Alam S., Marshall R. I. and Sasaki S.. 1996 *Tribol Int* **29** 487.
[6] Simma L. 2009, *Chem. Pet. Eng.* **45** 649.
[7] Brown L. and Joyce P. 2012 *J. Test Eval.* **41**104.
[8] Samadi A. and Shahbazkhani H. R. 2014 *Int. J. Cast Metals Res.* **27** 129.
[9] Diouf P. and Jones A .2010 *Metall Mat Trans A Phys Metall Mat Sci* **41** 603.
[10] Cao Y., Zeng F., Liu B., Liu Y., Lu J.Z., Gan Z.Y. and Tang H.P. 2016 *Mater. Sci. Eng. A* **654** 418.
[11] Mironov , Stankevich , Tatarinov A., Zemchenkov V. and Boiko I. 2015 *IOP Conf. Ser. Mater. Sci. Eng.* **96** 012016.
[12] Dourandish M., Simchi A., Hokamoto K. and Tanaka S., 2010 *Mater. Sci. Eng. A* **527** 449.
[13] Costa F. A. D., Silva J. F. D, Silva A. G. P. D. and Gomes U. U., Alves J. C. 2008 *Int J Refract Met Hard Mater* **26** 207.
[14] Maneshian M. H. and Simchi A. 2008 *J Alloys Compd* **463** 153.
[15] Nadhrah M., Sitirabiatull A. and Mahadzir, I.M. 2016 *Proc IEEE CPMT Int Electron Manuf Technol IEMT Symp* **11** 7761946.

[16] Fan X. P, Wang B. J., Ren X. Q. and Peng F. C. 2014 *Mater Sci Forum* **815** 297.
[17] Chen H. Y. and Chen C. 2009 *J Electron Mater* **38** 338.
[18] Yamamoto K., Akahoshi H., Kato T., Kawamura T., Koizumi M. and Satoh R. 2008 *IEEE Trans. Compon. Packag. Technol.* **31** 849.
[19] Mittal J. and Lin K. L. 2012 *J Mater Res* **27** 1142.
[20] Yu C., Yang Y., Chen J. S., Xu J., Chen J. S., Xu J. J., Chen J. M. and Lu H. 2014 *Mater Lett* **128** 9.
[21] Elsaka S. E., Hamouda I. M., Elewady Y. A., Abouelatta O. B. and Swain M. V. 2010 *Dent. Mater.* **26** 793.
[22] Arash V., Anoush K., Rabiee S. M., Rahmatei M. and Tavanafar S. 2015 *Scanning* **37** 294.
[23] Liu H. and Chen W. 2008 *Surf. Coat. Technol.* **202** 4019.
[24] Li Y., Tao Y., Ke D., Ma Y. and Han S. 2015 *Appl Surf Sci* **357** 1714.
[25] Li B. S., Lin A. and Gan F. 2006 *Rare Met* **25** 645.
[26] Zhang H., Guo T. W., Song Z. X., Wang X. J. and Xu K. W. 2006 *Xiyou Jinshu Cailiao Yu Gongcheng* **201** 5637.
[27] Sun H., Yazdi M. A. P. , Ducros C., Chen S. C., Aubry E., Wen C. K., Hsieh J. H., Samchette . and Billard A. 2017 *Mater Sci Semicond Process* **63** 295.
[28] Kelly P. J., Li H., Benson P. S., Whitehead K. A., Verran J., Arnell R. D. and Iordanova I. 2010 *Surf. Coat. Technol.* **205** 1606.
[29] Li J., Zhang H., Fan A. and Tang B. 2016 *Surf. Coat. Technol.* **294** 30.
[30] Wu Y. P., Leng Y. X., Gui X., Sun H., Huang N., Zhu S. F., Zhang P. C. and Bai B. 2009 *Mater Sci Forum* **610-613** 647.
[31] Luo X. T., Wei Y. K, Wang Y. and Li C. J. 2015 *Mater. Des.* **85** 527.
[32] Kumar M., Singh H. and Singh N. 2013 *Surf Eng* **29** 419.
[33] Chen Y. M., Yu G. P. and Huang J. H. 2002 *Surf. Coat. Technol* **155** 239.
[34] Yoo J. H., Ahn S. H., Kim J. G. and Lee S. Y. 2002 *Surf. Coat. Technol* **157** 47.
[35] Tang J. and Azumi K. 2011 *Surf. Coat. Technol* **205** 3050.
[36] HussainT., Mccartney D. G., Shipway P. H. and Zhang D. 2009 *J Therm Spray Technol* **18** 364.
[37] Ning X. J., Kim J. H., Kim H. J., Lee C. H. 2009 *Appl Surf Sci* **255** 3933.
[38] Jan V., Čupera J. and Cizek J. 2015 *Surf. Coat. Technol* **268** 216.
[39] Tsega M., Kuo D. H. and Dejene F. B. 2015 *Thin Solid Films* **589** 712.
[40] Gabbitas B. L., Ariff T. F. and Cao P. 2011 *Powder Metall.* **54** 488.
[41] Vahdat S. E. 2016 *Arch. Foundry Eng.* **16** 131.
[42] Zamora R., Hernandez-Ortega J. J., Faura F., Lopez J. and Hernandez J. 2008 *J Manuf Sci Eng Trans ASME* **130** 65.

Acknowledgments:
This research was supported by National Natural Science Foundation of China 51705310, China Postdoctoral Science Foundation of 2017M611555, and Marine Low Speed Engine Project of CDGC01-KT11

Analytical and Numerical Design Analysis of Concentric Tube Heat Exchangers – A Review

Karthik Silaipillayarputhur, Tawfiq Al Mughanam, Abdulaziz Al Mojil and Mohammed Al Dhmoush

King Faisal University, Saudi Arabia

Abstract. This paper considers an analytical and a numerical approach in the design of a concentric tube heat exchanger. Sensible heat transfer is considered in the analysis and the heat exchanger is developed for actual operating conditions in a chemical plant. The heat exchanger is a concentric tube heat exchanger where hot oil exchanges heat with hot water. Hot oil is in the inner pipe and the heating medium, hot water, is in the outer pipe (annular side) of the heat exchanger. An analytical model employing effectiveness-number of transfer units (ε-NTU) approach and log mean temperature difference (LMTD) approach were employed in the design of the concentric tube heat exchanger. In the design process, performance charts were developed for concentric tube heat exchanger. Performance charts describe the performance of the heat exchanger in terms of crucial dimensionless parameters. Performance charts help to select the right number of transfer units (NTU) for the given heat exchanger. Both parallel and counter flow configurations were considered for the design analysis. Likewise, a numerical model was also considered in the design of the heat exchanger. The results from the analysis are presented and compared. From the results it can be seen that both numerical and analytical approaches produce the exact same results. The designer certainly has the flexibility to choose an appropriate design methodology based on the available inputs and requirements.

1. Introduction

Heat exchangers (HX) are used in all process and manufacturing industries worldwide. Heat exchanger equipment transfers heat from a hot fluid to a cold fluid. Herein both fluids are separated by means of a solid wall. Typically, most heat exchangers are two-fluid heat exchangers though three-fluid heat exchangers are becoming popular. Heat exchangers can be classified in terms of flow and construction. In terms of flow, heat exchangers can be classified as parallel flow, counter flow and cross flow. In terms of construction, they can be classified as shell and tube, concentric tube and finned tube heat exchangers. The choice of a heat exchanger for a given application is dependent on the application itself, available resources, space, existing connections in the field, etc. Whatever may be the choice of a heat exchanger, it is very essential that a heat exchanger be designed such that it delivers the required heat transfer while occupying less space, being light weight, and yet be priced competitively.

In this project, a concentric tube heat exchanger is designed for heating hot oil. Hot oil is used for a certain process heating application and hot water is the heating medium. The heat exchanger will be designed such that the hot oil shall flow through the inner pipe and hot water through the outer pipe (annular side) of the heat exchanger.

A conventional ε-NTU approach, LMTD approach and a numerical approach are employed in design the concentric tube heat exchanger. Both parallel and counter flow configurations are considered in the design analysis. Herein, performance charts were developed to aid the design process. Performance charts describe the variation of heat exchanger effectiveness with respect to capacity rate ratio and number of transfer units (NTU). Since all parameters in performance charts are expressed in dimensionless basis, the developed performance charts are applicable for any system of units, inlet fluid temperatures, fluid flow rates, materials of construction and size. Each design approach has unique characteristics and these aspects are discussed in the subsequent sections of the paper.

There are numerous references available in the literature pertaining to heat exchanger performance modelling, and only the most pertinent are discussed. Kays and London [1] and Rohsenow [2] described both the logarithmic mean temperature difference and the methods in order to size and predict the performance of a heat exchanger. Domingos [3] presented a general method of calculating overall performance and intermediate temperatures of complex crossflow heat exchangers using the concept of effectiveness and a local energy balance. Pignotti and Shah [4] and Shah and Pignotti [5] discussed the tools developed previously (such as Domingos' method, the Pignotti chain rule, etc.,) to determine the relationship for highly complex heat exchanger flow arrangements. As compared with the present investigation, these studies pertained to quite different geometries such as cross flow and shell and tube heat exchangers. Furthermore, they did not address the design of optimal heat exchangers, which achieve the required task at the lowest cost while satisfying imposed constraints. Mott and Mills [6] and Genic et al. [7] are among those researchers who described optimization analysis based primarily on minimizing energy costs related to pumping of a fluid. Kovarik [8] described a technique to optimize a cross flow heat exchanger. The objective function which was employed included cost factors related to the heat exchanger size and pumping power, as well as the required heat transfer rate. Similarly Rao et al. [9] and Caputo et al. [10] proposed methods to minimize capital and operating costs for shell-and-tube heat exchangers while satisfying the required heat transfer duty. Silaipillayarputhur and Idem [11] considered a numerical approach in the design and optimization of a double pipe heat exchanger. Therein, the heat exchanger was optimized such that the capital cost and operation cost of the heat exchanger were minimized.

2. Nomenclature

Symbol	**Name**
A_o	Heat exchanger surface area (m^2)
C_{min}	Minimum capacity rate fluid (W/K)
C_{max}	Maximum capacity rate fluid (W/K)
C_r	Capacity rate ratio (dimensionless)
C_p	Specific heat (J/kg·K)
D_H	Tube hydraulic diameter (m)
d_{in}	Tube inside diameter (m)
d_{out}	Tube outside diameter (m)
h_i	Internal heat transfer coefficient (W/m^2K)
h_o	External heat transfer coefficient (W/m^2K)
$\dot{m}$	Mass flow rate (kg/s)

K	Thermal conductivity (W/m·K)
L	Tube length (m)
Nu	Nusselt Number (dimensionless)
NTU	Number of transfer units (dimensionless)
Pr	Prandtl Number (dimensionless)
Q	Rate of heat transfer (W)
Re	Reynolds number (dimensionless)
T	Temperature (K)
$\Delta T_{\ell m}$	log mean temperature difference (K)
V	Velocity (m/s)
R_f	Fouling factor
f	Darcy friction factor
f	Function, dimensionless
U_o	Overall heat transfer coefficient (W/m^{2}K)

Greek Symbols

ε	Heat exchanger effectiveness
ε_1	Surface roughness of pipe, m
μ	Dynamic Viscosity (N·s/m^2)
ρ	Density (kg/m3)
ΔT	Temperature Difference (K)
ΔP	Pressure drop in pipe (Pa)

3. Analytical Approach

This project considers the design of a concentric tube heat exchanger wherein the hot oil is in the inner pipe of the heat exchanger and hot water is in the outer pipe (annular side) of the heat exchanger. The hot oil is used for a certain heating application in the process plant. It is proposed to pre-heat the oil using the available "waste" hot water in the facility. It is desired to raise the temperature of 400 lbm/hr of hot oil from 90°F to 100°F by employing hot water that is available at 5000 lbm/hr at 150°F. The hot oil employed in the study is a heat transfer fluid, rated for open systems, having a viscosity grade of ISO 46. The thermophysical properties of the fluids employed in the heat exchanger are described in Tables 1 and 2 respectively.

IOP Conf. Series: Materials Science and Engineering **272** (2017) 012006 doi:10.1088/1757-899X/272/1/012006

Table 1. Hot Oil Thermal Properties.

ρ (kg/m^3)	$\dot{m}(kg/s)$	**C_P (J/kg-°C)**	**k (W/m-°C)**	**μ (N-s/m^2)**
761	0.0504	2558	0.133	0.00117

Table 2. Hot Water Thermal Properties.

ρ (kg/m^3)	$\dot{m}(kg/s)$	**C_P (J/kg-°C)**	**k (W/m-°C)**	**μ (N-s/m^2)**
1000	0.63	4180	0.65613	0.000446

The design constraints for the project are described in Table 3. The design constraints are heat transfer duty, heat exchanger diameter, and length. An additional design constraint, allowable pressure drop, will be considered while employing the numerical model. Employing this constraint in the conventional ε-NTU approach and LMTD approach will make the design process a little tedious. Therefore, in practical applications, this constraint is examined in terms of flow velocity. In real life applications, to keep the pressure drop within reasonable limits, it is a common accepted practice to keep the velocity of flow in the pipe to be around 3 m/s.

Table 3. Design Constraints.

Design Constraints	
Parameter	**Condition**
Discharge oil temperature	37°C
Max. heat exchanger external diameter	75 mm (2.5 in NPS)
Max. heat exchanger length	1.2 m (4 ft)
Allowable pressure drop, oil side*	500 Pa
* - for Numerical design analysis	

The assumptions employed in the analysis are that the heat excahnger is operating at steady state, the properties of the fluids and the heat exchanger wall remains constant, and the heat exchanger is operating adiabatically.

The rate of heat transfer or the heat transfer duty, being one of the design constraint can be determined as follows [12]

$$Q = \dot{m} c_p \Delta T \tag{1}$$

Here, the term ΔT refers to the absolute temperature difference between the fluid at inlet and at exit of the heat exchanger. Assuming the concentric tube heat exchanger to be well insulated it can be assumed that the rate of heat gained by the hot oil will be equal to rate of heat lost by the hot water such that

$$Q = Q_{oil} = Q_{hotwater} \tag{2}$$

The average specific heat and the mass flow rate of both the fluids rates are known. The required discharge temperature of the hot oil is known as well. Therefore, the rate of heat transfer and the discharge temperature of hot water can be readily determined by employing Equations (1) and (2).

The capacity rate ratio for the given arrangement may be given as [12]

$$C_r = \frac{C_{min}}{C_{max}} = \frac{(\dot{m} c_p)_{oil}}{(\dot{m} c_p)_{hotwater}} \tag{3}$$

Here, C_{min} corresponds to the capacity rate of minimum capacity rate fluid, which is hot oil and C_{max} corresponds to the capacity rate of maximum capacity rate fluid, which is hot water. The values for C_{min} and C_{max} can be readily computed as the pertinent information are available. The heat exchanger effectiveness may then be determined by employing the following equation [12]

$$\varepsilon = \frac{Q}{Q_{max}} = \frac{Q}{C_{min}\left[T_{Hotwater,inlet} - T_{oil,inlet}\right]} \tag{4}$$

Here, the rate of heat transfer Q is given by Equation (1). For a concentric tube heat exchanger subjected to parallel flow, the heat exchanger effectiveness may also be given as [12]

$$\varepsilon = \frac{1 - \exp\left[-NTU(1 + C_r)\right]}{1 + C_r} \tag{5}$$

Likewise, for a concentric tube heat exchanger subjected to counter flow, the heat exchanger effectiveness may also be given as [12]

$$\varepsilon = \frac{1 - \exp\left[-NTU(1 - C_r)\right]}{1 + C_r \exp\left[-NTU(1 - C_r)\right]} \tag{6}$$

Number of transfer units (NTU) is a dimensionless parameter that is widely used by process engineers and heat exchanger designers. NTU is a physically significant dimensionless parameter as it encompasses material characteristics, fluid characteristics, flow characteristics, heat exchanger size, fouling, etc. Using Equation (5) and (6), for a range of NTU varying between 0.1 and 10, and capacity rate ratios varying between 0 and 1 the heat exchanger effectiveness can be readily plotted and performance charts can be developed for the concentric tube heat exchanger. The following figures 1 and 2 describe the performance charts for concentric tube heat exchanger. Using the performance charts, the required NTU for the heat transfer duty can be determined.

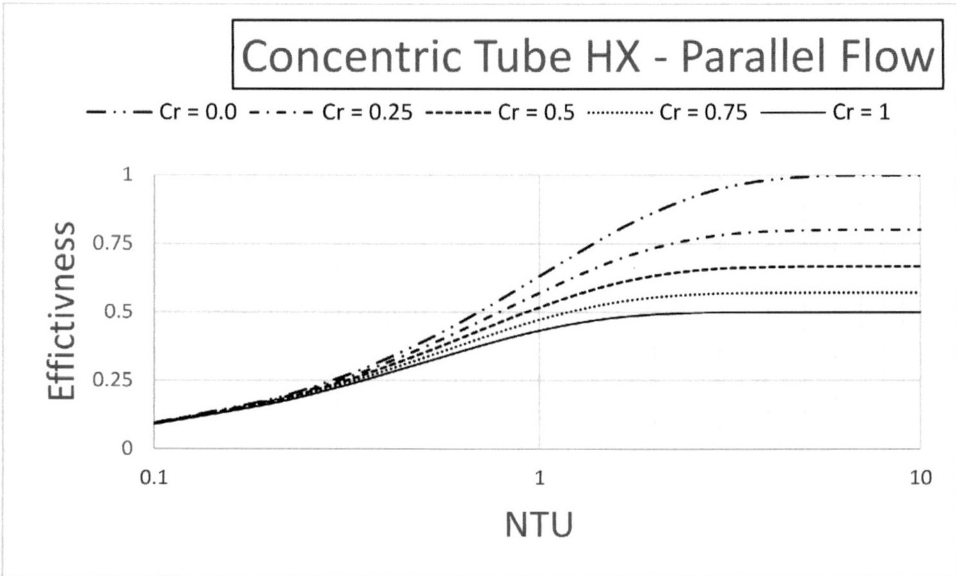

Figure 1. Performance of concentric tube heat exchanger subjected to parallel flow.

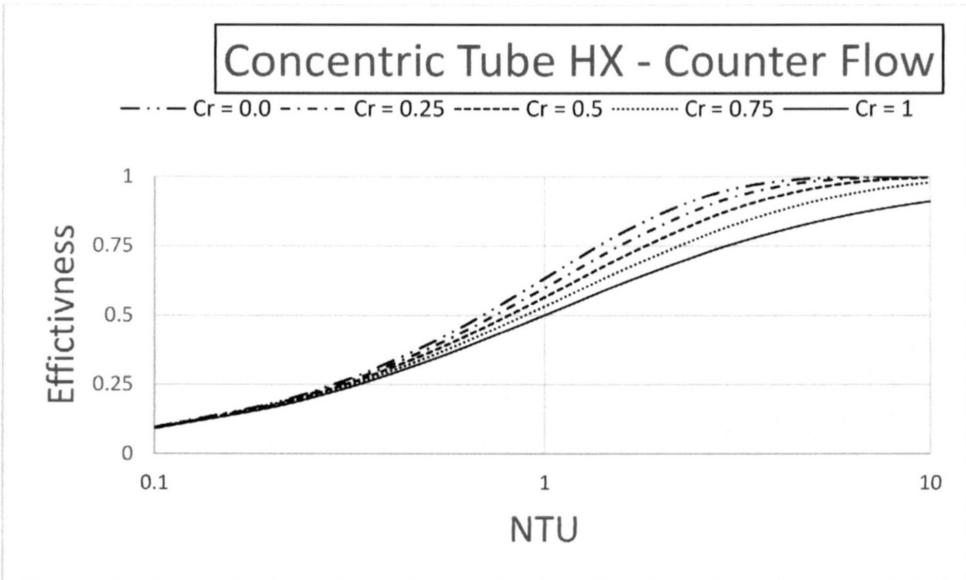

Figure 2. Performance of concentric tube heat exchanger subjected to counter flow.

NTU is a dimensionless parameter that accounts for material characteristics, flow characteristics, size, construction, fouling, etc. Therefore, NTU is a physically significant dimensionless parameter that is used by engineers during the design phase of the heat exchanger. Likewise, it must be recognized that higher the NTU, the more would the area, material, size, weight and cost. In keeping up with the competition, it is important to design the equipment such that it is compact, less weight, low cost and yet deliver the required heat transfer. Performance charts described in Figures 1 and 2 shall help in optimizing the heat exchanger during the development phase. In addition, it is a common belief that increasing the surface area enhances the rate of heat transfer. However, this is true only until a threshold limit and increasing the surface area (i.e., NTU) beyond that limit, adds unnecessary weight and cost. Figures 1 and 2 is certainly helpful in choosing the right surface area (i.e., NTU) for the given application.

The determination of internal and external heat transfer coefficients are described extensively in [11] and [12] and therefore they are not discussed herein.

Considering the fouling resistances, and conduction resistances, the overall heat transfer coefficient may be given as [12]

$$\frac{1}{U_o} = \frac{1}{h_o} + R_{fo} + \frac{r_{in,out}}{k_{pipematerial,in}} \ln \frac{r_{in,out}}{r_{in,in}} + \frac{r_{in,out}}{r_{in,in}} R_{fi} + \frac{r_{in,out}}{r_{in,in}} \frac{1}{h_i} \tag{7}$$

Recall that the required NTU for the heat exchanger was determined using performance charts. The NTU of a heat exchanger can also be described as [12]

$$NTU = \frac{U_o A_o}{C_{min}} \tag{8}$$

Hence, from Equation (8), the required surface area of the heat exchanger can be readily determined. The surface area of the heat exchanger can be described as

$$A_o = \pi d_{in,out} L \tag{9}$$

From Equation (9) the required length for the concentric tube heat exchanger can be determined.

If log mean temperature difference (LMTD) approach were to be used, the rate of heat transfer may be described as [12]

$$Q = U_o A_o \Delta T_{lm} \tag{10}$$

The computation of log mean temperature difference $\Delta T_{\ell m}$ for parallel and counter flows are described in detail in [12] and therefore not included herein.

For LMTD approach, the required heat transfer surface area is determined by employing Equation (10) and the required length of the concentric tube heat exchanger can be determined by employing Equation (9).

Using the equations described in this section a MATLAB model was developed to design the concentric tube heat exchanger.

4. Numerical Approach

This section of the document describes the governing equations required to design the heat exchanger using numerical approach. Only the hot oil side is considered in the analysis, as heating up of hot oil is the primary focus in the project. The dimensions of the inner pipe and the flow velocity of oil are determined such that the required heat transfer duty is satisfied for a given mass flow rate and for an allowable pressure drop. From Equation (1), for the given conditions, the temperature difference in the hot water between the inlet and discharge was found to be negligible. This means that the hot water temperature remains approximately a constant while flowing through the concentric tube heat exchanger. Hence, it is reasonable to assume that the heat exchanger wall is at constant temperature during the steady state operation of the heat exchanger.

Considering the heat exchanger to opearte adiabatically, an overall energy balance yields the heat transferred from the tube wall to the hot oil [11, 12]

$$Q = \dot{m}_{hotoil} c_{p,hotoil} \left(T_{hotoil,out} - T_{hotoil,in}\right) = \pi D_{H,inner} L h_i \Delta T_{\ell m} \tag{11}$$

The log-mean temperature difference is the mean temperature difference between the tube wall and the hot oil [11, 12]

$$\Delta T_{\ell m} = \frac{\left(T_w - T_{hotoil,in}\right) - \left(T_w - T_{hotoil,out}\right)}{\ell n\left(\frac{T_w - T_{hotoil,in}}{T_w - T_{hotoil,out}}\right)} \tag{12}$$

The empirical Dittus-Boelter correlation for turbulent convection heating can be given as [11, 12]

$$Nu = \frac{h_i D_{H,inner}}{k_{hotoil}} = 0.023 Re^{0.8} Pr^{0.4} \tag{13}$$

The Darcy friction factor represents the dimensionless pressure loss per unit length in a pipe and can be expressed in terms of the mean oil velocity [11, 12, 16]

$$f = \frac{\Delta P / L}{\frac{1}{2}\rho_{hotoil} V_{hotoil}^{2} / D_{H,inner}} \tag{14}$$

An expression from Haaland [11, 13, 15, 16] provides an explicit relation between the friction factor, pipe relative roughness, and the Reynolds number

$$\frac{1}{\sqrt{f}} = -1.8\log\left[\left(\frac{\varepsilon_1 / D_{H,inner}}{3.7}\right)^{1.11} + \frac{6.9}{Re}\right] \tag{15}$$

The Reynolds number may be defined as [11, 12]

$$Re = \frac{\rho_{hotoil} V_{hotoil} D_{H,inner}}{\mu_{hotoil}} \tag{16}$$

The mean oil velocity in the pipe is determined from the continuity equation [11, 16]

$$V_{hotoil} = \frac{4\dot{m}_{hotoil}}{\rho_{hotoil} \pi D_{H,inner}^{2}} \tag{17}$$

Upon rearranging, the energy conservation can be expressed in dimensionless as follows [11]

$$f_1 = \frac{\dot{m}_{hotoil} c_{p,hotoil}}{k_{hotoil} L} - 0.023\pi \left(\frac{\rho_{hotoil} V_{hotoil} D_{H,inner}}{\mu_{hotoil}} \right)^{0.8} \left(\frac{\mu_{hotoil} c_{p,hotoil}}{k_{hotoil}} \right)^{0.4} \frac{1}{\ln\left(\frac{T_w - T_{hotoil,in}}{T_w - T_{hotoil,out}} \right)} = 0 \quad (18)$$

Upon rearranging Equations (14) and (15), the following dimensionless equation can be obtained [11]

$$f_2 = \sqrt{\frac{\rho_{hotoil} V_{hotoil}^2 L}{2\Delta p D_{H,inner}}} + 1.8 \log \left[\left(\frac{\varepsilon_1 / D_{H,inner}}{3.7} \right)^{1.11} + 6.9 \frac{\mu_{hotoil}}{\rho_{hotoil} V_{hotoil} D_{H,inner}} \right] = 0 \quad (19)$$

Upon rearranging Equation (17), the following dimensionless equation can be obtained [11]

$$f_3 = \frac{\rho_{hotoil} V_{hotoil} D_{H,inner}^2}{\dot{m}_{hotoil}} - \frac{4}{\pi} = 0 \quad (20)$$

The quantities f_1, f_2, and f_3 represent simultaneous nonlinear algebraic equations whose numerical roots are to be determined by employing techniques such as Newton Raphson method, Secant method, etc. MATLAB software has an inbuilt solver for solving simultaneous nonlinear algebraic equations. By solving the three simultaneous non linear equations, the three unknowns, namely, the pipe diameter, length and pipe velocity can be determined.

5. Results

Based on the analytical (ε-NTU and LMTD) approach and numerical approach as discussed in the sections 3 & 4 respectivily, Matlab models were developed to design the concentric tube heat exchanger. The heat exchanger was subjected to the input conditions as described in section 3 and the design constraints as described in Table 3. The results from both models for parallel and counter flow concentric tube heat exchanger are described in Table 4.

Table 4. Description of Concentric Tube Heat Exchanger.

Design of Concentric Tube Heat Exchanger		
Item	**Inner Side**	**Annular Side**
Fluid	**Hot Oil**	**Hot Water**
Inlet Temp (°C)	32	65
Discharge Temp (°C)	37	64.8
Mass flow rate (kg/s)	0.0504	0.63
Heat Transfer (W)	721	721
Pressure drop, oil side (Pa)*	250	NA
Material	Carbon Steel, SMLS A53B	Carbon Steel, ERW A 53B
Diameter (in) - Analytical	1/2	1 1/4
Diameter (in) - Numerical	1/2	NA
Length (m) - (ε-NTU) approach, parallel flow	0.81	0.81
Length (m) - (ε-NTU) approach, counter flow	0.81	0.81
Length (m) - LMTD approach, parallel flow	0.81	0.81
Length (m) - LMTD approach, counter flow	0.81	0.81
Length (m) - Numerical approach	0.81	0.81
* - for numerical design analysis		

It can be clearly seen that both the analytical and numerical approach deliver the same results for the prescribed conditions. Though from performance charts, it can be clearly seen that the counterflow heat exchangers perform significantly better than the parallel flow heat exchangers, for the given application, there is absolutely no difference in the results between counter and parallel flow configurations. This is due to the prevailing uniform temperature of the hot fluid in the heat exchanger. The material selection for the heat exchanger was based on DuPont Engineering Standards as prescribed in [14].

6. Conclusions

This paper reviews the analytical and numerical approach in the design of a concentric tube heat exchanger. Therein, both parallel and counter flow configurations are analyzed. In the design process, performance charts were developed for concentric tube heat exchanger. Performance charts help the engineers to quickly determine the performance of the heat exchanger without performing tedious calculations. Likewise, performance charts help the engineers to determine the required NTU for the heat exchanger. NTU accounts for the physical characteristics of the heat exchanger, fluid and thermal properties, type of heat exchanger and fouling. Since NTU accounts for all the pertinent parameters, optimizing NTU is an essential item in the design of the heat exchanger. An over sized exchanger (having higher NTU) will certainly incur higher capital and operational costs and while an under sized heat exchanger (having lower NTU) shall certainly not deliver the required heat transfer. Performance charts developed in this paper will help the designers and engineers by providing this crucial information.

In addition, both analytical and numerical approaches are reviewed in this paper. Numerical approach accounts for tube side pressure drop and the heat exchanger can be designed such that this design criterion is satisfied. Analytical approach consisting of ε-NTU and LMTD methods do not explicitly satisfy this criterion. Likewise, employing the analytical approach can help to optimize the heat exchanger by choosing the right NTU for the given application. However, if the required discharge temperatures are unknown, it is quite tedious to model using LMTD method as the design process shall be iterative. Therefore, based on the available information and the required specifications, the designers can choose an appropriate method for the design of a concentric tube heat exchanger.

7. References

[1] Kays, W. M., and London A. L., Compact Heat Exchangers, 3rd Ed., McGraw-Hill, New York, 1984.
[2] Rohsenow, W. M., Heat Exchangers – Basic Methods, in Heat Exchangers Thermal-Hydraulic Fundamentals and Design, Hemisphere Publishing Corporation, Washington D.C., 1982.
[3] Domingos, J. D., Analysis of Complex Assemblies of Heat Exchangers, Int. J. Heat Mass Transfer, Vol. 12, pp. 537-548, 1969.
[4] Pignotti, A. and Shah, R. K., Effectiveness-number of transfer units relationships for heat exchanger complex flow arrangements, Int. J. Heat Mass Transfer, Vol. 35, No. 5, pp. 1275-1291, 1992.
[5] Shah, R. K. and Pignotti, A., Thermal Analysis of Complex Crossflow Exchangers in Terms of Standard Configurations, J. Heat Transfer, Vol. 115, pp. 353-359, 1993.
[6] Mott, J. E. and Mills, R. R., Computerized Design of a Minimum Cost Heat Exchanger, ASME Paper 72-HT-26, 1972.
[7] Genic, S. B., Jacimovic, B. M., and Genic, V. B., Economic optimization of pipe diameter for complete turbulence, Energy and Buildings, Vol. 45, pp. 335-338, 2012.
[8] Kovaric, M., Optimal Heat Exchangers, J. Heat Transfer, Vol. 111, pp. 287-293, 1989.

[9] Ramananda Rao, K., Shrinivasa, U., and Srinivasan, J., Synthesis of Cost-Optimal Shell-and Tube Heat Exchangers, Heat Transfer Engineering, Vol. 12, No. 3, pp. 47-55, 1991.
[10] Caputo, A. C., Pelagagge, M. P., and Salini, P., Heat Exchanger Design Based on Economic Optimization, Paper CIT06-0387, ENCIT 2006 -ABCM, Curitiba, Brazil, 2006.
[11] Silaipillayarputhur, K., Idem, S., "Design of a single tube fuel oil preheater", Heat Transfer Engineering, Volume 38, Issue 6, page 1396-1403, Taylor and Francis, USA.
[12] Bergman, T. L., Lavine, A. S., Incropera, F. P., and DeWitt, D. P., Fundamentals of Heat and Mass Transfer, 7th Ed., John Wiley, Hoboken, NJ, 2011.
[13] Haaland, S. E., Simple and Explicit Expressions for the Friction Factor in Turbulent Pipe Flow, Journal of Fluids Engineering, vol. 105, pp. 89-90, 1983.
[14] DuPont Technology Consulting; E.I. du Pont Nemours Company; Engg Standards; 2002 revision.
[15] Douglas, J. M., Conceptual Design of Chemical Processes, pp. 327-328, McGraw Hill, New York, 1988.
[16] Janna, W. S., Design of Fluid Thermal Systems, 2nd Ed., pp. 162-163, PWS Publishing Company, Boston, 1998.

IOP Conf. Series: Materials Science and Engineering **272** (2017) 012007 doi:10.1088/1757-899X/272/1/012007

Internal Flow Field Uniformity Study of Dust Collector for A Street Vacuum Sweeper Based on CFD

Yuan Xi[1,2], Yan Dai[1], Yonghou Xiao[1], Kai Cheng[2,*], Tao Xiao[3], Shixiang Zhao[4].

[1] Panjin Industrial Technology Institute of DUT,NO.2 Dagong Road, Panjin, China.
[2] School of Mechanical Science and Engineering, Jilin University, NO.5988 Renmin Street, Changchun,China.
[3] FAW Jiefang Automotive Co.,Ltd. , NO.3082 Jincheng Street,Changchun,China.
[4] Jilin Zijin Copper Industry Co., Ltd, NO.18 Village, Hunchun,China.

* E-mail: 2932532935@qq.com

Abstract. Internal flow field of dust collector for a street vacuum is studied by Computational Fluid Dynamics (CFD). The causes of uneven distribution of flow field were diagnosed by simulation, and an improvement measure was put forward. The result shows that flow ratce distribution of dust collector is uneven, the reason for which is the higher wind velocity at the Inlet. A baffle plate was added at the Inlet to improve the uniformity of internal flow field. The main technical parameters, maximum flow rate uneven amplitude Δk_i and overall flow rate uneven amplitude $\overline{\Delta k_i}$ and overall flow rate uneven amplitude, are decreased in this way. Δk_i is reduced from 0.3429 to 0.1308. While $\overline{\Delta k_i}$ is reduced from 0.1012 to 0.0402.

1. Introduction

City sanitation attracts more attention for the past few years. Sweeper is typically practiced to remove the dust accumulation from road surface [1]. Li-Ming Lo found higher pleat ratios is effective for particle collection efficiency [2]. Some studies used fabric filter bags [3] and ceramic filters [4] to improve operating conditions of regenerable filtering systems, but fewer studies focus on sweeper dust collector [5]. The sweeper dust collector is modeled and simulated by using CFD technology. The simulation results were analyzed and discussed, and the distribution regularities of internal flow field were finally summarized, which provides theoretical support for the design and development for sweeper dust collector.

2. Numerical simulation

2.1. Model and grid generation

Fig1 shows the actual picture of dust collector, geometric model and mesh model. Dusty gas enters from Inlet（the purple in Fig 1（b））. The cartridges, the green in Fig 1（b）, make dusty gas flow from outside to inside to collect the dust. The filtered air is finally discharged to atmosphere at the Outlet（the red in Fig 1（b））. Cartridge model was established as cylinder because of complicated pleated structure [5-6].The main size parameters are listed in Table1. Fig1(c) shows the mesh model.

Table1. The main size parameters.

Parameter	Dimensions (mm)
Geometric Model	800×800×1500 (length×width×height)
Inlet	300×230 (length×width)
Outlet	600×250 (length×width)
Cartridge length	660
Cartridge diameter	325

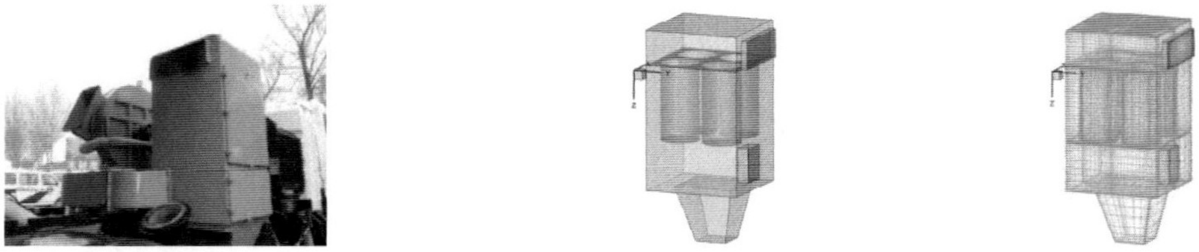

(a) Actual Picture (b) Geometric Model (c) Mesh Model

Figure 1. Schematic diagram of dust collector for a street vacuum sweeper.

2.2. Boundary conditions

The flow field can be simplified as a single-phase flow because of particle low concentration [7-9]. SIMPLE method, and second-order upwind interpolation scheme were chosen for pressure–velocity coupling. Inlet surface was treated as the velocity inlet, and its velocity was set as 5.95 m/s based on the KASDA-KV621 hot-wire anemometer measured values. Outlet surface was treated as the pressure out, and its pressure was set as the atmosphere based on connecting to the atmosphere.

Porous jump model was used to simulate porous medium of cartridge. The one-dimensional model provides the advantages of robustness and better convergence [10-11], and the static wall was applied to the rest of the dust collector shells. The main Face Permeability is 1.95×10^{-9} m^2 and Medium Thickness is 1 mm.

3. Parameters of flow field distribution

3.1. Unevenness

Uniformity is evaluated with American RMS standard, and the formula can be expressed as follows:

$$\delta=\left(\frac{1}{n}\sum_{i=1}^{n}\left(\frac{v_i-v_p}{v_p}\right)^2\right)^{1/2} \quad (1)$$

Where v_i is the each point velocity, v_p is the cross section average velocity, and n is the amount of points. The uniformity of flow field is qualified when δ is not more than 0.25 [12].

3.2. Flow rate distribution coefficient

K_i is flow rate distribution coefficient of each cartridge, and the formulas can be expressed as follows:

$$K_i=Q_i/\overline{Q} \quad (i=1,\ 2,\ 3,\ \ldots\ldots,\ n) \quad (2)$$

Where Q_i is the flow rate through a single cartridge. $\overline{Q}$ is the average flow rate.

3.3. Maximum and overall flow rate uneven amplitude

Δk_i is the maximum flow rate uneven amplitude, and $\overline{\Delta k_\xi}$ is overall flow rate uneven amplitude. The formulas can be expressed as follows:

IOP Conf. Series: Materials Science and Engineering **272** (2017) 012007 doi:10.1088/1757-899X/272/1/012007

$$\Delta k_i = k_{i\max} - k_{i\min} = Q^+ - Q^- \tag{3}$$

$$\Delta \overline{k_\xi} = \sum\left(\left|K_i - 1.0\right| / N\right) \tag{4}$$

Where $k_{i\max}$ is the maximum flow rate distribution coefficient, $k_{i\max}$ is the minimum flow rate distribution coefficient, Q^+ and Q^- are the maximum positive and negative deviation of flow rate distribution, respectively. N is the amount of cartridges. The smaller the Δk_i is, the better the uniformity of flow field is [7].

4. Modeling verification and analysis

4.1. Numerical modeling verification

Fig 2 shows cartridge number and location of the monitoring points (Row 1 &3). Cross section Z=400 and Z=600mm were selected. Numerical and experimental results, velocities of 16 monitoring points on Row 1and Row 3, were compared.

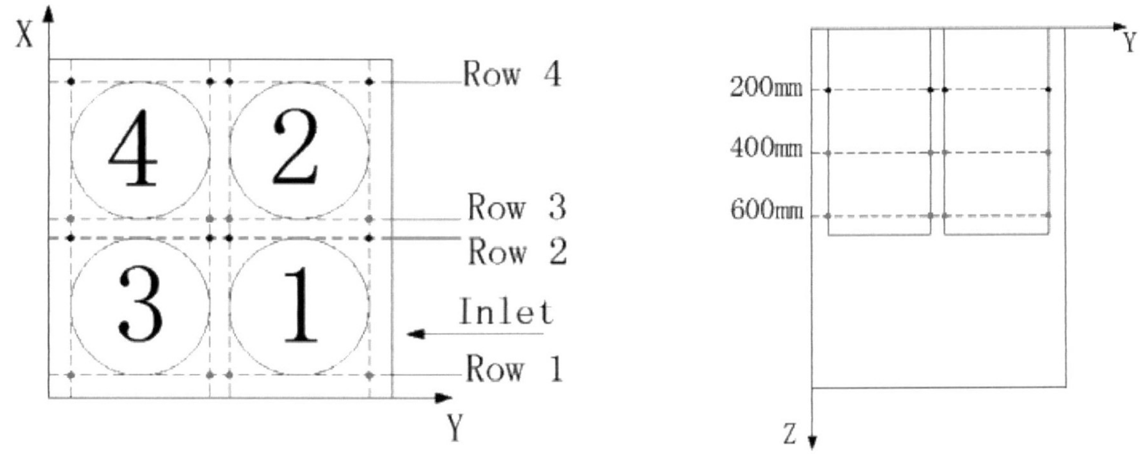

Figure 2. Distribution of monitoring points.

Fig 3 shows the comparison between numerical and experimental values. 4 points are selected because sensors` position effect the flow field. And maximum relative error of the comparison is 12.8%. The result shows that the simulation precision is high and reliable. The reasons which cause experiment errors are as follows: （1）Porous jump model was used; therefore the flow was treated as isotropic [11].（2）The sensors position have effects on the internal flow field.

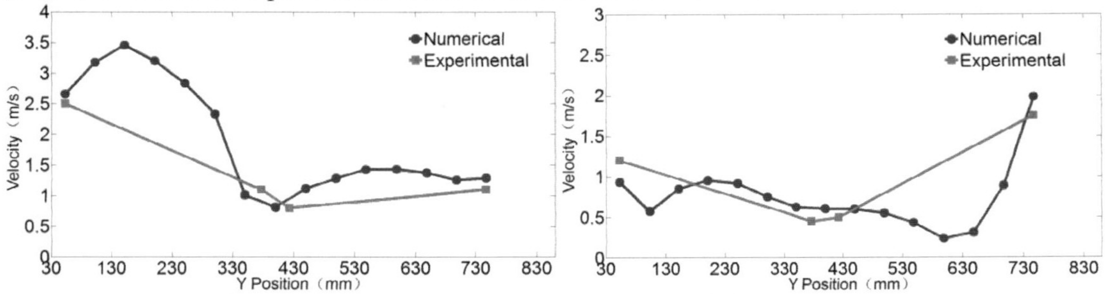

（a）Z=400mm, Row 1 （b）Z=400mm, Row 3

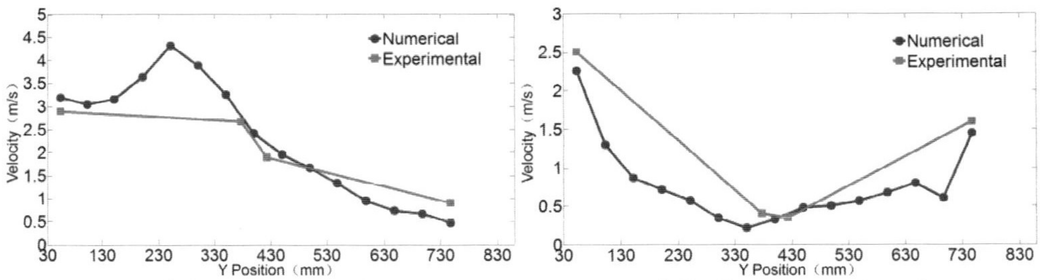

(c) Z=600mm, Row (d) Z=600mm, Row 3

Figure 3. Comparison between simulation and experimental results.

4.2. Analysis of the velocity field

As is shown in Fig 4 (a) (b) (c), the "climbing" phenomenon appears because airflow rises along the side walls gradually. This can be explained by the fact that the resistance is smaller between the cartridges and the side walls, and dusty air flows along the direction of smaller resistance. The velocities of filtered air reduce significantly because of the existence of cartridges.

4.3. Analysis of parameters of flow field distribution

According to the equation (1), the unevenness of cross section Z=200, Z=400, and Z=600mm can be obtained. The unevenness of Z-axis Section is shown in Table 2.

Table 2. Unevenness of Z-axis Section.

Unevenness	Z-axis Section (mm)		
	Z=200	Z=400	Z=600
δ	0.59	0.65	0.76

The distribution of internal flow field is uneven because of $\delta > 0.25$. The reasons for this phenomenon may lie in the fact that the big change of section area when dusty air enters from the Inlet, and dynamic - static pressure converts drastically, which leads to the disorder of local airflow. In order to improve the uniformity of the internal flow field, and service life of the cartridge; therefore the orifice or baffle plate should be added at the Inlet to achieve dynamic - static pressure conversion.

Flow distribution coefficient is shown in Table 3. Flow rate through cartridge 1 and cartridge 2 is larger, while flow rate through cartridge 3 and cartridge 4 is smaller. According to the formulas (3), (4) and (5), the maximum flow uneven amplitude Δk_i is 0.3429, and overall flow uneven amplitude $\overline{\Delta k_\xi}$ is 0.1012, Neither Q^+ =17.45% nor Q^- =-16.84%is in the error range of -15% to 15%. The results show that uniformity of flow field is poor and flow distribution is uneven.

Table 3. Flow distribution coefficient of the original model

Flow distribution coefficient	Cartridge number			
	1	2	3	4
K_i	1.0278	1.1745	0.9660	0.8316

4.4. An improvement measure

Considering interference between the sweeper and the dust collector, the structure of the dust collector should not be changed drastically. In order to reduce the velocities of Inlet, a baffle plate is employed at the Inlet. The final program is determined after several repeated simulations and optimizations. Actual picture of the improvement measure is shown in Fig 5 (a), and Fig 5 (b) shows the partial enlargement of baffle plate.

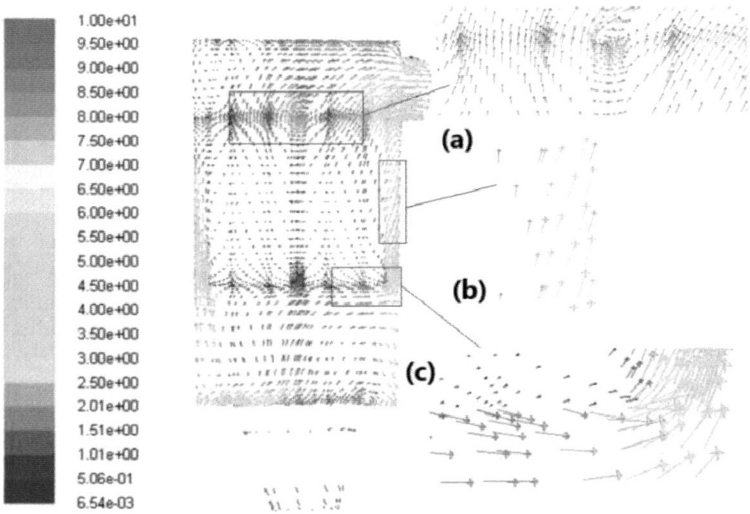

Figure 4. Velocity vector diagram of X=560mm.

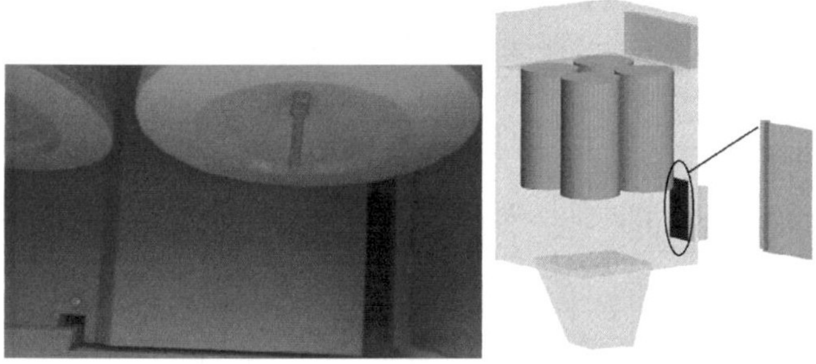

Figure 5. Improvement measure.

Flow distribution coefficient of the improved model is shown in Table 4. The maximum flow rate uneven amplitude is 0.1308, and overall flow rate uneven amplitude is 0.0402. Both Q^+ =6.63% and Q^- =-6.45% are in the error range of -15% to 15%. The results show that flow rate distribution is even. The intensity of jet flow becomes weaken because of the existence of the baffle plate, the distribution of internal flow field is even, and the erosion of cartridges is prevented.

Table 4. Flow distribution coefficient of the improvement model.

Flow distribution coefficient	Cartridge number			
	1	2	3	4
K_i	0.9841	1.0663	0.9355	1.0141

5. Conclusion

(1) The maximum flow rate uneven amplitude Δk_i is 0.3429, and overall flow rate uneven amplitude is $\Delta \overline{k_\xi}$ 0.1012. Neither $Q^+ = 17.45\%$ nor $Q^- = -16.84\%$ is in the error range of -15% to 15%. So the uniformity of flow field is poor and flow rate distribution is uneven.

(2) The main technical parameters are Δk_i =0.1308, $\Delta \overline{k_\xi}$ =0.0402, Q^+ =6.63%, and Q^- =-6.45% after adding a baffle plate at the Inlet. The improvement measure makes the flow filed more uniform without changing the structure drastically, which is economical and practical.

6. Acknowledgments

This work is supported by Natural Science Foundation of Liaoning Province of China(No.20170520354,20170520148) and the Special Fund Project of Panjin Industrial Technology Institute of DUT (NO.YJYZXZJ2017004).

7. References

[1] Khanal R, Furumai H, Nakajima F. 2014 *Journal of hazardous materials* **264** 53
[2] Li-Ming Lo, Shih-cheng Hu, Da-Ren Chen,David Y.H. Pui. 2010 *Powder Technology* **198** 75
[3] H.C. Lu, C.J. Tsai. 2017 *APPLIED SURFACE SCIENCE* **416** 639.
[4] HLF Magalhães, AGBD Lima, SRDF Neto, HG Alves, JSD Souza. 2017 *Advances in Mechanical Engineering* **9** 1.
[5] ZHENG Juan 2013 *Journal of Zhejiang Ocean University (Natural Science)* **32** 16
[6] Qing Zhang, Yun-lou Qian, Yuan-xia Bi 2014 *Chinese Journal of Environmental Engineering* **8** 2975
[7] Hai-ming Fu,You-jun Zhao 2010 *Journal of Central South University (Science and Technology)* **41** 799
[8] Le Cloirec P, Bellettre J, Subrenat A 2003 *Chemical engineering science* **58** 4965
[9] Chun-Wan Chen,Sheng-Hsiu Huang,Che-Ming Chiang,Ta-Chih Hsiao,Chih-Chieh Chen 2008 *Annals of Occupational Hygiene* **52** 207.
[10] Li-Ming Lo,Shih-Cheng Hu,Da-Ren Chen,David Y.H.Pui 2010 *Powder Technology* **198** 75
[11] Zhou R, Shen H, Zhao M 2012 *Energy Procedia* **16** 426
[12] Wang Chun, Wang dian-yin 2009 *Dust removal equipment manual*(Beijing：Chemical Industry Press), p 211

IOP Conf. Series: Materials Science and Engineering **272** (2017) 012008 doi:10.1088/1757-899X/272/1/012008

Interactive Image Recognition of Space Target Objects

Xianheng Zeng[1] and Hanxu Sun[1]

[1]Department of Automation,Beijing University of Posts and Telecommunications ,Beijing, China

Abstract. An interactive image pre-processing algorithm is proposed to solve the problem that the target object is not accurate and the calculation is time consuming when the space robot is in orbit. Firstly, the Grabcut algorithm is improved to identify the space object. The Gaussian filtering, edge detection, morphological processing and other methods are combined with Grabcut algorithm. Then the Hough transform algorithm is improved, and the straight line with similar slope is fitted, and the intersection of the straight line is obtained, which is the key point of the object to describe the spatial coordinates of the object. Through the experimental verification, the method proposed in this paper can meet the task of identifying the object of the space object, which can improve the accuracy and recognition speed of the object recognition and reduce the time taken to identify the space object.

1. Introduction

In the orbit operation of a space robot, there is a need to identify a non-cooperative object of space, that is, an object that identifies information that cannot provide a valid key point to the space robot. Need to use the robot vision to achieve the detection and positioning of objects. If you cannot know the model of the space object in advance, it will need to deal with a lot of useless information in the identification process, and a substantial increase in the processing load of space computer.

Scholar Xu Wenfu[1] et proposed relatively fixed structure of the spacecraft as a measurement object, using binocular vision based on non-cooperative target pose measurement program. The scheme needs to assume that the approximate shape of the part under test is known and then the subsequent identification and measurement process can be completed automatically. But this scheme cannot recognize unknown objects. Inaba[2] et proposed a model-based matching method that uses the camera parameters and pose estimation results to project the characteristics of the target model onto the 2D image plane to obtain the object pose information.

This paper first proposes to combine Grabcut algorithm with image processing method to identify objects to reduce the computational load of space object recognition tasks. Then, an improved Hough change algorithm is proposed for the space target objects to calculate the image coordinates. Finally, the image of three sets of spatial target objects is used to test the algorithm to verify the feasibility and operation of the improved algorithm.

The structure of this paper is organized as follows: Part 2 provides a brief introduction to the target object recognition algorithm. Part 3 describes the improved the Grabcut algorithm and the Hough transform algorithm. Part 4 is the experimental evaluations. At last section is conclusions.

2. Target object recognition algorithm

2.1. Grabcut interactive image segmentation algorithm

IOP Conf. Series: Materials Science and Engineering **272** (2017) 012008 doi:10.1088/1757-899X/272/1/012008

The Grabcut algorithm first requires the user to interactively select the foreground and background samples, establish a Gaussian Mixture Model (GMM) for the foreground and background regions, and then obtain the segmented energy weights.

Scholars Achanta[3] et proposed a simple linear Iterative Clustering (SLIC) method and proved that it was significantly improved in the processing of ultra-pixel, and SLIC produced a lower block boundary error rate. But it cannot be a good solution to the foreground space object`s grey is similar to the background.

3. Target recognition extraction algorithm for space objects

3.1. Improved Grabcut algorithm to extract spatial objects

The improved Grabcut algorithm steps are shown in the figure 1:

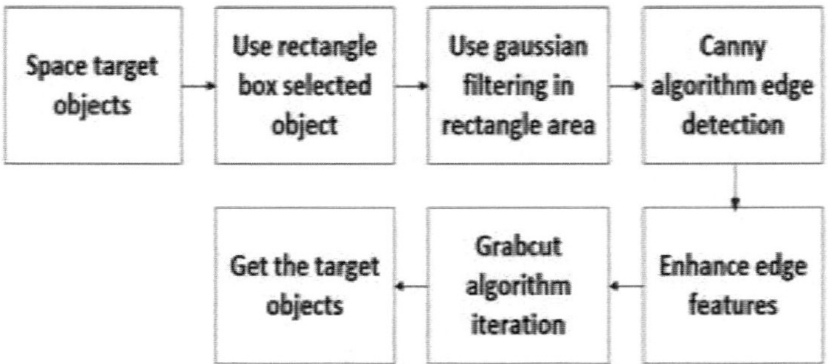

Figure 1. Improved Grabcut algorithm flow.

Gaussian filtering is a good way to get a better image edge in order to get a better image edge, a signal-to-noise ratio (SNR) image (reactive real signal)[4]. A Gaussian filter with a standard deviation of different sizes taken in the x-axis and y-axis directions is used as a smoothing filter.

Secondly, the Canny algorithm is used to detect the edge of the selected area. This method has a strong ability to suppress noise, is currently the most widely used edge detection method. The figure 2 shows the result which used Canny algorithm to detect the object.

(a) (b)

Figure 2. Canny algorithm edge detection.

Then, according to the grey value of the pixel in the rectangle area, use morphological processing to enhance the characteristic information of the object target. As a result of Canny algorithm, some features are not very obvious edge will be eliminated, so here is mainly used to expansion to enhance the edge of the object characteristics. The processing effect as shown in the figure 3, through the expansion method, can not only repair the break edge of the object, but also facilitate the next step of calculation.

IOP Conf. Series: Materials Science and Engineering **272** (2017) 012008 doi:10.1088/1757-899X/272/1/012008

(a) (b)

Figure 3. Morphological processing.

We use 2 * 2 structural elements for expansion[5]. After several experiments, we found that structural elements should not exceed 5 * 5. The main reason for the operation is that scheme aims at the target object which has slight edge, and the big structural elements seriously affect the structural characteristics of the edge.

After the morphological operation, each pixel point f inside the rectangular area is traversed to determine whether the grey value g is greater than the threshold value δ, where the threshold δ is taken as 50, taking into account the Canny algorithm after detection, the edge of the pixel grey Values are greater than or equal to 50. If there is a pixel`s grey value $g \geq \delta$ in the region, then $g=0$, otherwise $g=255$. The purpose of this operation is to distinguish the inner region of the target object from the background, since the grey value g of the spatial background area is usually small, so the grey value of the spatial target region can be improved, and the number of iterations of the Grabcut will reduce. The effect is as follows in figure 4:

Figure 4. Enhanced edge feature.

Finally, the Grabcut algorithm is calculated for the rectangle area to remove the interference factor in that area, leave only the space object within the region. After the previous steps, the calculation speed of Grabcut algorithm has been improved obviously, and the convergence speed is accelerated and the number of iterations is reduced. Iterations can achieve good results in two times.

Through the improved Grabcut algorithm, it can get the image of the space object and separate it from the background. This kind of processing is helpful to calculate the image coordinates of the space object, and further facilitate the further position and attitude coordinate solution. The figure 5 shows the comparison result between original picture and the picture which was processed by the algorithm.

(a) (b)

Figure 5. Algorithm before and after the comparison.

3.2. Improved Hough transform detect object edge information

The steps of extracting the key points on the line: first set the abscissa vector to x_i. And then through the slope of the straight line and the coordinates of the linear end point to calculate the vertical axis of the vector y_i, that is, each abscissa corresponding to the two straight lines on the vertical axis. In the next step, calculate the absolute delta of the difference between each ordinate. Set the difference threshold rho, when delta <rho, the resulting point of the vertical and horizontal coordinates of the straight line intersection of the vertical and horizontal coordinates, that is, the line intersect at that point.

In order to reduce the error, take the average of two points, because the coordinates of the points in the image represent the rows and columns of the points, respectively. Because the coordinates are all integers, so the nearest intersection of the coordinates of the nearest rounded.

The improved algorithm uses adaptive threshold segmentation instead of binarization method for segmentation[6]. The edge of the line structure is extracted by the Canny operator, and then the centre line is extracted by the column grey centre method[7]. Finally, the Hough function is used to fit the straight line.

As shown below in the figure 6, there are only two effective lines in those lines, so it is necessary to filter these lines to merger, establish the average slope to the final slope, and determine an endpoint to calculate the linear equation. Select the line comparison centre as the endpoint.

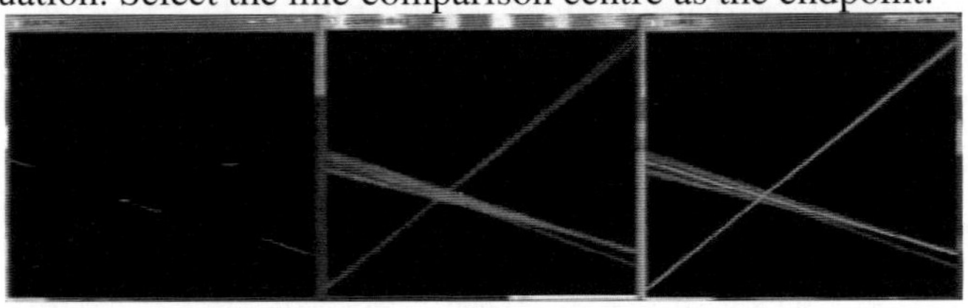

(a) (b) (c)

Figure 6. Improved Hough Transform detection results.

4. Experimental results and analysis

In this paper, we choose the more common operation objects in space, carry out algorithm test, use the same computer, run the original algorithm and improve the algorithm, compare the effect of object recognition. Experimental equipment: CPU: Intel T5870 2GHz, RAM: 4GB, Windows 10 operating system, using C ++ programming language and Opencv2.4.9, IDE: Microsoft Visual Studio 2015.

As shown in the figure 7, the experimental images are common to the space object operating objects: solar sails, space operating parts, unit antenna modules. The resolution of the pictures is 800 * 600. As the space robot vision system to use more grey scale operation, so we use greyscale images. The experimental images are shown below. The first column is the original image of the space objects, the second column is the images which are processed by the original algorithm, and the third column is the images which are processed by the improved algorithm.

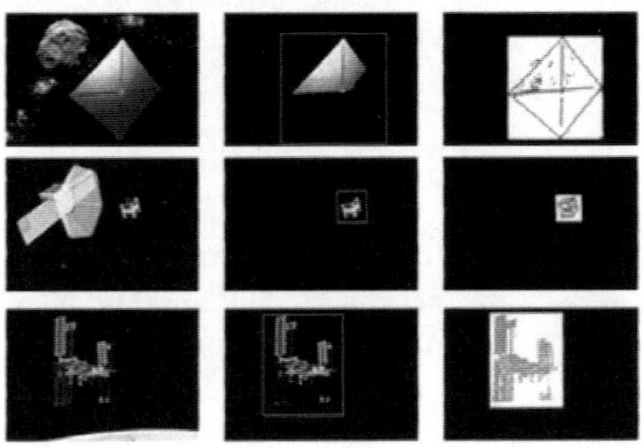

(a) (b) (c)

Figure 7. Comparison of improved algorithm and original algorithm.

Observe the above three sets of pictures, unmodified algorithm is not good to deal with the effect, in dealing with the first set of pictures, there have been over-segmentation, seriously affected the need to identify the structure of the space sail, unable to get a complete object contour. And the processing results of the second picture and the third picture are also not ideal. The original algorithm can not adapt to this situation.

Observation of the second column and the third column of the picture we can see that the improved algorithm to deal with the effect of space object has been improved.

The experiment also compares the iterative times of the algorithm and the convergence time of the algorithm, and analysis the effect of the improved algorithm. The data is shown in the following table 1:

Table 1. Improved algorithm and the original algorithm processing speed comparison

Image	Resolution	Number of iterations		Convergence time /s	
		Original algorithm	Improved algorithm	Original algorithm	Improved algorithm
1	800*600	6	2	14.857	4.708
2	800*600	4	1	9.905	2.328
3	800*600	5	2	12.412	5.024

From the above table 1, we can see that the improved algorithm can speed up the convergence speed, thus reducing the number of iterations and shortening the convergence time. After testing, Gaussian filtering in image processing, canny algorithm edge detection, morphological expansion and enhancement of target characteristics, the total time is less than 2s. Even with the time taken for image processing, the time of the improved algorithm reduce significantly.

For the image obtained by the improved Grabcut algorithm proposed in this paper, the contour extraction of the object is carried out by using the improved Hough linear detection. After testing, the extracted key points of the pixel coordinates (x_i, y_i) of the error around 5 pixels, and the detected straight lines and key points can describe the position of the space object in the image.

5. Conclusions

This paper presents an improved Grabcut algorithm, and an improved Hough transform algorithm[8]. Combining the two algorithms enables the task of identifying spatial objects. The operators only need to select the space object through the host computer to operate, according to the identification system can get the spatial object in the image of the contour feature and the key point of the image coordinates.

The experimental results show that the improved algorithm can improve the accuracy of the recognition in Xu Wenfu`s algorithm. At the same time effectively reduce the Grabcut algorithm to calculate the amount of processing, to reduce the use of identification time[9]. A solution to the object recognition of space object is proposed, which paves the way for the space manipulator to grasp the object and track the object. Based on the spatial object recognition algorithm proposed in this paper, the efficiency of space robot can be improved.

6. References

[1] Xu Wenfu et Space robot to capture non-cooperative target measurement and planning method [J]. Robotics, 2010, 32 (1): 61-69.

[2] Inaba N，Oda M，Asano M. Rescuing a stranded satellite in space-experimental robotic capture of non-cooperative satellites ［J］. Transactions of the Japan Society for Aeronautical and Space Sciences，2006，**48**(162) : 213-220.

[3] Radhakrishna A， Shaji A， Smith K， et al.SLIC superpixels， Technical Report 149300[R].EPFL，2010.

[4] Rother C, Kolmogorov V, Blake A. "Grabcut": Interactive foreground extraction using iterated graph cuts. ACM Trans. on Graphics, 2004, 23(3): 307–312.

[5] Rincen J, Illingworth J, Kittler J. A formal definition of the Hough Transform properties and relationships. JMathematical Imaging and Vision, 1992, 1:153-148

[6] Bergen J R, Shvayster H. A probabilistic algorithm for computing Hough Transform. [J] Algorithm, 1991, 12: 639-656

[7] Mikhailov L,Tsvetinov P.Evaluation of services using a fuzzy ana — lytic hierarchy process[J].Applied Soft Computing,2004,5(1):**23-33**.

[8] Brent E. Tweddle, Alvar Saenz-Otero and David W. Miller. Design and Development of a Visual Navigation Testbed for Spacecraft Proximity Operations[C]. AIAA SPACE 2009 Conference & Exposition, Californi, 2009: 6547-6561.

[9] Tom Drummond, Roberto Cipolla. Real-time visual tracking of complex structures [J]. IEEE Transactions on Pattern Analysis and Machine Intelligence, 2002, 24(7): 932-946.

IOP Conf. Series: Materials Science and Engineering **272** (2017) 012009 doi:10.1088/1757-899X/272/1/012009

The FEM Simulation on End Mill of Plastic Doors and Windows Corner Cleaning Based on Deform-3D

Guoping Li, Zhenyong Huang and Xiaohui Wang

School of Mechanical Engineering, University of Jinan, Jinan 250022

Abstract. In the plastic doors and windows corner cleaning process, the rotating speed, the feed rate and the milling cutter diameter are the main factors that affect the efficiency and quality of the of corner cleaning. In this paper, SolidWorks will be used to establish the 3D model of end mills, and use Deform-3D to research the end mill milling process. And using orthogonal experiment design method to analyze the effect of rotating speed, the feed rate and the milling cutter diameter on the axial force variation, and to get the overall trend of axial force and the selection of various parameters according to the influence of axial force change. Finally, simulate milling experiment used to get the actual axial force data to verify the reliability of the FEM simulation model. And the conclusion obtained in this paper has important theoretical value in improving the plastic doors and windows corner cleaning efficiency and quality.

1. Introduction

Corner cleaning as a main technology in plastic doors and windows production process, the quality of its corner cleaning has a direct impact on the quality and the strength. But in the actual inner countersink slot cleaning, the corner cleaning is incompletely, the cutting edge crack and the serious abrasion in cutter, these problems will affect the quality of the plastic doors and windows. And there are many factors that lead to this phenomenon, such as cutting speed, feed rate, etc[1].

Based on that, the effects of rotating speed, feed rate and milling cutter diameter on milling axial force were analyzed by using orthogonal design method in this paper. First, use SolidWorks to establish the 3D model of end mills, and use Deform-3D to research the end mill milling process. Then, using orthogonal experiment design method to analyze the effect of rotating speed, the feed rate and the milling cutter diameter on the axial force variation, to get the overall trend of axial force and the selection of various parameters according to the influence of axial force change. Finally, use simulate milling experiment to get the actual axial force data to verify the reliability of the FEM simulation model[2].

2. End Mill and Workpiece Model Creatio

2.1. End Mill and Workpiece Model Creation

Plastic doors and windows profiles use PVC resin as the main raw material, with a certain percentage of stabilizers, colorants, fillers, UV absorbers and other extrusion molding. Main performance parameters of the plastic doors and windows profile as shown in Table 1.

Table 1. Performance parameters of the profile.

Project		Target
Thermal Expansion Coefficient, 10^{-6}/K	≥	5
Heat Transfer Coefficient, $W/m^2 \cdot K$	≥	1.91
Elastic Modulus, Mpa	≥	3000
Bending Modulus of Elasticity, Mpa	≥	100
Melting Point, ℃	≥	210
Poisson's Ratio		0.40
Density, $\rho/(kg \cdot m^3)$		1385
Specific Heat Capacity, $c/(J \cdot kg^{-1} \cdot K^{-1})$		0.8—1.5

The inner countersink slot cleaning is shown in figure 1. The inner countersink slot cleaning was used to hold the glue strips, and the effect of cleaning will affects the tightness. Use SolidWorks to establish the profile model. And the model of the profile is shown in figure 2.

Figure 1. The inner countersink slot cleaning

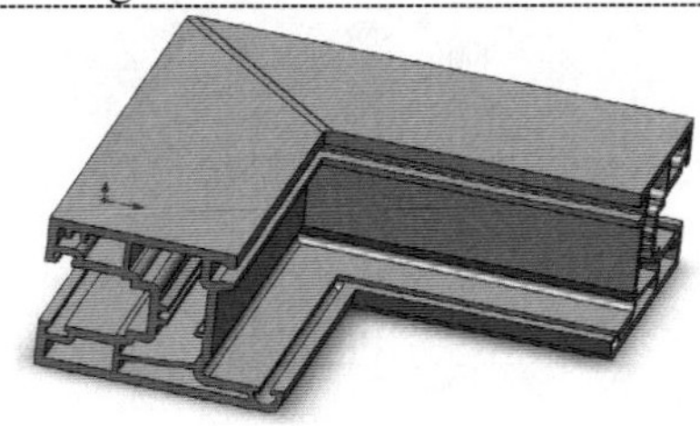

Figure 2. Profile model

Use SolidWorks to establish solid model of end mill according to the geometrical parameters of the end mill. The model of the end mill is shown in figure 3.

Table 2. The geometrical parameters of the end mill.

Diameter	4	5	6
Full Length	83	92	101
Blade Length	24	24	24
Number of Teeth	4	4	4
Rake Angle	0°	0°	0°
Clearance Angle	12°	12°	12°
Helix Angle	30°	30°	30°

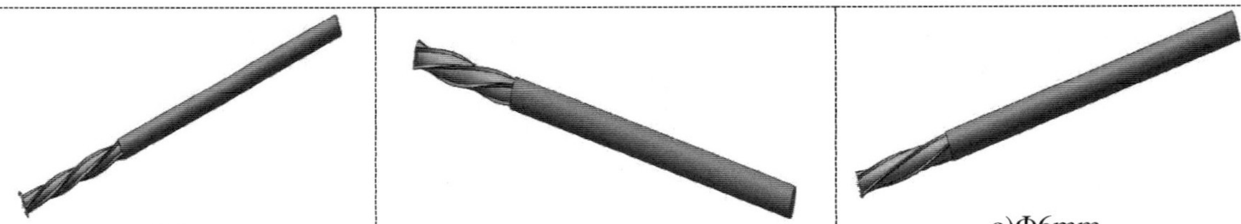

a)Φ4mm b)Φ5mm c)Φ6mm

Figure 3. The model of the end mill.

It is necessary to adjust the position relations between the end mill and the workpiece in SolidWorks to reduce the computation time in the process of simulation. Milling simulation model is shown in figure 4.

2.2. Material Selection and Meshing

The end mill material is cemented carbide, and the end mill is set to a rigid type. PVC as the workpiece material was selected, and set the workpiece as a plastic type. Deform-3D uses adaptive meshing method to divide the grid, which can prevent the excessive distortion of the division unit and affect the simulation accuracy. At the same time, in order to reduce the computation time of the simulation, the meshing method of the local mesh should be adopted to make the mesh away from the milling area become thinner gradually when the mesh is divided. End mill relative grid division method, the number of grid 20000[3], but the profile adopt the absolute mesh method, the ratio of the maximum mesh size to the minimum mesh size is set to 2, the minimum cell side length is 0.6 mm[4], the ratio of the maximum mesh size to the minimum mesh size is set to 4, the minimum cell side length is 0.2 mm[5]. The meshed model is shown in figure 5.

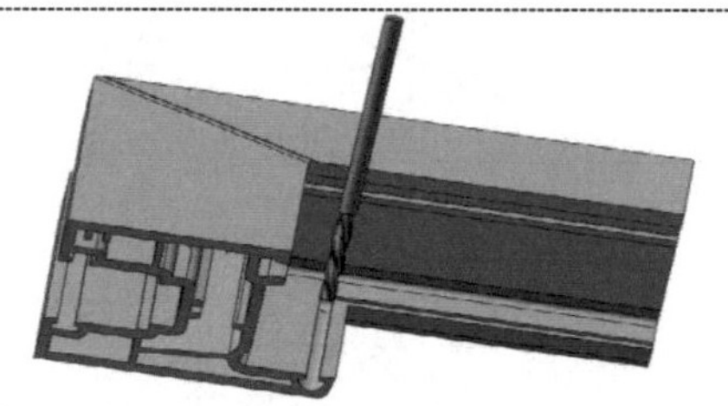

Figure 4. Milling simulation model.

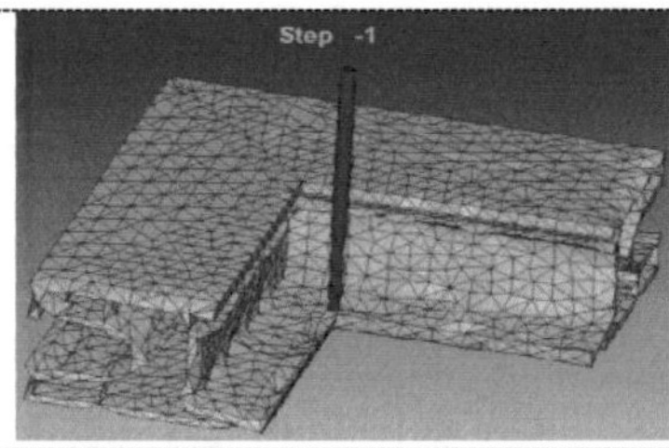

Figure 5. The meshed model.

2.3. Simulation Parameters and Boundary Conditions

Set the milling depth was 3mm, the tool wear parameter a and b are 1e^5 and 855. In the simulation control window, set the number of steps was 1000, the storage increment is stored once every 25 steps, and the time step is 0.0002s[6].

In the boundary condition setting, set the speed of the workpiece side in the X, Y, Z axis in three directions were 0, to limit the movement of the workpiece. The tool feeds in the -Z direction and rotates around the Z axis. At the same time, set the tool and workpiece surface and all external heat transfer, ambient temperature, the end mill and the workpiece temperature is defined as 20℃, the heat transfer coefficient is 45N/sec/mm/C, the convection coefficient is 0.02N/sec/mm/C. The contact condition of the workpiece is shear friction, and the friction coefficient is 0.6. Its tool wear Usui's model [7]:

$$w = \int apVe^{-b/T}dt$$

In this formula, w is the wear volume; p is the contact surface pressure; V is the sliding speed; dt is the time increment; a, b are the experimental calibration coefficient, a is 1.0e^5 and b is 855 according to experience.

3. Simulation Analysis of End Mill

3.1. Simulation Experiment

In this part, the orthogonal experiment design method is used to study the effect of milling parameters on axial force. Then, select milling cutter diameter, feed speed and spindle speed as the three factors, the test factors and level as shown in Table 3.

Table 3. Factor level table.

Status	Factor		
	A:Rotating speed(r/min)	B:Feed Rate(mm/r)	C:Diameter(mm)
Level	1000	0.1	4
	2000	0.2	5
	3000	0.3	6

3.2. Axial Force Analysis

From the table 3.1, it can be seen that there are three factors (A is rotation speed, B is feed rate and C is diameter) and three levels in the table. The orthogonal table L9 (34) is selected and the plan is discharged according to L9 (34), namely, nine groups of experiments were performed. The axial force variation curve of the end mill in each test condition is shown in figure 6.

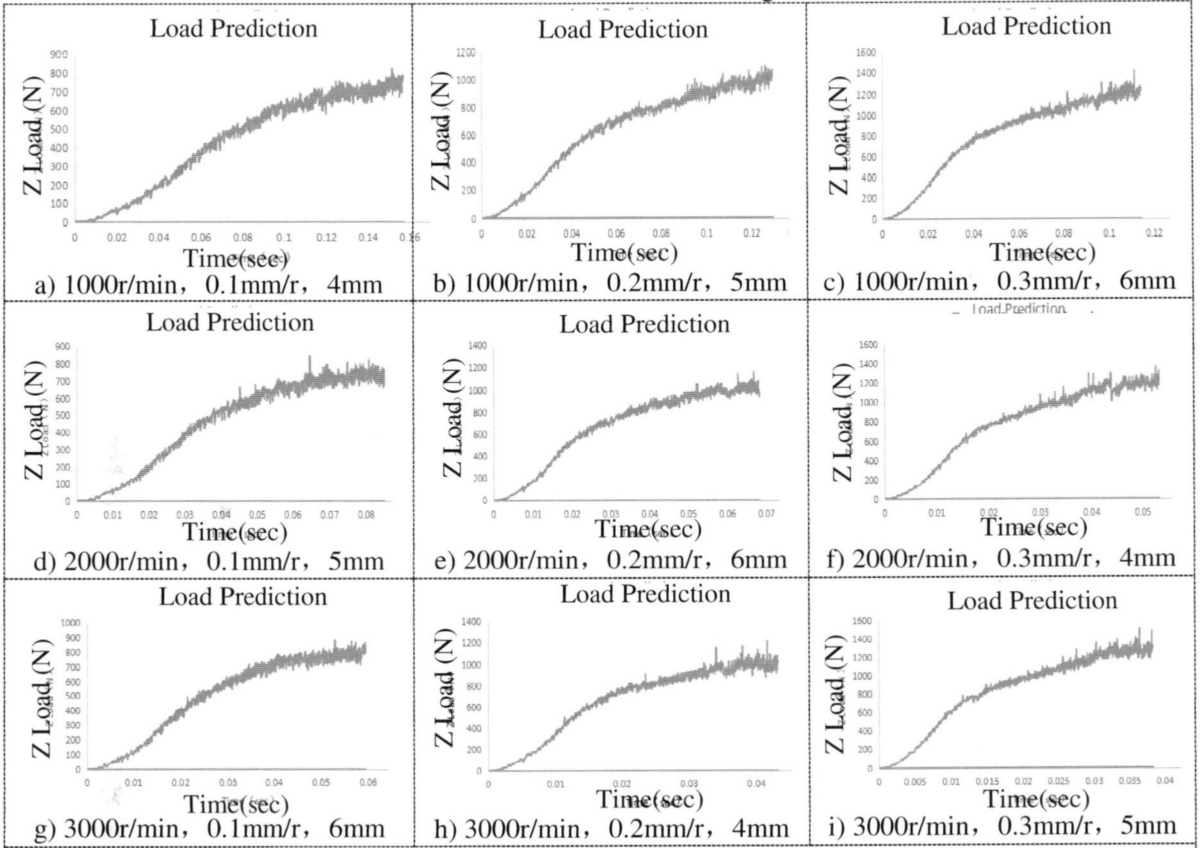

Figure 6. The axial force variation curve of the end mill in each test condition.

In figure 6, the axial force of the end mill in the test condition is gradually increased as the milling blade and the profile begin to contact with each other in the initial milling force variation curve. The axial force will reach the maximum end mill when the milling cutter edge all involved in the clearance of plastic doors and windows profile milling. Owing to the re-division of the mesh in Deform-3D lead to discontinuities in the simulation process or chip breakage, the curve will fluctuate up and down when the axial force tends to be stable.

According to the simulation results, take the average of the steady-state data to complete Table 4.

In Table 4, the three sets of data in row K1are the sum of the corresponding axial force of the end mill in the test where the first level of factors A, B and C. the three sets of data in row K2are the sum of the corresponding axial force of the end mill in the test where the second level of factors A, B and C. And the three sets of data in row K3 are the sum of the corresponding axial force of the end mill in the test where the third level of factors A, B and C. But beyond that, k1, k2and k3 rows of three sets of data for the K1, K2and K3 rows corresponding to the three groups of data obtained by dividing 3 respectively. The range is the difference between the maximum value and the minimum value of k1, k2, and k3 in the same column. According to Table 4, the influence of each factor on the axial force is shown in figure 7.

Table 4. Orthogonal test simulation results

Factor / Group	1 *A*	2 *B*	3 *C*	Axial force（N）
1	1	1	1	747.23
2	1	2	2	1007.57
3	1	3	3	1227.22
4	2	1	2	736.95
5	2	2	3	1025.41
6	2	3	1	1197.37
7	3	1	3	796.19
8	3	2	1	996.05
9	3	3	2	1282.87
K_1	2982.02	2280.37	2940.65	
K_2	2959.73	3029.03	3029.03	
K_3	3075.11	3707.46	3048.82	
k_1	994.01	760.12	980.22	
k_2	986.58	1009.68	1009.68	
k_3	1025.04	1235.82	1016.27	
Range	38.46	475.7	36.05	
Optimal Solution	A_1	B_1	C_1	

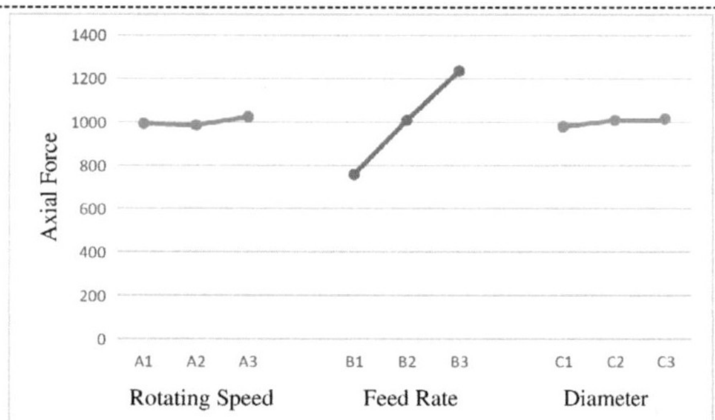

Figure 7. The influence of each factor on the axial force.

It can be seen from figure 7 that with the spindle speed increases, making the cutting edge of the blade edge thickness increases, and ultimately the overall increase in the axial force of the trend. But the growth trend is not significant, and it shows that the spindle speed has little effect on the milling force. The axial force increases rapidly when the feed rate increases, which is due to the increase in feed lead to elastic deformation profiles, resistance and friction are also increases. The axial force increases when the diameter of the milling cutter increases, namely, the diameter of the milling edge increases, and the area involved in the milling increases, so that the axial force increases as the deformation and friction increase accordingly. It can be seen that the feed rate of the end mill has the greatest influence on the axial force, and the sort of each influencing factor as follows: Feed Rate>Rotating speed> Diameter.

Generally speaking, the range of the line is different, which shows that the level of the factors change the impact of the test index is different. The larger the range is, the greater the effect of the change of the level of the factor on the test index, and the column of the range of maximum deviation is the main factor to be considered in this article. It can be seen from the table 4 that the three ranges are 38.46, 475.7, 36.05, obviously, the second column (factor B) corresponds to the maximum range, namely, the level of factor B changes in the test index of the most significant factor. So, factor B is an important

factor to be considered in this article. The axial force of the end mill should be reduced in the corner cleaning process to reduce the excessive tool axial force caused by tool wear and chipping phenomenon. The average axial force corresponding to three levels of factor B are 760.12N, 1009.68N and 1235.82N. It can be seen that the axial force corresponding to the first level is the smallest, so, the first level of factor B is the best; Analogously, the value of the first level of factor A and the value of the first level of factor C are the best. Therefore, B1A1C1 is the best scheme.
In order to improve the corner cleaning efficiency, it is important to increase the speed of the end mill for the third factor A, namely, speed is3000r/min, so, scheme B1A3C1can be choosed. At the same time, the large diameter of the end mill can be choosed to increase the milling area, namely, scheme B1A3C2or B1A3C3 also can be choosed. But the scheme B1A3C1and B1A3C2did not appear in the nine groups of tests. In order to validate the results of scheme B1A3C3, B1A3C1and B1A3C2, the milling force simulation tests can be carried out on solutions B1A3C1 and B1A3C2. The axial force variation curves of the two schemes are shown in figure 8.

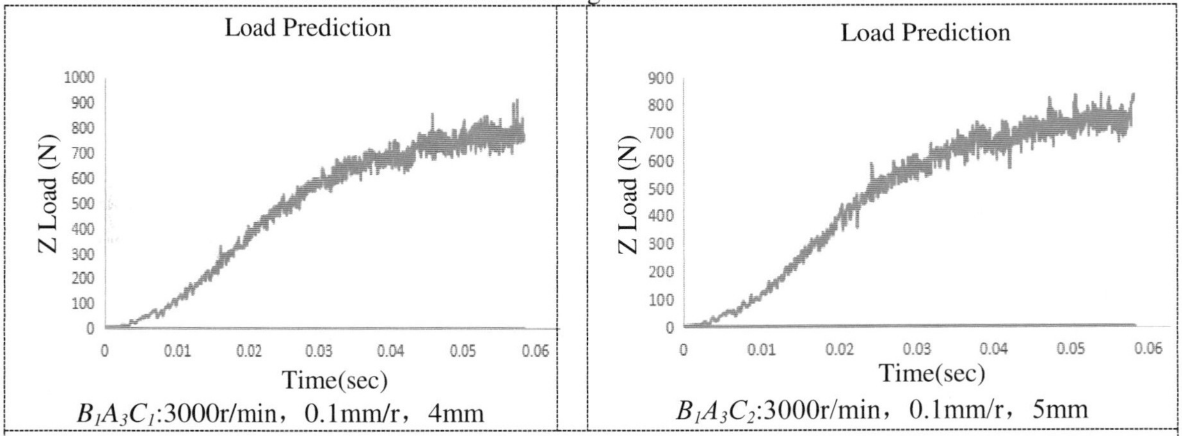

Figure 8. The axial force variation curves of the two schemes.

According to the figure 8, the results of the axial force comparison of the three schemes was completed as shown in Table 5.

Table 5. The results of the axial force comparison of the three schemes.

Group \ Factor	1 *A*	2 *B*	3 *C*	Axial Force（N）
7	3	1	3	796.19
10	3	1	1	779.29
11	3	1	2	762.25

It can be seen from the Table 5 that the axial forces of the 10th group and the 11th group are all smaller than those of the 7th group, which shows that the two schemes (B1A3C1 and B1A3C2)are better than B1A3C3, and B1A3C2is the best. That is to say, feed rate can be selected 0.1mm/r, speed can be selected 3000r/min, and the diameter can be selected 5mm.

4. Simulation Results Verify

In this article, the axial force test is carried out on 11 experiments in Table 4 and Table 5, to verify the accuracy of the simulation results in section 3.2. The force measuring device included: Kistler9257B three-way dynamic piezoelectric dynamometer, Charge amplifier, DEWE-50-FW-16 cutting force measurement system, Dewesoft65 computer processing system, YCM-V116B CNC, End mill and Profile. The system connection is shown in figure 9.

Milling force tests were carried out on 11 test schemes, and five sets of experimental data were collected for each scheme to reduce the errors caused by the experiment. The mean error of five sets of experimental data for each protocol was compared with the simulation data, and the results in Table 6 were completed.

Table 6. Comparison of experimental data and simulation data.

Group	Experimental Data（N）	Simulation Data（N）	Relative Error（%）
1	614.28	747.23	-17.79
2	862.99	1007.57	-14.35
3	1067.62	1227.22	-13.01
4	660.42	736.95	-10.38
5	922.72	1025.41	-10.01
6	1090.95	1197.37	-8.89
7	681.98	796.19	-14.34
8	885.79	996.05	-11.07
9	1180.07	1282.87	-8.01
10	680.37	779.29	-12.69
11	634.34	762.25	-16.78

From Table 6, the histogram of the experimental data and simulation data is plotted in figure 10.

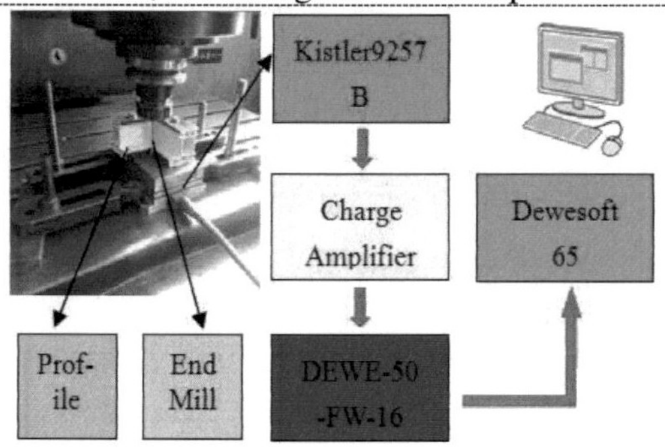

Figure 9. The force measuring device.

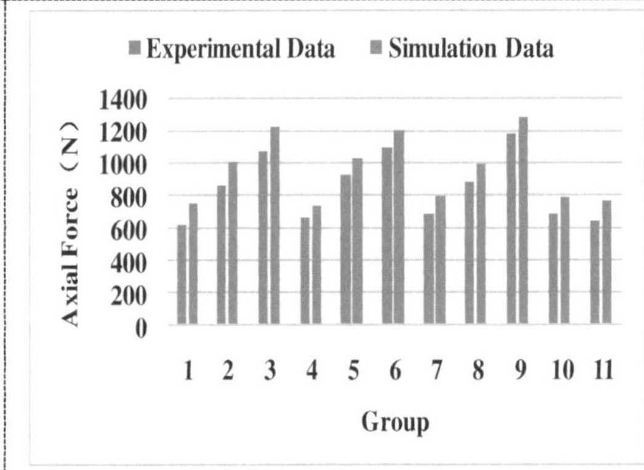

Figure 10. The histogram of the experimental data and simulation data.

It can be seen from Table 6 that the relative error between the simulation data and the experimental data is within 20%. The effect of data read is the difference between the actual processing environment and simulation processing environment, as well as the installation of the experimental device is difficult to guarantee accuracy. From the figgure 10, it shows that the experimental data are in agreement with the simulation data, which indicates that the simulation data of the axial force of the end mill can be referenced by the finite element simulation method.

5. Conclusion

In this article, the SolidWorks were used to establish three-dimensional model of three different diameter end mills, and use Deform-3D to research the end mill milling process. And using orthogonal experiment design method to analyze the milling blade diameter, feed rate and spindle speed on the axial force influence, to get the overall trend of axial force and the selection of various parameters on the influence of axial force change. Finally, simulate milling machining used to get the actual axial force data to verify the reliability of the FEM simulation model.

(1)Through the milling process simulation analysis, the axial force variation curve of end mill is obtained, namely, the axial force will increase gradually when the cutting edge and the profile begin to come into contact in the initial stage. The axial force of the end mill will reach the maximum value when the milling edge is all involved in the corner cleaning of the profile, and then, the axial force of the end mill will tend to be stable with the milling of the milling cutter. And it shows that the feed rate of the end mill has the greatest influence on the axial force, and the sort of each influencing factor as follows: Feed Rate>Rotating speed> Diameter.
(2)Through the analysis and the comparison, it shows that the reasonable value of the parameter in the corner cleaning process of end mill can be selected as feed rate is 0.1mm /r, speed is 3000r/min and diameter is 5mm.
(3)From the conclusions, it can be known that the simulation test data of the axial force of the end mill with the finite element simulation method are relatively reliable, namely, the data can be referenced.

6. References

[1] Zhou Yanli,Wang Xiaofei,Yu Yongchang. Numerical Simulation of GCr15 Steel in Equal Channel Angular Extrusion Based on DEFORM-3D[J].Foundry Technology,2017,(02):402-404.
[2] Gu Meilin, Li Ning, Shi Wen, Mu Jincheng. Finite Element Analysis of Tool Wear Based on Deform-3D for Spiral Taps[J].Tool Engineering,2017,(04):65-68.
[3] Xu Minggang,Zhang Zhen,Ma Xiaolin,Huang Wenyong. Simulation Analysis of Micro Texture Tool Cutting Nature Based on DEFORM-3D[J].Modular Machine Tool & Automatic Manufacturing Technique,2016,(03):44-47.
[4] Xu Minggang,Huang Wenyong,Zhang Zhen.Micro texture knives simulation analysis and cutting experiment research based on DEFORM-3D[J].Mach Tool Hydrau,2016,(24):50-56.
[5] Zhang Rongrong,Zhao Xianfeng,Li Changhong,Zhang Haining. Analysis of the residual Stress in the Turning Cutting Based on Deform-3D[J].Modular Machine Tool & Automatic Manufacturing Technique,2016,(05):22-24+28.
[6] He Aidong. Study on FE Simulation of Chip Formation during Machining Vit1 by DEFORM-3D[J].Coal Mine Machinery,2016,(08):89-91.
[7] Zhu Chao, Bai Haiqing. FEM Simulation of Drilling Process by DEFORM-3D[J].Coal Mine Machinery,2015,(06):286-289.

Order Selection for General Expression of Nonlinear Autoregressive Model Based on Multivariate Stepwise Regression

Jinfei Shi, Songqing Zhu, Ruwen Chen

Nanjing Institute of Technology, Nanjing, China, 211167

Abstract. An order selection method based on multiple stepwise regressions is proposed for General Expression of Nonlinear Autoregressive model which converts the model order problem into the variable selection of multiple linear regression equation. The partial autocorrelation function is adopted to define the linear term in GNAR model. The result is set as the initial model, and then the nonlinear terms are introduced gradually. Statistics are chosen to study the improvements of both the new introduced and originally existed variables for the model characteristics, which are adopted to determine the model variables to retain or eliminate. So the optimal model is obtained through data fitting effect measurement or significance test. The simulation and classic time-series data experiment results show that the method proposed is simple, reliable and can be applied to practical engineering.

1. Introduction

Time series analysis technology is a kind of system identification method which can establish models based on the inherent law of data with no need for system inputs, so it has important applications in natural and social science field of industrial process control, economy and biomedical engineering, etc.[1]-[5].

Determining model's order plays a very important role in time series analysis process. Insufficient order of the model tends to the losing of valuable information to decrease model's tracking and prediction ability of data sequence. Meanwhile, excessive order makes a complex model and huge computing quantity.

At present, order determination for the linear models based on the stationary assumptions have many research and lots of classic criterions are applied in engineering such as AIC criterion, residual test, minimization of residual sum of squares or final prediction error and other information criterions[6][7]. But for nonlinear systems, due to their multifarious characteristics, there are no universal order determinations or goodness judgment strategies.

The General Expression of Nonlinear Autoregressive (GNAR) model is a novel time-series model, which can be applied to system identification, data tracking and system prediction, etc.[8][9]. In this paper, an order selection method based on multiple stepwise regressions is proposed for GNAR model which converts the model order selection into variable selection for multiple linear regression equation. Firstly, the partial autocorrelation function is adopted to define the linear term in GNAR model. And the linear result is set as the initial model, and then the nonlinear terms are added gradually. Statistics are chosen to study all the variables and their improvements are adopted to determine the model variables to retain or eliminate. Finally, the optimal model is obtained through data fitting effect measurement or significance test.

This remainder paper is organized as follows. Section 2 introduces the definition of GNAR model, Section 3 discusses the order selection steps based on multivariate stepwise regression for linear and nonlinear terms in GNAR model, Section 4 describes its application and finally Section 5 presents the conclusion.

2. Definition of GNAR model

The expression of GNAR model can be expressed as:

$$\begin{aligned} w_t &= \sum_{i_1=1}^{\infty} \alpha_{i_1} w_{t-i_1} + \sum_{i_1=1}^{\infty}\sum_{i_2=1}^{\infty} \alpha_{i_1,i_2} w_{t-i_1} w_{t-i_2} + \cdots + \sum_{i_1=1}^{\infty}\cdots\sum_{i_p=1}^{\infty} \alpha_{i_1,\cdots,i_p} \prod_{\tau=1}^{p} w_{t-i_\tau} + a_t \\ &= \sum_{j=1}^{\infty}\sum_{i_1=1}^{\infty}\cdots\sum_{i_j=1}^{\infty} \alpha_{i_1,\cdots,i_j} \prod_{\tau=1}^{j} w_{t-i_\tau} + a_t \end{aligned} \tag{1}$$

where w_{t-i} represents the system observed data at the moment of t-i, i=0, 1, 2, …; $\alpha_{i_1}, \cdots, \alpha_{i_1,i_2}, \cdots$ are model parameters; a_t is white noise with zero mean.

When modeling in engineering, the model order should not be infinite, so Eq.(1) is rewritten as:

$$\begin{aligned} w_t &= \sum_{i_1=1}^{n_1} \alpha_{i_1} w_{t-i_1} + \sum_{i_1=1}^{n_2}\sum_{i_2=1}^{n_2} \alpha_{i_1,i_2} w_{t-i_1} w_{t-i_2} + \cdots + \sum_{i_1=1}^{n_p}\cdots\sum_{i_p=1}^{n_p} \alpha_{i_1,\cdots,i_p} \prod_{\tau=1}^{p} w_{t-i_\tau} + a_t \\ &= \sum_{j=1}^{p}\sum_{i_1=1}^{n_j}\cdots\sum_{i_j=1}^{n_j} \alpha_{i_1,\cdots,i_j} \prod_{\tau=1}^{j} w_{t-i_\tau} + a_t \end{aligned} \tag{2}$$

in which p is the polynomial order; $\sum_{i_1=1}^{n_1} \alpha_{i_1} w_{t-i_1}$ is the first order linear term, α_{i_1} (i_1=1,2,…, n_1) are linear coefficients; $\sum_{i_1=1}^{n_2}\sum_{i_2=1}^{n_2} \alpha_{i_1,i_2} w_{t-i_1} w_{t-i_2}$ is the second order nonlinear term, α_{i_1,i_2} (i_1, i_2=1,2,…, n_2) are the second order nonlinear coefficients; $\sum_{i_1=1}^{n_p}\cdots\sum_{i_p=1}^{n_p} \alpha_{i_1,\cdots,i_p} \prod_{\tau=1}^{p} w_{t-i_\tau}$ is the p order nonlinear term, $\alpha_{i_1,\cdots,i_p}$ (i_1, …, i_p =1,2,…, n_p) are the p order nonlinear coefficients; n_j(j=1,2,…, p) are the memory steps of every linear or nonlinear term. The model (2) can be denoted as GNAR (p; n_1, n_2,…, n_p) in abbreviation.

3. Order selection method for GNAR model based on multivariate stepwise regression

3.1. Order selection method

The expansion of Eq. (2) is as follows:

$$\begin{aligned} w_t = {} & \alpha_1 w_{t-1} + \alpha_2 w_{t-2} + \cdots + \alpha_{n_1} w_{t-n_1} + \\ & \alpha_{1,1} w_{t-1}^2 + \alpha_{1,2} w_{t-1} w_{t-2} + \cdots + \alpha_{n_2,n_2} w_{t-n_2}^2 + \\ & \alpha_{1,1,1} w_{t-1}^3 + \alpha_{1,1,2} w_{t-1}^2 w_{t-2} + \cdots + \alpha_{n_3,n_3,n_3} w_{t-n_3}^3 + \\ & \vdots \\ & + \alpha_{\underbrace{1,\cdots,1}_{p}} w_{t-1}^p \cdots + \alpha_{\underbrace{n_p,\cdots,n_p}_{p}} w_{t-n_p}^p + a_t \end{aligned} \tag{3}$$

Let $W_t = [w_t, w_{t-1}, w_{t-2}, \cdots, w_{t-N+1+p}]^{\mathrm{T}}$, $\psi = \left[\alpha_1, \cdots, \alpha_{n_1}, \alpha_{1,1}, \cdots, \alpha_{n_2,n_2}, \cdots\cdots, \alpha_{n_p,\cdots,n_p}\right]^{\mathrm{T}}$, then GNAR expression for a time series with the length of N is:

$$W_t = \begin{pmatrix} w_{t-1} & \cdots & w_{t-n_1} & w_{t-1}^2 & \cdots & w_{t-n_2}^2 & w_{t-1}^3 & \cdots & w_{t-n_3}^3 & \cdots & w_{t-n_p}^p \\ w_{t-2} & \cdots & w_{t-n_1-1} & w_{t-2}^2 & \cdots & w_{t-n_2-1}^2 & w_{t-2}^3 & \cdots & w_{t-n_3-1}^3 & \cdots & w_{t-n_p-1}^p \\ \vdots \\ w_{t-N+p} & \cdots & w_1 & w_{t-N+p}^2 & & w_1^2 & w_{t-N+p-1}^3 & & w_1^3 & \cdots & w_1^p \end{pmatrix} \Psi + a_t \tag{4}$$

Then the order selection for GNAR model is to determine the variable matrix in the right of Eq. (4).

Regard the column vector in this variable matrix as regression variables, and set $X_1 = [w_{t-1}, w_{t-2}, \cdots, w_{t-N+p}]^{\mathrm{T}}$,

$\cdots,$ $X_{n_1} = [w_{t-n_1}, w_{t-2}, \cdots, w_{t-n_1-1}]^{\mathrm{T}},$ $X_{1,1} = [w_{t-n_1}, w_{t-2}, \cdots, w_{t-n_1-1}]^{\mathrm{T}}, \cdots,$ $X_{n_2,n_2} = [w_{t-n_2}^2, w_{t-n_2-1}^2, \cdots, w_1^2]^{\mathrm{T}}, \cdots,$

$X_{\underbrace{n_p,\cdots,n_p}_{p}} = [w_{t-n_p}^p, w_{t-n_p-1}^p, \cdots, w_1^p]^{\mathrm{T}}$, then:

$$W_t = \begin{pmatrix} X_1 & X_2 & X_{1,1} & \cdots & X_{n_2,n_2} \cdots & X_{\underbrace{n_p,\cdots,n_p}_{p}} \end{pmatrix} \Psi + a_t \tag{5}$$

Hence, the order selection for GNAR model is converted to select the variable X_1, X_2, $X_{1,1}$ ... in multiple linear regression equation (5). The forward or backward selection or stepwise regression methods based on AIC, BIC, residual mean square, or *Cp*, can be adopted to resolve variable selection problem [10].

3.2. Order selection for linear term

The coefficients in time series model generally do not have definite physical meaning, but they reflect some inherent characteristics of the system. The coefficients can be considered the weights of each component in the model. Stable system has a relatively stable weight function; especially the coefficients of the linear terms in the model have the robust characteristic [9]. Hence, the memory step of the linear part in GNAR model can be determined by Partial autocorrelation function (PACF) [11]. Take the GNAR(2; 3, 2) shown in Eq. (6) for example. 100 data sequence is generated by simulation, shown in Figure 1.

$$\begin{aligned} &w_t = -0.1w_{t-1} - 0.3w_{t-2} + 0.6w_{t-3} + 0.05w_{t-1}w_{t-1} - 0.04w_{t-1}w_{t-2} - 0.06w_{t-2}w_{t-2} + a_t \\ &a_t \sim \mathrm{N}(0,1) \quad t = 1, 2, \cdots \end{aligned} \tag{6}$$

The partial autocorrelation function of this data is obtained, as shown in Figure 2. As can be seen from the Figure 2, within the 95% confidence interval, partial autocorrelation of significance lowers greatly with the delay step of 3. Thus, the rudimentary memory step of the linear part can be set as 3.

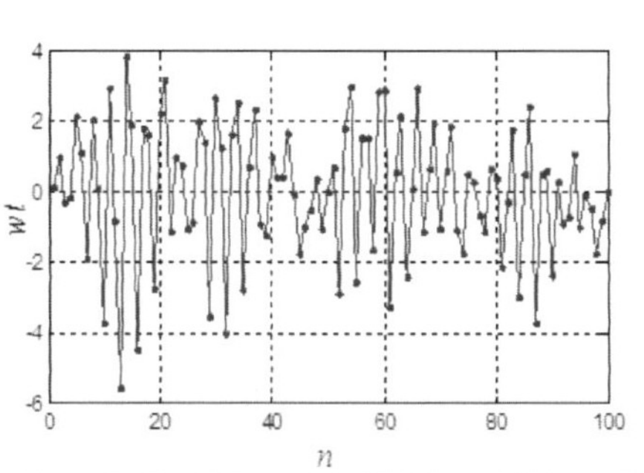

Figure 1. Simulation data sequence of GNAR(2; 3, 2).

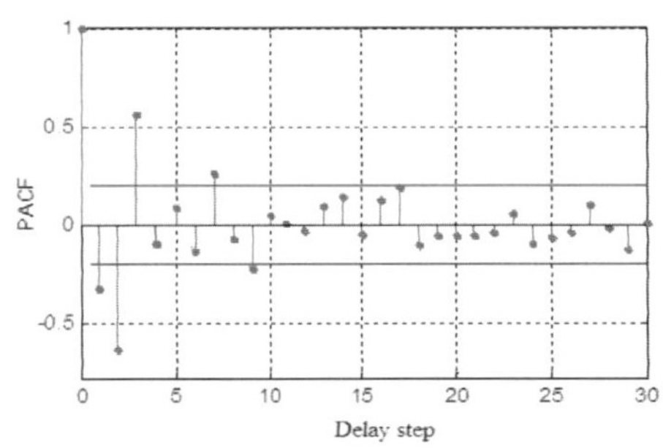

Figure 2. Partial autocorrelation function of the simulation data sequence.

3.3. Order selection for nonlinear term

The order and memory steps for nonlinear terms in GNAR model can be selected through stepwise regression method. The rudimentary linear model is set as the initial model and then the nonlinear terms $w_{t-1}w_{t-1}$, $w_{t-1}w_{t-2}$, $w_{t-1}w_{t-3}$...are introduced gradually. Statistics are chosen to study both the new introduced and originally existed variables, and their improvements are adopted to determine the model variables to retain or eliminate.

Still take the model of Eq. (6) for example. Least Squares parameter estimation method is adopted. Statistics magnitudes, AIC, BIC and residual mean square value shown in Eq. (7) to Eq. (9), are chosen to make a judgment.

$$\mathrm{AIC} = N\ln(\mathrm{SSE}/N) + 2p \tag{7}$$

$$\mathrm{BIC} = N\ln(\mathrm{SSE}/N) + p(\ln N) \tag{8}$$

$$\mathrm{RMS} = \mathrm{SSE}/(N-p) \tag{9}$$

where, N represents the sample data length; p represents the variable number in regression equation; SSE represents the sum of squared residuals in modeling, and $\mathrm{SSE} = \sum (w_t - \hat{w}_t)^2$.

The results of the variable selection for the simulation data are listed in Table 1. It can be learned from the table that GNAR(2; 3, 1) (No 2 in Table 1) is the preferable model according to the statistics of AIC and BIC and GNAR(2; 3,2) (No 4 in Table 1) to RMS. The models whose AIC values vary between 2 can be regarded as equivalent [2], so model 2, model 3 and model 4 are all suitable to describe this group of data.

Table 1. Results of the variable selection for the simulation data

No	Variable	p	SSE	AIC	BIC	RMS	R^2
1	w_{t-1}, w_{t-2}, w_{t-3}	3	129.833	32.108	39.924	1.338	0.646
2	w_{t-1}, w_{t-2}, w_{t-3}, $w_{t-1}w_{t-1}$	4	122.628	28.398	38.819	1.277	0.666
3	w_{t-1}, w_{t-2}, w_{t-3}, $w_{t-1}w_{t-1}$, $w_{t-1}w_{t-2}$	5	122.221	30.066	43.092	1.287	0.667
4	w_{t-1}, w_{t-2}, w_{t-3}, $w_{t-1}w_{t-1}$, $w_{t-1}w_{t-2}$, $w_{t-2}w_{t-2}$	6	119.888	30.139	45.770	1.275	0.673
5	w_{t-1}, w_{t-2}, w_{t-3}, $w_{t-1}w_{t-1}$, $w_{t-1}w_{t-2}$, $w_{t-1}w_{t-3}$, $w_{t-2}w_{t-2}$	7	119.847	32.105	50.341	1.289	0.673
6	w_{t-1}, w_{t-2}, w_{t-3}, $w_{t-1}w_{t-1}$, $w_{t-1}w_{t-2}$, $w_{t-1}w_{t-3}$, $w_{t-1}w_{t-4}$, $w_{t-2}w_{t-2}$	8	119.842	34.101	54.942	1.303	0.673
7	w_{t-1}, w_{t-2}, w_{t-3}, $w_{t-1}w_{t-1}$, $w_{t-1}w_{t-2}$, $w_{t-1}w_{t-3}$, $w_{t-1}w_{t-4}$, $w_{t-1}w_{t-5}$, $w_{t-2}w_{t-2}$	9	119.842	36.100	59.547	1.317	0.673
8	w_{t-1}, w_{t-2}, w_{t-3}, $w_{t-1}w_{t-1}$, $w_{t-1}w_{t-2}$, $w_{t-2}w_{t-2}$, $w_{t-2}w_{t-3}$	7	119.575	31.877	50.114	1.286	0.674
9	w_{t-1}, w_{t-2}, w_{t-3}, $w_{t-1}w_{t-1}$, $w_{t-1}w_{t-2}$, $w_{t-2}w_{t-2}$, $w_{t-2}w_{t-3}$, $w_{t-2}w_{t-4}$	8	119.323	33.667	54.508	1.297	0.675
10	w_{t-1}, w_{t-2}, w_{t-3}, $w_{t-1}w_{t-1}$, $w_{t-1}w_{t-2}$, $w_{t-2}w_{t-2}$, $w_{t-2}w_{t-3}$, $w_{t-2}w_{t-4}$, $w_{t-2}w_{t-5}$	9	118.806	35.232	58.679	1.306	0.676

After fitting the given data by GNAR model, we'd better measure model fitting effect for the data or test the model significance further more.

Determination coefficient R^2 can be used to evaluate the efficacy of the fitting; if the model fits the data well, R^2 should close to 1[10].

$$R^2 = 1\text{-}SSE/SST \quad (10)$$

Determination coefficients of the ten models in Table 1 are shown in Figure 3. R^2 varies gently after the fourth model which means that GNAR(2; 3,2) fits the data best and has the less variables. The result is in conformity with the nature of this group of data.

Analysis of variance for model 2, model 3 and model 4 is carried out to test regression effect through F test, and their statistical quantities of F significant are 8.206×10^{-22}、5.188×10^{-21}、1.402×10^{-20} in turn which are far less than the significance level 0.05. So the three model regression effect is remarkable.

From the above, it is feasible to select the GNAR model order based on multiple stepwise regressions. And the process of the order selection method is summarized as shown in Figure 4.

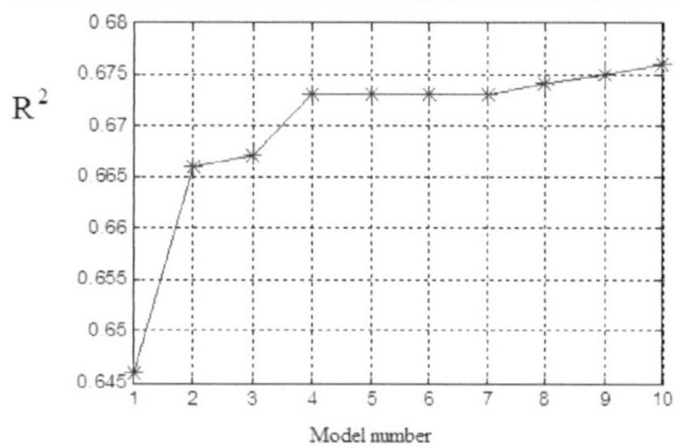

Figure 3. Determination coefficients of the ten models in Table 1.

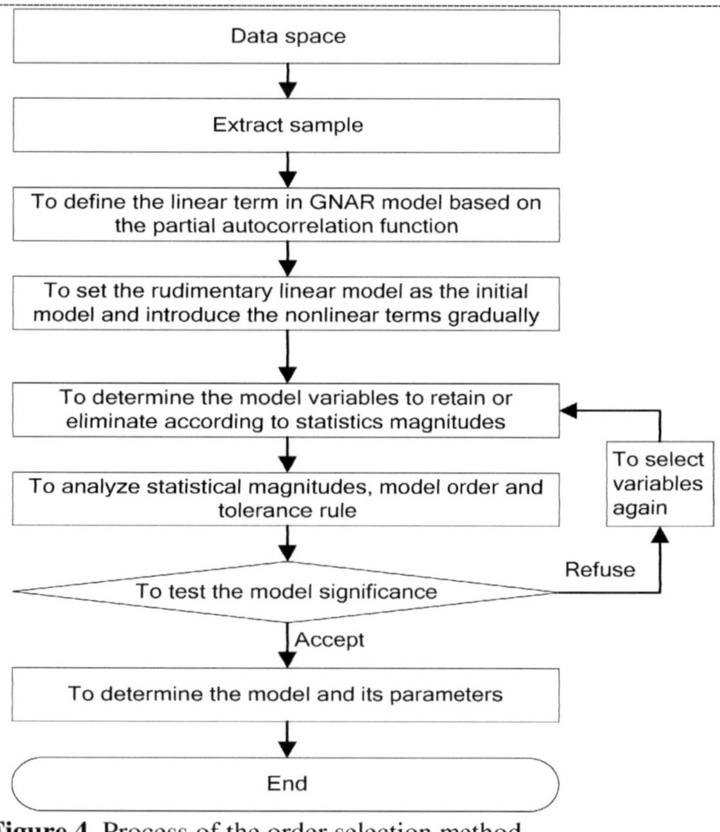

Figure 4. Process of the order selection method.

4. Application for classical data

The order selection method proposed is applied to model and analyse the observations sequence of sunspots from 1700 to1987. According to the steps shown in Figure 4, GNAR model is established based on the 200 sample data from 1700 to 1899.

Figure 5 shows the partial autocorrelation function of sunspots sequence from which the linear memory step length can be judged 2.

The Statistic index of variable selection for Sunspots modelling is presented in Table 2. From the table, it can be seen that BIC statistics increases the punishment for model parameters increasing, so the variable selection result stops with GNAR (2, 2, 2); yet the choice of AIC, RMS and R^2 is GNAR (2, 2, 7).

Forecasting is the important application of time series analysis technology, so we do the forecasting experiment for sunspot data from 1900 to 1987.

Forecasting error square sum named P_error is presented in Table 3. GNAR(2;2,7) has the minimum P_error which confirms that GNAR(2;2,7) is the most suitable model for the group of temporal data.

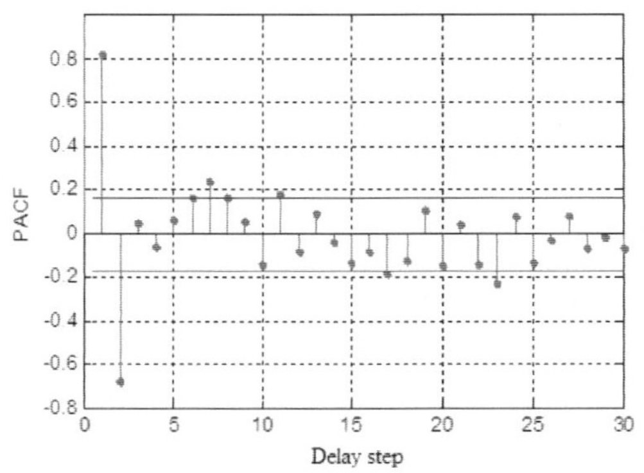

Figure 5. The partial autocorrelation function of sunspots sequence.

Table 2. The Statistic index of variable selection for Sunspots modeling

No	Model	AIC/10^3	BIC/10^3	RMS	R^2
1	GNAR(1;2)	1.244	1.251	498.132	0.432
2	GNAR(2;2,1)	1.236	1.246	477.153	0.458
3	GNAR(2;2,2)	1.221	1.237	437.706	0.508
4	GNAR(2;2,3)	1.218	1.245	425.935	0.527
5	GNAR(2;2,4)	1.217	1.257	415.347	0.547
6	GNAR(2;2,5)	1.211	1.267	393.328	0.583
7	GNAR(2;2,6)	1.207	1.283	375.529	0.614
8	GNAR(2;2,7)	1.198	1.297	349.092	0.655
9	GNAR(2;2,8)	1.232	1.357	398.915	0.623
10	GNAR(3;2,7,1)	1.279	1.450	479.942	0.587
11	GNAR(3;2,7,2)	1.281	1.475	475.315	0.610

Table 3. Forecasting error square sum

No	Model	P_error/10^4
1	GNAR(1;2)	4.242
2	GNAR(2;2,1)	5.087
3	GNAR(2;2,2)	4.833
4	GNAR(2;2,3)	4.667
5	GNAR(2;2,4)	4.385
6	GNAR(2;2,5)	4.018
7	GNAR(2;2,6)	3.652
8	GNAR(2;2,7)	3.372
9	GNAR(2;2,8)	4.107
10	GNAR(3;2,7,1)	4.552
11	GNAR(3;2,7,2)	4.411

5. Conclusion

In this paper, an order selection method based on multiple stepwise regressions is proposed for GNAR model. The process of order selection steps are found out, and the simulation data and the classic time-series data experiment results show that the method proposed is simple, reliable and can be applied to practical engineering.

For a given data set, choosing different statistics perhaps result in different models. Time series data reflect characteristics of the system; therefore, after finding the matching model through some statistics, we should understand the data structure further and analyze the application background to find the optimal model. This is also our future research direction.

Acknowledgments

This work is supported by the National Nature Science Foundation of China (51305194, 51405221), Qing Lan Project (2016) and Jiangsu province science and technology support project (BE2014142). The authors also gratefully acknowledge the helpful comments and suggestions of the reviewers, which have improved the presentation.

References

[1] Fan J Q, Yao Q W. Nonlinear Time Series [M]. Beijing: Higher Education Press, 2011.
[2] Yang Shuzai, Wu Ya, Xuan Jianping. Time series analysis in engineering application (the 2nd Edition)[M]. Wuhan: Huazhong University of Science and Technology Press, 2007.
[3] Kalid Yunus, Peiyuan Chen, Torbjörn Thiringer. Modelling spatially and temporally correlated wind speed time series over a large geographical area using VARMA [J]. IET Renewable Power Generation, 2017, 11(1): 132-142.
[4] Pomsak Deekumpa, Itsared Chanjarean, Kissana Audjai. Adaptive time series forecasting model with associate rules from time series application to business and financial[C]. 2017 International Conference on Digital Arts, Media and Technology, 2017, 320-324.
[5] Junpeng Zhan, C. Y. Chung, Elemer Demeter. Time Series Modeling for Dynamic Thermal Rating of Overhead Lines[J]. Power Systems, 2017, 32(3): 2172-2182.
[6] Yao Xiaodong, Huang Yiqiao, Ma Xiaobo, et al. Thermal Error Modeling and Real-Time Compensation of CNC Machine Tools Based on Time Series Method[J]. Journal of Shanghai Jiaotong University, 2016, 5(5): 673-679.

[7] Zhang Liangli, Zhu He, Wu Chaozhong, et al. Risk Estimation to Vehicles Collision at Intersection Based on ARMA Prediction Model [J]. Journal of Transportation Systems Engineering and Information Technology, 2015, 15(5): 239-245.
[8] Chen Ruwen, Wang Yuguo, Zhan Shishi. Identification Based on GNAR Model for Vehicle Suspension Anti-vibration Performance [J]. Automotive Technology, 2016, (4): 31-35.
[9] Chen Ruwen. The General Expression for Linear and Nonlinear Auto-regressive Time Series Model and Its Engineering Application [M]. Nanjing: Southeast university press, 2011.
[10] Samprit Chatterjee, Ali S Hadi. Regression Analysis by Example [M]. Beijing: China Machine Press, 2016.
[11] Ruey S Tsay. Analysis of Financial Time Series [M]. Beijing: Posts and Telecom Press, 2012.

IOP Conf. Series: Materials Science and Engineering **272** (2017) 012011 doi:10.1088/1757-899X/272/1/012011

Prediction and Optimization of Key Performance Indicators in the Production of Stator Core Using a GA-NN Approach

M. Rajora[1], P. Zou[2], W. Xu[3], L. Jin[3], W. Chen[3], and S.Y. Liang[1]

[1]George W. Woodruff School of Mechanical Engineering, Georgia Institute of Technology, Atlanta, Georgia, USA
[2]Mechanical Engineering College, Donghua University, Songjiang District, Shanghai, China
[3]Shanghai Electric Group Co., Ltd. Central Academe, Shanghai, China

Abstract. With the rapidly changing demands of the manufacturing market, intelligent techniques are being used to solve engineering problems due to their ability to handle non-linear complex problems. For example, in the conventional production of stator cores, it is relied upon experienced engineers to make an initial plan on the number of compensation sheets to be added to achieve uniform pressure distribution throughout the laminations. Additionally, these engineers must use their experience to revise the initial plans based upon the measurements made during the production of stator core. However, this method yields inconsistent results as humans are incapable of storing and analysing large amounts of data. In this article, first, a Neural Network (NN), trained using a hybrid Levenberg-Marquardt (LM) – Genetic Algorithm (GA), is developed to assist the engineers with the decision-making process. Next, the trained NN is used as a fitness function in an optimization algorithm to find the optimal values of the initial compensation sheet plan with the aim of minimizing the required revisions during the production of the stator core.

1. Introduction

Even though, over the years, the bounds of what manufacturing industries can accomplish has increased significantly, with the rapid development of global industry, manufacturing industries are facing many challenges, such as the higher complexity and flexibility of problem as well as increasing human labor cost. These problems create bottleneck to the traditional manufacturing system due to the engineer's limitations on handling uncertainties, complexity and memorizing large amount of data [1]. The recent advancement in computing has enabled researchers to develop many artificial intelligence techniques that have enabled the transfer of the years of experience stored in the minds of the human operators to computers, thereby transferring the decision-making burden form man to machine. Furthermore, optimization techniques have enabled engineers to find optimal process parameters that can help them the achieve the desired outputs.

The intelligent techniques have showed great results when used for creating prediction models in the absence of physics-based models and also for obtaining optimal process parameter combinations for optimization problems. Rajora et al. [2] used NNs, *k-means clustering* algorithm, and GA to create a forward prediction model and obtain multiple optimal solutions for the process of electrochemical micro-machining (μECM). Tansel et al. [3] used a cluster of NN to represent the friction stir welding operation. Rao and Pawar [4] used ABC, SA, and harmony search algorithms to minimize the production cost, maximize the production rate, and minimize the surface roughness of a grinding process. Ji et al. [5] used NN to model the process of minimum quantity lubrication (MQL) turning and then utilized GA to optimize its input process parameters. Lin et. al [6] developed a forward prediction model between various inputs of turning operation and the surface roughness and cutting force. Benardos and Vosniakos [7] created an ANN to predict the surface roughness in computer numerically controlled (CNC) face milling. Mahesh et. al [8] used GA for the optimization of cutting parameters to minimize the surface roughness. Kumar et. al [9] used a hybrid Taguchi-ANN for predicting surface roughness during machining of titanium alloys. Ji et. al [10] used ANNs to predict surface residual stress in MQL face turning.

Though intelligent techniques have gained widespread usage in the traditional manufacturing industries, an area where traditional human decision-making process is still relied upon is in the production of stator cores. In this paper, NNs have been used to help the engineers with the decision-making process. The NNs are trained using a hybrid Genetic Algorithm (GA) – Levenberg Marquardt (LM) algorithm with the aim of increasing their prediction accuracy. Next, the trained NNs are used as the fitness function in an optimization model, using GA, to obtain optimal initial compensation sheet plan. The rest of the paper is organized as follows: the production of stator cores and decision making involved in their production is introduced. Next, the NN prediction model and optimization model are introduced. The results obtained when the trained NN is used to model the production of stator core process are presented followed by the optimization results. Lastly, some conclusions are drawn based on the results obtained.

2. Description of the Problem

A turbogenerator stator is comprised of two major components: 1) the stator windings, and 2) the stator core. A stator core is built up as laminations are placed side-by-side, to make a complete circular or ringed layer. The next layer is laid, offsetting each layer like a brick or cinder block wall. Along its back, the stator core is held in place by rails, which are bolted to so-called guide bars. Once the stacking is complete, the stator core is heated and pressed with a defined force. The process of stacking, heating and pressing needs to be repeated six times in the production to ensure the total height of the stator core reaches the desired value. To achieve uniform pressure distribution among the stator lamination layers, engineers align different shapes and sizes of compensation sheets between the layers of stator lamination per the measurement result of the lead extrusions, as seen in Fig. 1.

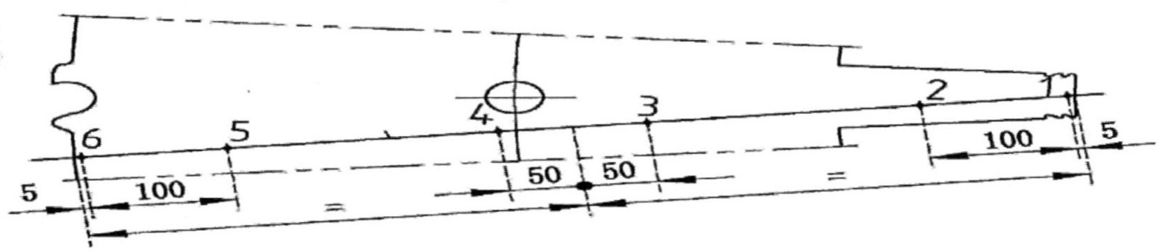

Figure 1. Measurement points of the lead extrusion.

In this study, there were nine different types and sizes of compensation sheets available that could be inserted among the stator lamination. Before the production of the stator laminations, experienced engineers make an initial guess on the number of each type of compensation sheets to be added after each pressure application process. Next, pressure is applied to the stator core for the first time and the value of four key performance indicators (KPIs), given by Equation (1) – (4) are measured. Based on the value of the four KPIs, the engineers use their experience to revise the initial compensation sheet assignment plan by either assigning more or fewer of each type of compensation sheets. Then, based on the revised compensation sheet assignment plan, the different compensation sheets are inserted at the eight location and the pressure is applied for the second time. The above process is repeated until the pressure application process is completed.

$$Average_1 = \frac{1}{8}\sum_{i=1}^{8}[avg(thickness_5^i + thickness_6^i) - avg(thickness_3^i + thickness_4^i)] \quad (1)$$

$$Deviation_1 = \sqrt{\frac{1}{7}\sum_{i=1}^{8}[avg(thickness_5^i + thickness_6^i) - Average_1)]^2} \quad (2)$$

$$Average_2 = \frac{1}{8}\sum_{i=1}^{8}[avg(thickness_1^i + thickness_2^i) - avg(thickness_3^i + thickness_4^i)] \quad (3)$$

$$Deviation_2 = \sqrt{\frac{1}{7}\sum_{i=1}^{8}[avg(thickness_1^i + thickness_2^i) - Average_2)]^2} \quad (4)$$

During the production of the stator cores, the engineers also aim to create a correct initial compensation sheet assignment plan based on their years of experience. Additional complexities are added to the process of creating an optimal initial compensation sheet assignment plan since there is an upper limit on how many times each compensation sheet can be used during the production of the stator core.

Under the current settings of the production of stator cores, coming up with the final compensation sheets assignment plan in which no additional adjustments are required is very crucial to help reduce the dependency on the experienced engineers and improve the efficiency. Therefore, it is important to create a prediction model that can accurately map from the inputs to the key performance indicators (KPIs) in the production of stator core and a corresponding optimization model the can create an optimal initial compensation sheet assignment plan which deviates very little from the adjusted compensation sheet assignment plan while meeting all the constraints of the process.

3. Modelling

3.1. Prediction Model

To map from the inputs to the outputs of the stator lamination production process NNs were utilized. Instead of Gradient Descent (GD), Levenberg-Marquardt (LM) algorithm was used for training as it has shown to outperform GD in a variety of problems. To overcome the drawback of gradient based algorithms (only local convergence), it was used in conjunction with GA, a metaheuristic algorithm. During the training procedure, GA was first used to search the function space for the optimal or sub-optimal weight and bias values. These values were then used as a starting point for the LM algorithm, which then fine-tuned the weight and the bias values to improve the results.

Since the pressure application was performed in six stages, five different NN models were created. Each NN had 13 inputs, four corresponding to the value of the KPIs obtained after a pressure application and the other nine corresponding to the number of each type of compensation sheet to be added per the initial compensation sheet plan before the next pressure application. The nine outputs were the revised compensation sheet plan based upon the experience of the engineers.

3.2. Optimization Model

The optimization objective in this study was to minimize the number of incorrectly predicted compensation sheets to be added before each force application. The four inputs corresponding to the KPI values were used as reference inputs (their values were not changed during optimization) while the nine inputs corresponding to the number of each type of compensation sheet to be added before each pressure application were the design variables. The NNs created in the previous step were used as the fitness function during the optimization process and GA was used to find the optimal values of the design variables.

4. Results

4.1. Training Results

Due to the lack of actual production data, 20 simulated data sets were generated with the help of experienced engineers. Of the 20 data sets available, 14 of them were used for training, three for validation, and three for testing the trained NN. The best NN structure, selected through trial-and-error, for each NN was a 13-12-9 structure. For each trained NN, its prediction accuracy was measured by observing how many outputs for each testing data set it could predict correctly. These results are shown in Figure 2, each column for a NN represents the number of outputs predicted correctly for a testing data set by that NN.

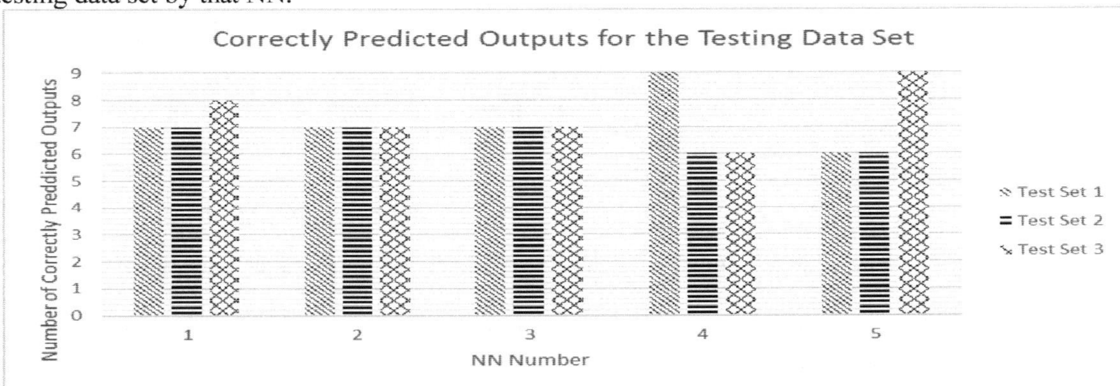

Figure 2. Prediction results obtained using the trained NNs.

4.2. Optimization Results

Once the NNs had been trained sufficiently, they were used as the fitness function for the GA optimization process. To test the optimization model, three different sets of four KPI value for the reference inputs were generated for each pressure application and given as inputs to the NNs. Next, the optimization algorithm was used to find the optimal values of the initial compensation sheets. The optimization objective was to minimize the difference between the initial compensation sheet plan and the adjusted compensation sheet plan, so that the optimized initial compensation sheet plan could be used in the production with little to no adjustment. The corresponding optimization result are calculated based on Equation (5) and are shown in Table 1.

$$\%\text{ difference} = \frac{\text{number of incorrectly predicted compensation sheets}}{\text{total compensation sheets predicted}} * 100\% \qquad (5)$$

Table 1. Optimization results obtained

	Percentage of difference using Settings A (%)	**Percentage of difference using Settings B (%)**	**Percentage of difference using Settings C (%)**
Simulation 1	26.7	26.7	26.7
Simulation 2	22.2	26.7	33.3
Simulation 3	33.3	33.3	33.3

Based on the results, the results obtained using setting A provides a slight improvement comparing to the results obtained by setting B and C. However, in general, the percentage of difference between the initial plan and the adjusted plan is around 22.2-33.3%, which may due to the relatively high number of optimization variables.

5. Conclusion

In this study, a prediction and optimization approach was proposed to solve the problem in the production of stator core that the assignment and adjustment of compensation sheets between each pressure application. In the proposed approach, a NN model was first trained with simulated training data with the aim of capturing the decision-making capabilities of experienced human engineers. Then, a GA based optimization model, which utilized the built NN as the fitness function, was created to minimize the difference between the initial assignment plan and the adjusted one.

The results of the prediction model showed that the NN trained using the hybrid LM-GA algorithm had a prediction accuracy of 90% indicating that the NN model can capture the decision-making capabilities of experienced human engineers. However, the optimization result shows that the algorithm is unable to find an optimal plan and the adjusted one was generally around 30%. Future research will be focused on improving the optimization process to narrow down the difference between the initial plan and the adjusted one. Possible solutions include combining simulated and historical data and selecting alternative inputs for the prediction model that might be better suited for the optimization purpose.

6. References

[1] S.Y. Liang, M. Rajora, P. Zou, Intelligent Manufacturing System for Next Generation Factories. Advances in Intelligent Systems Research. (2015); In print

[2] Rajora, M., Zou, P., Yang, Y.G., Fan, Z.W., Chen, H.Y., Wu, W.C., Li, B. and Liang, S.Y., 2016. A split-optimization approach for obtaining multiple solutions in single-objective process parameter optimization. SpringerPlus, 5(1), pp.1424

[3] Tansel, I.N., Demetgul, M., Okuyucu, H. and Yapici, A., 2010. Optimizations of friction stir welding of aluminum alloy by using genetically optimized neural network. The International Journal of Advanced Manufacturing Technology, 48(1-4), pp.95-101

[4] Rao, R.V. and Pawar, P.J., 2010. Grinding process parameter optimization using non-traditional optimization algorithms. Proceedings of the Institution of Mechanical Engineers, Part B: Journal of Engineering Manufacture, 224(6), pp.887-898.

[5] Ji, X., Shih, A.H., Rajora, M., Shao, Y.M. and Liang, S.Y., 2015. Optimization of Tensile Residual Stress on Machined Surface in MQL Turning. In Applied Mechanics and Materials (Vol. 713, pp. 209-212). Trans Tech Publications

[6] Lin, W.S., B.Y. Lee, and C.L. Wu., 2001. Modeling the Surface Roughness and Cutting Force for Turning. Journal of Materials Processing Technology 108(3), pp. 286-293.

[7] Benardos, P.G, and Vosniakos, G.C., 2002. Prediction of Surface Roughness in CNC Face Milling Using Neural Networks and Taguchi's Design of Experiments. Robotics and Computer-Integrated Manufacturing 18, pp. 343-354.

[8] Mahesh, G., Muthu S., Devadasan, S.R., 2015. Prediction of Surface Roughness of End Milling Operation Using Genetic Algorithm. International Journal of Advanced Manufacturing Technology 77.1, pp. 369-381.

[9] Kumar, S., Batish, A., Singh, R., Singh, T.P., 2014. A Hybrid Taguchi-artificial Neural Network Approach to Predict Surface Roughness during Electric Discharge Machining of Titanium Alloys." Journal of Material Science and Technology 28(7), pp. 2831-2844.

[10] Ji, X.; Shih, A.H., Rajora, M., Shao, Y., Liang, S.Y., 2014. A Hybrid Neural Network for Prediction of Surface Residual Stress in MQL Face Turning." Applied Mechanics & Materials 713-715, pp. 209-212.

IOP Conf. Series: Materials Science and Engineering **272** (2017) 012012 doi:10.1088/1757-899X/272/1/012012

Research and Design on a Product Data Definition System of Semiconductor Packaging Industry

Jinfei Shi[1], Qingyao Ma[2], Yifan Zhou[3] and Ruwen Chen[1]

[1] Nanjing Institute of Technology, Nanjing, China, 211167
[2] Huaihai Institute of Technology, Lianyungang, China, 222005
[3] Southeast University, Nanjing, China, 211189

Abstract. This paper develops a product data definition (PDD) system for a semiconductor packaging and testing company with independent intellectual property rights. The new PDD system can solve the problems such as, the effective control of production plans, the timely feedback of production processes, and the efficient schedule of resources. Firstly, this paper introduces the general requirements of the PDD system and depicts the operation flow and the data flow of the PDD system. Secondly, the overall design scheme of the PDD system is put forward. After that, the physical data model is developed using the Power Designer15.0 tool, and the database system is built. Finally, the function realization and running effects of the PDD system are analysed. The successful operation of the PDD system can realize the information flow among various production departments of the enterprise to meet the standard of the enterprise manufacturing integration and improve the efficiency of production management.

1. Introduction

The production process is complex in a semiconductor packaging and testing company, because the order quantity is large and the order also depends on the individual demand of different customs. Customized packaging requirements in the production-manufacturing process need a flexible MES system. Consequently, it is necessary for the semiconductor industry to develop, design, and research the MES system. Nowadays, many scholars study the relevant technologies of MES in semiconductor industry. For example, Hwa Gyoo Park et al [1] put forward an object-oriented simulator for manufacturing execution systems in the perspective of architecture, H. Wada [2] and B. Li [3] applied an autonomous agent approach in the computer for manufacturing execution control systems in modeling and Paul Valckenaers et al [4] presented schedule execution in autonomic manufacturing execution systems to expand the function module realization of MES. The PDD system is an important component of the MES. In order to solve problems in semiconductor packaging and testing, the authors used information integration and fusion theory to research and design a PDD system. The system could improve the management level and enhance the ability to cope with the challenge of globalization.

2. Design of the PDD system

The core of the semiconductor packaging industry demand is the special customization. The key to the research and design of the PDD system of MES is the component system layer. The mapping mechanism is applied to the semiconductor packaging industry, including materials, equipment, personnel, product characteristics, production process data, WIP, and finished goods status.

IOP Conf. Series: Materials Science and Engineering **272** (2017) 012012 doi:10.1088/1757-899X/272/1/012012

Consequently, it achieves the integration and fusion of information and knowledge based on the function construction, encapsulation, interface, and management. After that, basic information is integrated by database access technologies. Finally, component modelling, enterprise modelling support, and workflow management are used on the craftwork of customized products and the requirements of potential customers. The evolution and appreciation of information are achieved by using data warehouse, on-line analytical processing, and data mining technology [5].

2.1. General requirements of the system

The PDD system is mainly about using massive collected information, i.e., materials, equipment, personnel, product characteristics. The PDD system provides instructions on the back-end process of production. Requirement analysis of the PDD system is actually about investigating, summarizing, and clarifying the requirements of customers. The entire requirements of the PDD system consist of two major parts. One is maintenance on the Product Internal Device (referred to as PID); the other is maintenance on the Main Bill of Material (referred to as MBOM). The requirement analysis of the PDD system is listed in Figure 1.

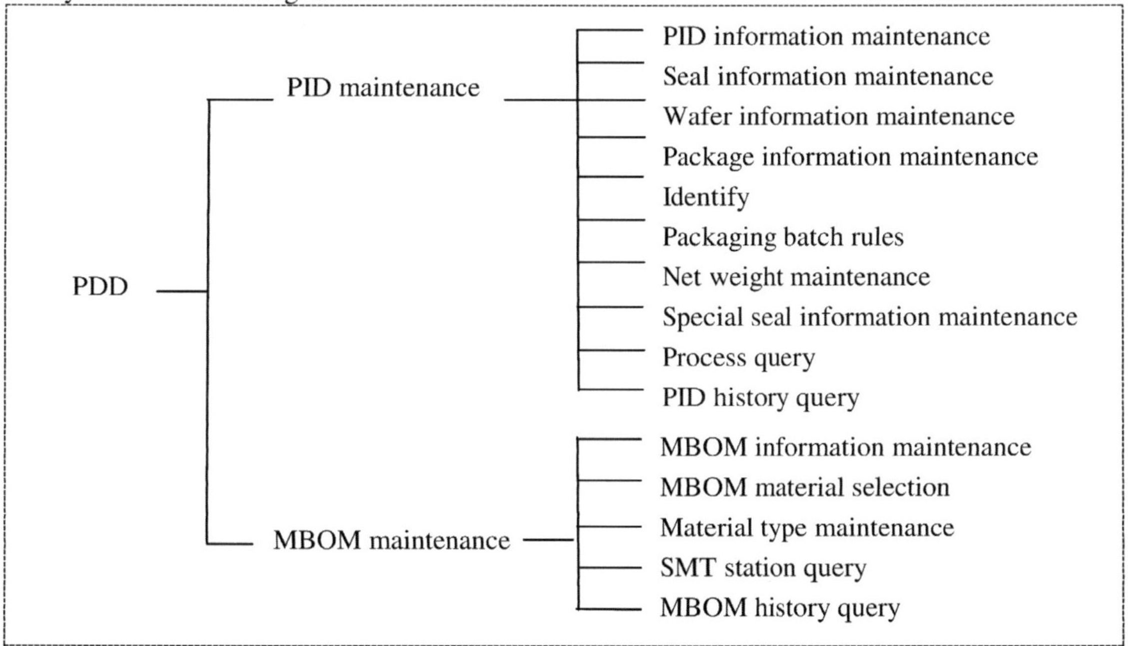

Figure 1. The requirement analysis of the PDD system.

The maintenance of PID includes PID information maintenance, seal information maintenance, wafer information maintenance, package information maintenance, packaging batch rules identification, net weight maintenance, special seal information maintenance, process query, and PID history query. The maintenance of the MBOM includes modules such as MBOM information maintenance, MBOM material selection, material type maintenance, SMT station query, and MBOM history query. Through the analysis of the requirements, the overall package operation flow and data flow of the system are summarized as Figures 2 and 3, respectively.

The operation flow of the system is as follows, a wafer from the front-end process is cut into smaller wafers (Die) by dicing process. The cut wafers is then glued to the island of the corresponding basal laminas (lead frame). After that, through the use of ultrafine metal (gold, tin, copper and aluminium) wire or conductive resin, the Bond Pad of the wafer is connected to the corresponding Lead of the basal lamina. The required circuit is also formed. In the next step, plastic housing is used for encapsulation and protection of individual wafers. After the completion of the package, a series of operations called finished product tests are performed, including incoming, testing, and packaging.

IOP Conf. Series: Materials Science and Engineering **272** (2017) 012012 doi:10.1088/1757-899X/272/1/012012

The data flow of the PDD system is to edit the data from the web collection platform, and the data are transferred to the production line system after identification. In the first step, the data flow from the web collection platform to the PDD system. During that step, staffs edit and filter data from the web collection platform. The second step identifies the edited data of the PDD system. This step transfers data from the PDD system to the production line system. So far, the entire data from the outside flow into the production line through the PDD system, and guide the production packaging line.

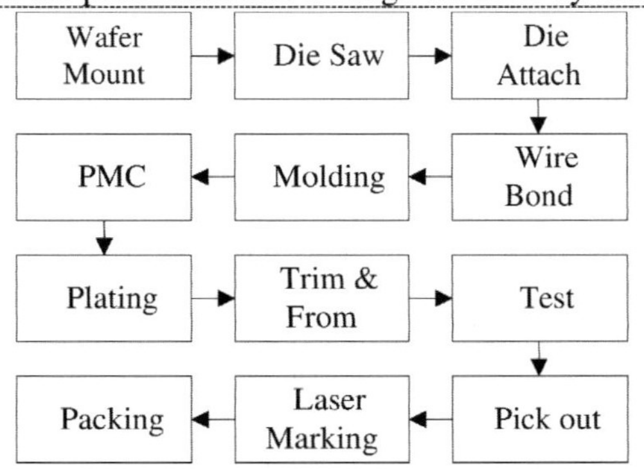

Figure 2. Flow chart of the overall package operation.

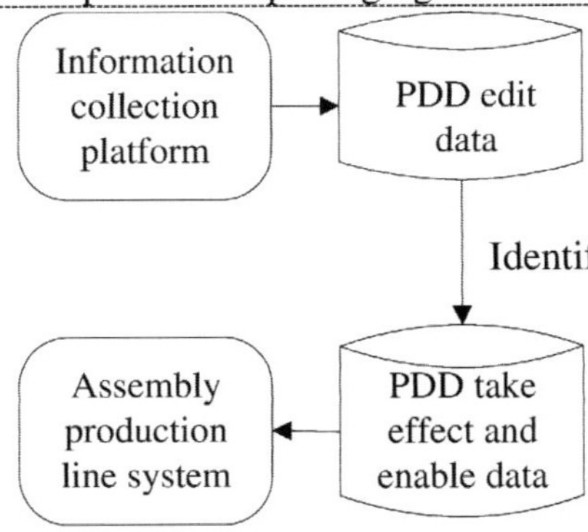

Figure 3. Flow chart of the PDD system data.

2.2. Overall structure of the system

Compared with the old framework, this system adopts the three-layer framework model of Client/Server, which is more flexible and hierarchically arranged. The working principle is to reduce the system operating burden by assigning tasks to the client and server [6]. The development of client and server programs is different. The user program is generally in the client-side, which is used for completing the specific operation of users. The server side is to manage the database, and share information with users. The three-layer framework divides the whole business applications into three parts: interface layer, business logic layer, and data access layer, as shown in Figure 4.

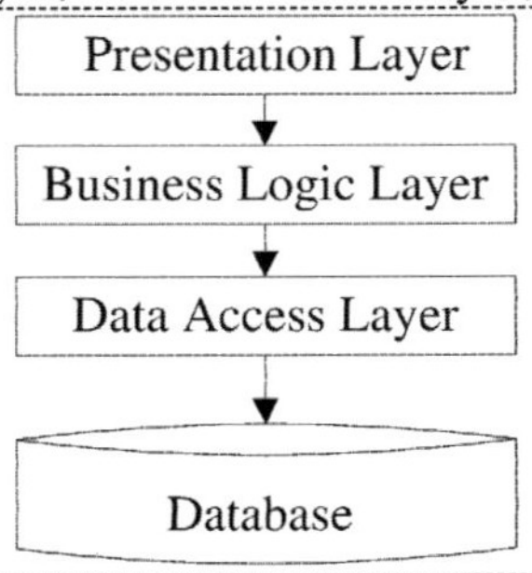

Figure 4. Diagram of three - tier C / S system framework.

2.3. Database Design of the System

The PDD system is connected with many modules, and its internal structure is complex. Therefore, the workload of database design is heavy. The Power Designer 15.0 tool is adopted to model and design the database, which can regulate the design task and shorten the design cycle. It provides strategies for standard database design and operational analysis for software development management by integrating several standard data modelling techniques and main-stream developing platform. It also provides users with the tools of selected model examples, which can help users to create the conceptual and physical models of the database.

Generally, there are two methods to design the conceptual structure of the database system, which are the centralized model method and the view integration method [7]. The view integration method is used in this paper. Firstly, the system is designed from the local to the whole. The PDD system is complex and has many branches. The database system is divided into separate modules for design, and then sharing among these modules is achieved. Through the analysis of the requirements, the Power Design15.0 tool is used to model the database and establish the E-R data model diagram.

The logical structure of the database system is designed, after the conceptual structure is identified, which turns the entity relation data model (E-R model) into a logical model identifiable to the database. The logical model could be expressed in many ways. The most popular and practical one is the relational scheme model. The detailed steps are as follows [8].

(1) The relational model is established through conversion and mapping of entity classes in the E-R model.
(2) Each link in the E-R model is converted. If the relationship is more than one pair, it is necessary to add a set of additional relational patterns.
(3) The attributes in the E-R diagram can also be converted to the attributes of the relationship.

The design of physical data model of PDD system is to do SQL execution of the conceptual data model by the Power Designer. Then, the physical data model of the relation based list data structure is established in the Oracle database [9]. The physical data model that is designed for the relational database management system is abstract to users. There are two main goals of design: improving the performance of the database system and enhancing the database system storage space efficiency. After the internal characteristics of the backstage Oracle 11g database are fully understood, the physical design of the database system considers the following four aspects: anti-normalization processing, data storage structure design, data storage path, and storage location design. System configuration is optimized taking into account of time efficiency, space efficiency, and maintenance cost of the PDD system database [10]. The anti-normalization process is to return a number of tables that have been normalized to the last specification level to improve the efficiency of the PDD system. Three factors are taken into consideration while designing the data storage structure. They are access time, storage space utilization, and maintenance costs. There are inner restricted relationships among the three [10]. Data storage location design is to storing data files of system tables and user tables in different space on the disks drive in order to divide the storage of data and index. The data storage path usually creates an index to improve the searching speed for tables and aggregate data. As for system configuration optimization, the Oracle11g database allocates parameters for storage provided by the user which gives the reasonable default values. However, sometimes, certain parameters should be set according to the actual needs of the system and the specific environment.

3. Implementation of the PDD system

3.1. Implementation of functions of the system

The function of the PDD system is the same as its roles, which is to serve MES. The PDD system is designed for improving the adaptability and timelines of the production plan and increasing the information flow about the bottom control activities. It is also designed for filling the gaps between management plan and production control. The PDD system is designed based on the development of the MES system.

The function interface is about settings for PID maintenance. It sets the marking, packing, special note, and curing profile through PID. The interface can be divided into two parts. One is used as the browser of property of PID itself. The other is used for tabbing. Ways of paging include four parts: departs, marking, packing, special note and curing profile. Each page is set with the Edit button in the browse block respectively. Then the data is maintained by the pop-up editing interface. Among them, the state of PID could be "UnApproved", "Active", or "Frozen". These states are specified manually in the PID editing interface. When the state of the PID is designated as “Frozen”, information like marking, packing, special note, curing profile information cannot be edited. Meanwhile, when a PID is selected, related information or data will show up i. Every time new PID information is queried, the data will be cleared first. Then corresponding results will show up according to the query. In addition, the states of PID are marked with different colures in the query results, such as “Frozen” (red), “Active”(green), “UnApproved” (yellow).

The function interface is mainly used to the maintain BOM and sub BOMs. It also completes the job of adding, modifying, deleting, or querying BOM. BOMs have three kinds of state. After creation, with the default version of V1, the state is “Created”. Then you can modify and maintain BOM information. When released, the state is changed into “Active”. At this moment, all the BOM information is not allowed to modify. In addition, when released, among all versions of BOM, only one could be “Active”. That is, during the release, the current version turns into “Active ”and the rest versions become “Frozen”.

3.2. Running effects of the system

The PDD system designed and developed in this paper has been already in stable operation in an enterprise. It not only improves the information level of the production department, but also greatly improves the work efficiency. The system can reduce the manual records and cut down errors in operation. In previous manual operation, disoperation often occurred after shifting, because the latter one has no idea whether the previous one has done certain operation or not. With this system, however, employees can query all the previous operations and then operate accordingly. The disoperation rate is reduced by more than 25%. Moreover, the utilization rate of the automation equipment of the enterprise is improved significantly. Employees can find unoccupied machines to realize reasonable allocation. The work efficiency is thus increased by 20% approximately.

4. Conclusion

The system orients to semiconductor packaging integration information of MES. With the development of computer technology and network communication technology in China, it uses advanced production management concept, especially the design of PDD system of MES. In this way, it realizes the integration, sharing, application, evolution and appreciation, of the information on management level and lower levels for producing. It also results in unifying of logistics, capital flow, and information flow. The practical application shows that the system contributes much to the integration, interaction, and sharing of information among different departments of the semiconductor packaging enterprises.

5. References

[1] Hwa Gyoo Park, Jong Myung Baik and Sang Bong Park. A development of object-oriented simulator for manufacturing execution systems. Computers & Industrial Engineering. 1999, 37:239-242.

[2] H. Wada, S. Okada. An autonomous agent approach for manufacturing execution control systems. Integrated Computer-aided Engineering, 2002, 9(3): 251-262.

[3] B Li, Z. H. Chen, C. Ying. Research on reconfigurable manufacturing execution system. In International Conference on Intelligent Mechatronics and Automation, Chengdu, 2004, 157- 161.

[4] Paul Valckenaers, Hendrik Van Brusse, Paul Verstraete, Bart Saint Germain, Hadeli. Schedule execution in autonomic manufacturing execution systems. Journal of Manufacturing Systems, 2007, 26(2): 75-84.

[5] Xing Yingjie, Yang Hua. Application of web - based ERP / MES system in Liaoning special steel Co., Ltd [J]. Computer Integrated manufacturing Systems. 2004.10(5): 553-554.

[6] SeungWoo Lee, SoJeong Nam, and Jai-Kyung Lee. Real-time data acquisition system and HMI for MES. Journal of Mechanical Science and Technology, 2012, 26(8): 2381-2388.
[7] SUE Cheng. Manufacturing execution system based on the soda ash industry. In Intemational Conference on Manufacturing Science and Technology(ICMST 2011), Singapore, 2011, 383-390.
[8] Yu Lei . My view on the application of database technology in the era of big data[J]. Digital technology and Application. 2016(11).
[9] Wenqian Cao, Shaohong Jing, Xiaohong Wang. Research on manufacturing execution system for cement industry. In 3rd IEEE Conference on Industrial Electronics and Applications (ICIEA 2008), Singapore, 2008, 1614-1618.
[10] Huang Fei. Research & Realization of production management system in a semiconductor manufacturing [D]: Department of Mechanical Engineering Southeast University, 2004. 3:23-27.

A Multi-Component Automated Laser-Origami System for Cyber-Manufacturing

Woo-Hyun Ko[1], Arun Srinivasa[2] and P. R. Kumar[1]

[1]Department of Electrical and Computer Engineering, Texas A&M University, 188 Bizzell st, College Station, TX 77843, USA
[2]Department of Mechanical Engineering, Texas A&M University, 188 Bizzell st, College Station, TX 77843, USA

E-mail:prk@tamu.edu

Abstract. Cyber-manufacturing systems can be enhanced by an integrated network architecture that is easily configurable, reliable, and scalable. We consider a cyber-physical system for use in an origami-type laser-based custom manufacturing machine employing folding and cutting of sheet material to manufacture 3D objects. We have developed such a system for use in a laser-based autonomous custom manufacturing machine equipped with real-time sensing and control. The basic elements in the architecture are built around the laser processing machine. They include a sensing system to estimate the state of the workpiece, a control system determining control inputs for a laser system based on the estimated data and user's job requests, a robotic arm manipulating the workpiece in the work space, and middleware, named Etherware, supporting the communication among the systems. We demonstrate automated 3D laser cutting and bending to fabricate a 3D product as an experimental result.

1. Introduction

Advanced technologies have led to many innovations in the manufacturing industry such as productivity enhancement by automation technology, facility enlargement by information communication technology, and quality improvement and product diversity by manufacturing process technology [1][2][3]. Employing their innovations, the manufacturing industry is proceeding towards producing customized and high quality products at low costs.

3D printing technology is emerging to lead the manufacturing industry. However, additive-type 3D printing has some shortcomings like long process times, unhealthy powder additives, and limit to product sizes. In order to resolve these issues, a laser origami technology has been introduced [4]. It rapidly manufactures customized 3D creatures by using a laser to cut, bend or fold sheet material. However, manually operated bending processes require a human operator to perform and monitor the whole process. As the switching between cutting and bending processes is performed by a human operator, he or she can be exposed to hazards caused by the laser. Moreover, it takes significant amount of time to find the proper intensity of a laser beam and feed rate of a laser head by trial and error. Also, the experience level of the operator can affect the quality of the product due to poor intensity and feed rate. It can be difficult to guarantee the quality of the product with manually bending.

On the contrary, the proposed feedback-based automatic laser fabricating system can not only assure the safety of a human operator, but it can also reliably repeatedly reproduce high-quality products.

Even more, considering the scalability of the automatic system, massive and customized production with high-quality is enabled.

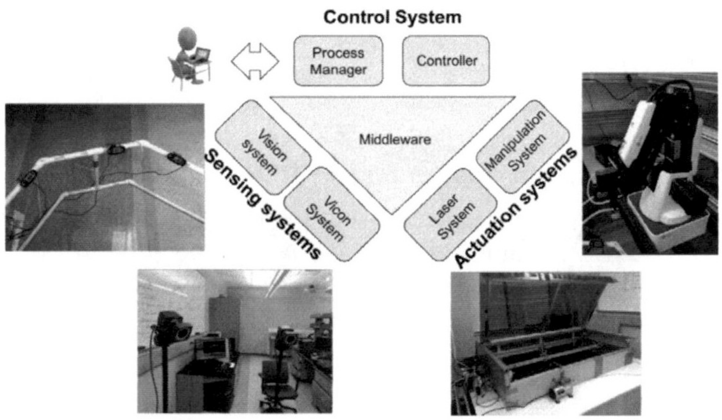

Figure 1. The overview of an automated laser origami system based on middleware.

We propose an automated laser origami system for cyber manufacturing. Figure 1 shows the overview of an automated laser origami system based on middleware. A sensing system with cameras monitors the status of the manufacturing process and a control system controls the feed rate of a laser head, the intensity of a laser beam and the position of a robotic arm based on the feedback information from the sensing system to operate cutting and bending processes, automatically. Middleware enables laser, sensing and control systems, which are connected over a network, to communicate with one another by sending and receiving messages. It can also support component-based development which offers simple quick deployment, easy extension, and maintenance. We validate the proposed system by experimenting with laser cutting and bending in a real laser origami system.

2. Related works

Balkcom et al. [6] extended some well-known necessary conditions for origami patterns to be flat-foldable, such as Kawasaki's theorem, Maekawa's theorem, and Meguro's theorem, for automatic planning and folding manipulation, and presented the world's first origami folding robot through a video. However, all processing was open-loop so that the number of foldings was limited due to the accumulated errors in the frictional insertion of the paper into the slot and in the positioning and grasping of the paper. Namiki and Yokosawa [7] achieved a feedback-based robotic origami folding in a real system. They extracted dynamic motion primitives from a valley folding process and experimentally demonstrated valley folds twice in a row in real time by using 3D visual and force-torque information and a physical simulation model. On the other hand, there was still a limit to the kind of a material that could be only a flexible sheet, like a paper. Tanaka et al. [8] analyzed the difficulty of the origami folding tasks by decomposing them into subtasks, and defined the levels of difficulty for each subtask. Based on observation of manipulation by humans, they designed a four-fingered origami folding robot and achieved folding of the Tadpole. However, their system was able to manipulate only a material similar to paper, and required online teaching by humans (i.e., semi-automatic) to generate the trajectories of the hands to fold a paper. Mueller et al. [4][5] suggested a prototyping system which could manufacture a 3D object by using a regular laser cutter. A laser was defocused to heat a desired region on the surface of a workpiece until it was bent or stretched to produce a 3D object. They proposed three design elements by using LaserOrigami and presented various 3D creations such as a card holder, a suspender, and a pen holder, which were manufactured by LaserOrigami without manual assembly. However, the open-loop process required an operator to supervise the bending process to achieve the desired angles.

3. Etherware

Conventional manufacturing machines have usually been operated and monitored by onsite supervisors and there have been difficulties in changing or expanding their system configurations. On the other hand, middleware offers extensibility and remote controllability to cyber manufacturing. Customized cyber-manufacturing needs heterogeneous manufacturing machines to cater to the convenience and requirements of customers. Component-based middleware enables the management of various kinds of devices which allows dynamic system configuration. Employing a remote control, supervisors do not need to stand by machines to manage and monitor them. They can operate the whole system with graphical user interfaces in real time based on the streams of sensing information and control commands over a network.

We have implemented the system over Etherware which is a middleware developed in our prior work [10]. Etherware has been developed to support a rapid, reliable and evolvable application development. Figure 2 describes the architecture of Etherware.

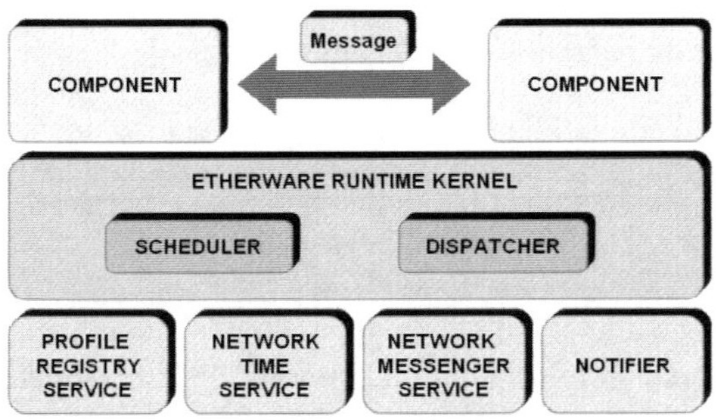

Figure 2. The architecture of Etherware.

The kernel of Etherware is designed with micro-kernel architecture which consists of a scheduler and a dispatcher. A scheduler manages messages requested by components and a dispatcher executes the messages transferred by the scheduler. There are three Etherware services: Profile Registry Service (PRS), Network Time Service (NTS), and Network Messenger Service (NMS), which provide communication networking among components. PRS allows semantic naming of components. One component can send a message to another component with the profile name of the destination component instead of its IP address. NTS deals with the clock differences among different computing devices to synchronize their events. It calculates the time offsets of the computers and translates the timestamp of the received message into local time. NMS manages the network connectivity among computers and handles message delivery over the network. It uses an XML format to transfer messages so that a developer can easily define type and contents of a message to be sent.

4. Middleware-based system integration

Current machine tool monitoring technologies are mostly limited to using static or well-accessed wireless sensors. Recent advances in mobile communications and computer vision algorithms offer a radically new approach for real-time process control and quality assurance. We have built a multi-component manufacturing system centered around a laser processing machine. The network architecture for the manufacturing system offers considerable advantages for custom manufacturing systems. For the system integration, Etherware was used to implement the networking among the individual system components. The middleware enables control over a communication network and computations can be carried out elsewhere in the cloud with seamless migration.

Figure 3 shows the feedback-based laser origami system. All systems are connected over Ethernet and communicate with each other by Etherware which supports message-based communication. The kernel of Etherware, which is run in each computer, executes components, and the profile registry service of Etherware provides a semantic naming service with the profile table mapping IP addresses of computers and IDs of components which are assigned by the kernel to the profile names of components. For example, a Vision Sensor component can send a message containing the bending angle to a Controller component by using only its name 'Controller', instead of the IP address of the computer which the component is running on. Networking among the system components enables the scalability of the system, Etherware supports component-based application developments and facilitates not only the development of the individual system components, but also system integration, management, and maintenance.

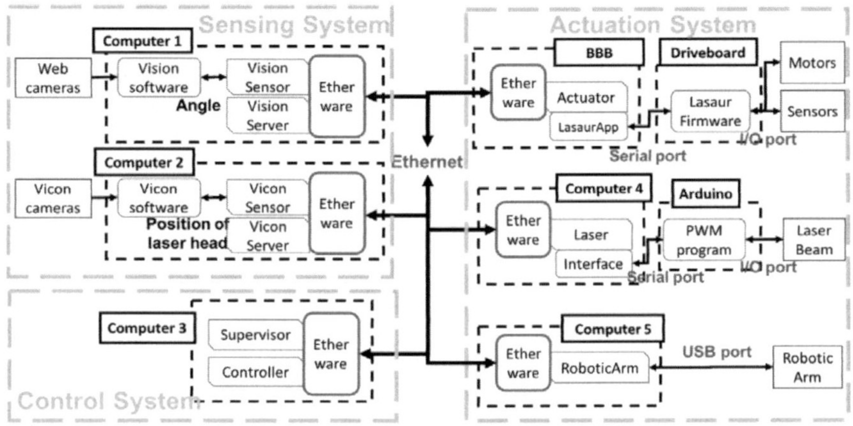

Figure 3. The system configuration of the automatic laser origami control system based on middleware.

The proposed automatic manufacturing system consists of three subsystems: a sensing system, a control system, and an actuation system. The sensing system observes the status of the process and provides observed information to a control system. For the sensing system, we have used two individual vision systems consisting of a multi-camera vision system which finds the bending angle of the workpiece during the bending procedure, and a Vicon system which finds the position and velocity of the laser head. Vision and Vicon sensor components obtain observation data from image processing programs, and Server components deliver them to a Controller component.

The control system has a supervisor as a high-level process manager and a bending controller. The supervisor receives the list of tasks from an operator and sends each task to a bending controller or an actuator system. With the exception of bending tasks all tasks such as cutting tasks or motion tasks of a robotic arm are executed by the actuator system directly because they do not need any feedback. When the current task is a bending task, it is executed by a bending controller. Once the bending controller receives the desired angle, it determines Gcode-type commands describing the intensity of the laser beam, and the feed rate of the laser head, based on the feedback from the sensing system, and sends them to the actuator system.

The actuator system plays the role of moving the laser head, regulating the intensity of the laser beam and manipulating the robotic arm. It executes the sequence of gcodes received from the high-level process manager and the bending controller to manufacture a 3D product from a sheet material by cut-bend-fold operations.

5. Feedback control scheme for laser bending

A feedback-based laser bending control system has shortened manufacturing lead time. Since manual operation for laser bending uses constant intensity of laser beam and feed rate of laser head throughout the whole bending process, it demands much trial and error to find the best heating policy manually. Especially, in the case of the changes of some parameters such as kind of work sheet or length of a bending line, a new heating policy will need to be found by trial and error. Feedback-based control system automatically deals with those issues. Based on the latest observation data, a control policy determines control inputs such as the feed rate of laser head and intensity of laser beam every control period, so that it not only accomplishes fast bending, but also prevents overheating which can degrade the quality of the bending process.

For laser bending, the feedback system controls the feed rate of the laser head and the intensity of the laser beam. A PID controller determines control inputs based on the desired angle from a supervisor component and the previous bending angle from the sensing system.

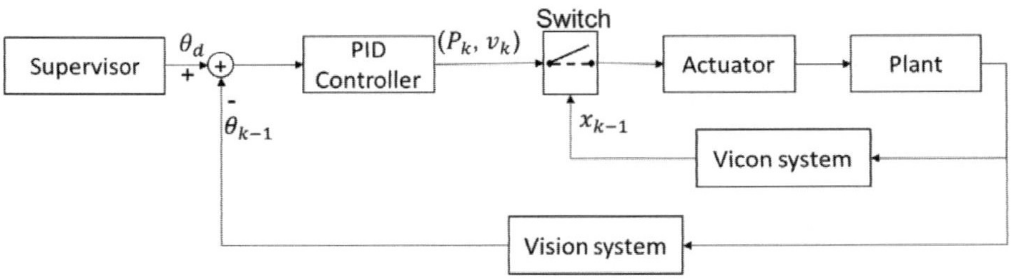

Figure 4. The control scheme for laser bending.

Figure 4 shows the control scheme with a PID controller for laser bending, where θ_d is the desired bending angle, θ_{k-1} is an observed bending angle at time $k-1$, and P_k and v_k are the intensity of a laser beam and the feed rate of a laser head, respectively, at time k. Let $e_{k-1} = \theta_d - \theta_{k-1}$ denote the error of the bending angle at time $k-1$. The PID controller for calculating the desired derivative of the intensity of the laser beam ΔP_k is designed as

$$\Delta P_k = k_{P_P} \cdot e_{k-1} + k_{I_P} \cdot \sum_{i=0}^{k-1} e_i \cdot dt_i \quad + k_{D_P} \cdot \left(\frac{e_{k-1} - e_{k-2}}{dt_{k-1}}\right) - P_{k-1}, \tag{1}$$

where k_{P_P}, k_{I_P} *and* k_{D_P} are the proportional, integral and derivative gains, respectively. The intensity of the laser beam at time k is determined as

$$P_k = \begin{cases} P_{k-1} + P_{step}, & if\ \Delta P_k > P_{step} \\ P_{k-1} + P_{step}, & if\ \Delta P_k < 0 \\ P_{k-1}, & otherwise \end{cases}, \tag{2}$$

where P_{step} is the step size for increasing or decreasing the intensity of the laser power. In order to design a controller for the feed rate of the laser head, let $\dot{e}_{k-1} = (e_{k-1} - e_{k-2})/dt_{k-1}$ denote the derivative of the error of the bending angle. The derivative portion of the controller for calculating a desired derivative feed rate is designed as

$$\Delta v_k = k_{d_v} \cdot (\dot{e}_{k-1}) + v_{base} - v_{k-1}, \tag{3}$$

where k_{d_v} is a derivative gain and v_{base} is the minimum feed rate. The feed rate of the laser head at time k is determined as

$$v_k = \begin{cases} v_{k-1} + v_{step}, & if\ \Delta v_k > v_{step} \\ v_{k-1} + v_{step}, & if\ \Delta v_k < 0 \\ v_{k-1}, & otherwise \end{cases}, \tag{4}$$

where v_{step} is the step size for increasing or decreasing the feed rate of the laser head. The control inputs P_k and v_k are delivered to an actuator component every control interval. The control interval is determined as

IOP Conf. Series: Materials Science and Engineering **272** (2017) 012013 doi:10.1088/1757-899X/272/1/012013

$$dt_k = \frac{2 \cdot L_{bending}}{v_{k-1}}, \tag{5}$$

where v_{k-1} is the previous feed rate of the laser head and $L_{bending}$ is the length of a bending line. This means that the control interval is varying during bending control process. When the previous feed rate of the laser head is slow, the current control interval is long. Conversely, a fast previous feed rate makes the control interval short. A flexible interval enables the control decision timing to adapt to the status of the bending. For example, when the angular velocity of bending is zero, the bending line is heated up radically with slow feed rates of the laser head. When the unfixed segment of the part starts falling down, the bending angle is controlled more frequently with fast feed rate of the laser head. This allows achievement of more precise bending, even mid-air stopping at 60 degrees.

The switch obtains the position data of the laser head from a Vicon system to check if the back and forth movement of the laser head is completed. When the laser head returns back to the starting position, the next control input is delivered to the actuator. According to the actuator's output signals, the laser system bends the material by the movement of the laser head and the heat from the laser beam. When the bending angle reaches the desired angle, the bending process is terminated.

6. Experiment

The experiment of evaluation validates the performance of the proposed feedback-based control system for cyber-manufacturing. Existing laser bending processes require an operator to manually repeat the sending commands to heat the bending line or stop the bending process while monitoring the status of the bending angle. This has several inherent issues such as the separation of the bending process from cutting processes, the possibility of overheating caused by human errors, the dependency of the process performance on the skill level of the operator, and repeated trial and error experimentation to obtain better performance. On the other hand, the proposed feedback-based control system eliminates the need for an operator to carry out the repetitive tasks. Moreover, it results in uniform and proper heating, and minimizes the deformation of the bending line by controlling the heat output of the laser and the velocity of the laser head in real time. It results in high performance and quality of the bending process. The performance of the proposed system is compared with that of a manually bending system.

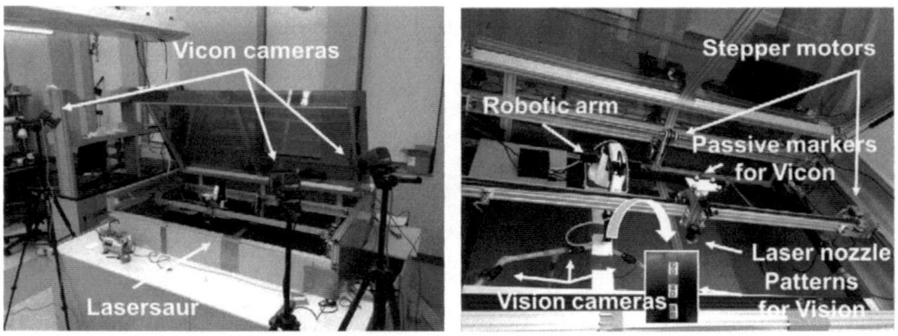

Figure 5. The overview of the experimental testbed.

6.1 Testbed

The figure 5 shows a feedback-based laser origami control system which controls a laser system to manufacture a 3D product by cut-bend-fold operations with a sheet material. The cyber-manufacturing system consists of a vision system estimating the state of the workpiece, a Vicon system observing the position and velocity of a laser head, a control system determining Gcode-type control inputs to a laser system based on the estimated data and user's job requests, a robotic arm manipulating the workpiece, and a laser system processing the workpiece with an incremental forming tool and a laser source.

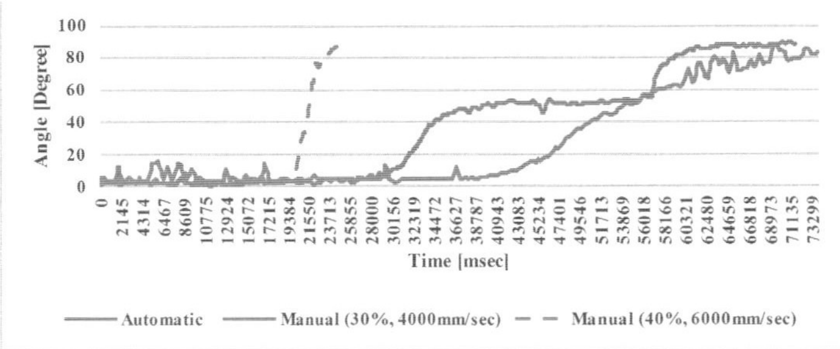

(a) 90 degree bending

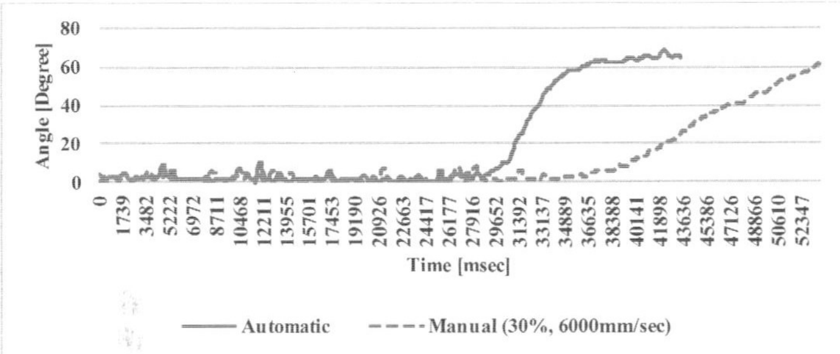

(b) 60 degree bending

Figure 6. The comparison of the performances of automatic bending and manual bending from the viewpoint of completion time.

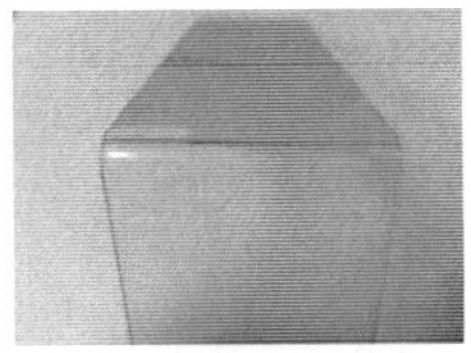

(a) Automatically bent

(b) Manually bent (30%, 4000 mm/sec)

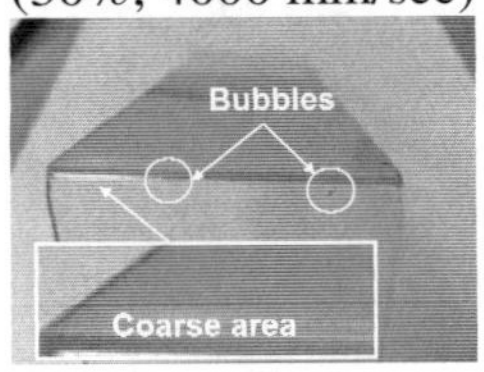

(c) Manually bent (40%, 6000 mm/sec)

Figure 7. The comparison of the performances of automatic bending and manual bending from the viewpoint of quality of bending.

The laser system is Lasersaur [11]. Its maximum laser power is 120W. It has two stepper motors to move a laser head with the maximum speed 6000 mm/min on the work area 1220 x 610 mm. The resolution of the movement of a laser head is less than 0.1 mm.

The robotic arm is Dobot magician [12]. It has 4 degrees of freedom. It can move the end effector along the x, y and z axes in the work area which have dimensions, 340 x 300 x 400 mm, and rotate up to 270 degrees. The payload is 500 grams.

A multi-camera vision system uses three web cameras whose resolution is 640 by 480 pixels. The cameras detect two dimensional barcode pattern patches which are attached on the surface of the workpiece. The vision program calculates the angle between the two planes, defined by the patterns, every 150 milliseconds. Moreover, we used three Vicon cameras, which are LED cameras, to detect the shape of a group of small balls on the top of the laser head every 100 milliseconds. The information is used for the control system to determine when the control commands should be executed because of the varying control period during the bending procedure.

6.2 Automatic cutting and bending

We experimentally demonstrate that the proposed system noticeably improves the performance of the process and the quality of the product. Figure 6 and figure 7 compare the performances of those manually bent versus those automatically bent, with respect to completion time and quality of bending. Figure 6 describes the graphs of the bending angles according to time. The dimension of work piece used in the experiment was 355 x 70 x 1 mm. The length of the bending line was 100 mm, which was longer than the width of the work piece to avoid overheating at both end sides by laser. The focal distance from the laser nozzle to the work piece for bending was 150 mm. For the manual bending, we used two pairs of static intensities of the laser beam and feed rates of the laser head because the nonlinearity and uncertainty of the bending model make it very hard to find the optimal combination of various intensities and feed rates manually. They are 30% with 4000 mm/sec, and 40% with 6000 mm/sec. For the automatic bending, the parameters of the PID controller were tuned by trial and error due to the nonlinearity of the system model. The gains of the PID controller for controlling laser intensity, k_{P_P}, k_{I_P} *and* k_{D_P} are 0.1, 10 and 0.002, respectively. The step size of increasing or decreasing the laser intensity, P_{step} is 10%. For controlling the velocity of the laser head, the derivative gain, k_{d_v} the minimum feed rate, v_{base} and the step size of the feed rate, v_{step} are set as 2000, 3000 and 1000, respectively.

In figure 6(a), when we used the pair of 40% and 6000 mm/sec, the bending angle, shown by the big dotted line, dramatically reached the desired degree with 8 back and forth laser heatings. However, figure 7(c) shows that radical heating by the strong laser beam degraded the quality of the bent surface by causing some bubbles and coarse area. Likewise, although figure 7(b) shows that the manual bending with 30% intensity and 4000 mm/sec feed rate achieved high quality of bending surface, its completion time for bending was too long, which is indicated by the small dotted line shown in figure 6(a), with 20 back and forth laser heatings. Otherwise, the automatic bending, shown by the solid line in figure 6(a), accomplished an earlier completion time than manual bending with 30% and 4000 mm/sec as well as better quality of bending than manual bending with 40% and 6000 mm/sec in figure 7(a).

The high performance of the automatic bending is also shown in figure 6(b). Without a rotation machine, in order to reach a mid-air angle which is not 90 degrees, an operator needs to keep the checking current bending angle and stop the back and forth heating as quickly as possible when the current bending angle is near the desired angle. If the feed rate is set fast to minimize the response time, it also causes a delay of completion time. Figure 6(b) shows that the completion time of the automatic bending is much faster than one of manual bending with 30% and 6000 mm/sec, to reach 60 degrees.

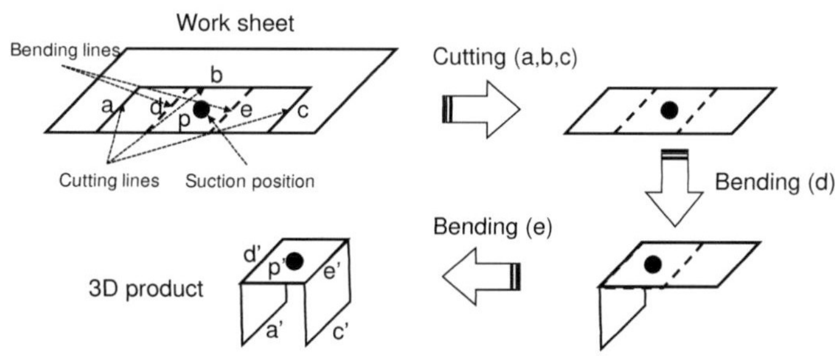

Figure 8. The procedure for 3 cutting and 2 bending processes.

We also demonstrate that the proposed automatic laser manufacturing system performs cut-bend operations effectively. An operator mounts a work sheet on the robotic arm and executes the gcode file which includes cutting and bending tasks. The system automatically manages and controls the whole process to fabricate a 3D-shaped creature. We used an acrylic sheet whose size is 355 x 279 x 1 mm. Figure 8 shows the entire procedure of cutting and bending.

An automatic laser fabricating system makes a simple 3D creation by performing 3 cutting and 2 bending processes. In figure 8 (a), 'p' is the holding position of the robotic arm, 'a', 'b' and 'c' are lines for cutting, and 'd' and 'e' for bending. At first, a laser system cuts the three cutting lines and then, bends along the two bending lines. For cutting, the robotic arm lifts the acrylic sheet up to the focal length of cutting. After cutting the three lines, the outer piece of the sheet is removed. As the bending process requires a long focal length of the laser to melt the bending line, the robotic arm lowers the sheet from the laser nozzle. The desired bending angle is 90 degrees for the both bending processes. During the bending process, a control system controls the bending angle to reach the desired value by sending control inputs such as the intensity of the laser beam and the feed rate of the laser head in real time. When the temperature of the heated region is high enough to melt the workpiece, the unattached portion of the workpiece is pulled downward by gravity, and multiple cameras measure the angle of the falling part and send the information to the control system as feedback. After the first bending process, the robotic arm rotates the workpiece to place the other side in the bending position. Once all cutting and bending processes are completed, we obtain the 3D product as in the Figure 8. The automatic two-sided bending experiment can be seen in [13].

7. Conclusion

We have proposed a feedback-based laser origami control system for cyber-manufacturing. We have established a sensing system and a control system around a laser system to operate the laser cutting and bending processes automatically to make 3D objects. We also integrated the robotic arm with a laser control system to switch between cutting and bending processes automatically due to their different focal distances. We experimentally demonstrate that our proposed automatic laser system successfully produces a 3D creation by cut-bend operations with short processing time and high quality of bending. Future work will include the anticipatory control based on the bending model to obtain fast and accurate bending control. We plan to design a laser bending model for an acrylic sheet by referring the analytic model, which one of the laser-bending literatures suggested to predict bending angle based on the knowledge about the thermo-mechanical process of laser bending [9]. Also, the temperature information around the area heated by a laser will be used to build a bending model to predict the future states of the bending angle. The anticipatory control is expected to reduce not only the lag effect of initial heating up, but also the inertia effect of final cooling down.

8. References

[1] P Wright, Cyber-physical product manufacturing, *Manufacturing Letters*, 2014. 2(2): p. 49-53.
[2] M M Herterich, F Uebernickel and W Brenner, The Impact of Cyber-Physical Systems on Industrial Services in Manufacturing, *Procedia CIRP*, 2015, 30: pp. 323-328.
[3] J Lee, B Bagheri and H-A Kao, Cyber-Physical Systems architecture for Industry 4.0-based manufacturing systems, *Manufacturing Letters*, 2015, 3(0): pp. 18-23.
[4] S Mueller, B Kruck and P Baudisch, LaserOrigami: Laser-Cutting 3D Objects', *CHI* 2013, April 27-May 2, 2013, Paris, France, pp. 2586-2592.
[5] S Mueller, LaserOrigami: laser-cutting 3d objects, 2013. Available from: https://www.youtube.com/watch?v=arjRtCjI9AQ.
[6] D J Balkcom and M T Mason, Introducing robotic origami folding, *Proceedings on IEEE International Conference on Robotics and Automation*, April 26 - May 1, 2004, New Orleans, USA
[7] A Namiki and S Yokosawa, Robotic Origami Folding with Dynamic Motion Primitives, *IEEE /RSJ International Conference on Intelligent Robots and Systems*, Sept 28 - Oct 2, 2015, Hamburg, Germany, pp. 5623-5628.
[8] K Tanaka, Y Kamotani and Y Yokokohji, Origami folding by a robotic hand, *IEEE/RSJ International Conference on Intelligent Robots and Systems*, Oct. 29 - Nov. 2, 2007
[9] A Eideh, U S Dixit and R Echempati, A Simple Analytical Model of Laser Bending, *5th International and 26th All India Manufacturing Technology, Design and Research Conference*, Dec. 12-14, 2014, IIT Guwahati, Assam, India
[10] G Baliga, S Graham, S Lui and P R Kumar, Etherware: domainware for wireless control networks Object-Oriented Real-Time Distributed Computing, *Proc., 7th IEEE International Symposium*, pp. 155-162, 2004.
[11] Lasersaur, http://www.lasersaur.com/
[12] Dobot magician robotic arm, https://www.dobot.us/
[13] The automatic two-side laser bending, Available from: https://youtu.be/Ufeu0zgYzng

Acknowledgments
This material is based upon work partially supported by NSF under Contract Nos. ECCS-1547075, CNS-1646449 and Science & Technology Center Grant CCF-0939370, and the U.S. Army Research Office under Contract No. W911NF-15-1-0279.

IOP Conf. Series: Materials Science and Engineering **272** (2017) 012014 doi:10.1088/1757-899X/272/1/012014

System of error detection in the manufacture of garments using artificial vision

J J Moreno[1,3], A Aguila[1,4], E Partida[1,4], C L Martínez[1,3], O Morales[1,3] and R Tejeida[2,3]

[1]ESIME-Zacatenco,
[2]EST, Instituto Politécnico Nacional (IPN), México.
[3]SNI, Research Professor.
[4]IPN, Thesis Student.

Abstract. A computer vision system is implemented to detect errors in the cutting stage within the manufacturing process of garments in the textile industry. It provides solution to errors within the process that cannot be easily detected by any employee, in addition to significantly increase the speed of quality review. In the textile industry as in many others, quality control is required in manufactured products and this has been carried out manually by means of visual inspection by employees over the years. For this reason, the objective of this project is to design a quality control system using computer vision to identify errors in the cutting stage within the garment manufacturing process to increase the productivity of textile processes by reducing costs.

1. Introduction

In the textile industry, quality control is required in manufactured products and this has been carried out manually by means of visual inspection by employees over the years, observing the product, analysing it, classifying it and discovering its imperfections. However, people may overlook several errors because of various external factors, in addition to the human visual inspection has drawbacks such as distraction and fatigue of the operators to be a repetitive task or they are simply very slow when doing the inspection, which generates losses of product replacement or a decrease in profits when selling the products at lower prices. It is necessary to evaluate methods to increase the efficiency of these processes, focusing on the use of automatic machines to find small details that those in charge of the inspection cannot observe, so that the quality control is a competitive advantage in the textile industry [1]. A system of quality control is proposed using computer vision, to increase the precision in the inspection of the manufactured products and thus to achieve speed in the process of inspection and to increase the production. As for the device being implemented for processing, it is an embedded system Raspberry Pi 3, which has an operating system based Linux, Raspbian, in which an application compiled in OpenCV or free software to make it possible the implementation of complete computer vision system [2,3]. The Raspberry Pi 3 system has been chosen due to its hardware performance and the relative ease of using Raspbian for the intended implementation; some other development boards, such as Arduino, would force a more complex computer vision program with no apparent benefit compared to Raspberry Pi 3, in turn, the low consumption and small dimensions of the Raspberry Pi 3 make it ideal for its handling, assembly and operation, compared to a conventional computer.

2. Historical Framework

With the success of the technological revolution or second industrial revolution in the late nineteenth and early twentieth century, mankind has always sought to adopt technology to streamline its processes. Thus, with the advance of the creation of digital images that emerged in the decade of the twenties and continues to the present day, computer vision took a lot of interest in researchers and entrepreneurs. The idea of obtaining information through the analysis of an image revolutionized the way of seeing many processes, one of them, the process of detecting errors in quality control, which began to take increasing importance. During the time of the craft era in the nineteenth century, products had been sought to meet the quality attributes that the end-user desired, with the passage of time and increase in production volume, the piece-by-piece inspection became impossible to perform, in consequently, Harold F. Dodge and Harry Roming, quality researchers, implemented acceptance sampling, thus only inspecting several products within the total. However, this solution was not sufficient to obtain the total quality, although the costs for errors in the products were reduced significantly, still exist errors that represented costs for the companies. Technology applied to quality was necessary, so computer vision systems became important in quality control posts for the most delicate processes. From this background, a large variety of research and applications have been made on the use of computer vision and image processing for error detection. On the other hand, having food, clothing and a home to live are human physical needs. Clothing has been taking on more and more meaning for humanity, to the point of becoming a representation of us, our personality and our way of life. It is for this reason that the manufacture of garments has evolved notably from periods such as the Neolithic period. Currently the design and manufacture of garments not only covers the most basic needs of users, but also offers technologies ranging from waterproof fabrics to wearables, smart devices that can be worn. Being necessary even in the design and construction of automobiles and airplanes. Textile production has become essential. Therefore, the textile industry is one of the largest industries in the world at the sales level only behind industries such as tourism and information [4].

3. Theoretical Framework

3.1 Basic Manufacturing Process in the Textile Industry

In order to carry out the fabrication process of textile products, various steps are represented by stages, Figure 1.

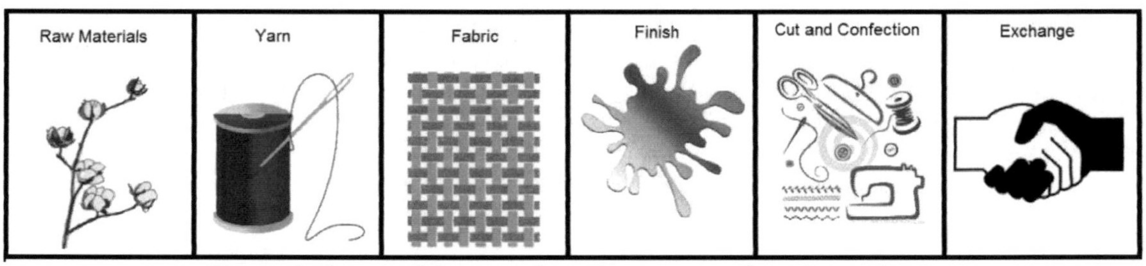

Figure 1. Stages of the manufacturing process of textile products.

Raw Materials: These can be divided into natural fibres and synthetic fibres, such as cotton, wool, silk and nylon, polyester, rayon, respectively. Naturally, with the exponential growth of processes and the world population, the need for textiles has grown in the same way. It is believed that an economically stable person consumes between 20 and 25kg of textile products annually. This has created the need to produce more and more synthetic fibres to meet the needs.

Yarn: Process through which the fibres are either natural or synthetic are transformed into fine, resistant, flexible and extremely long thread that allows to create fabrics.

Fabric: This is a process that has not changed since the origins of the first garments. However, it is also the most complex due to the large number of fabrics that exist. It consists of 4 basic steps: Tissue

organization, positioning of tissues in parallel, Drawn, and Twisting. The process at its most basic level tries to organize the fibres strategically according to the type of fabric desired and to interlace them in such a way that they maintain a consistency throughout the process.
Finish: This process is responsible for 3 main tasks: 1) Remove impurities that may appear on the fabric. 2) Dyeing and stamping determined by the use that will be given to the fabric. 3) Modifications in the surface of the fabric, either to improve the physical properties of the product which translates into greater comfort or simply aesthetic improvements. *Cut and confection*: It is the last stage of the manufacturing process of a textile product, in this, the fabrics treated in the previous stages are transformed into a final product, passing through sub-stages, such as: Cutting: The goal of this stage is to turn the fabric into pieces with specific shapes to be joined later through the sewing stage. Depending always on the type of technology used the cutting stage can be divided into 7 basic steps:1) A cut design is established for the parts of the product to be processed. 2) An optimization pattern is created to reduce waste as much as possible. 3) Prepare the fabric to cut. 4) Raw cuts are made that will serve for a next fine cut of contours 5) A mark is made on the pieces that will aid in fine cutting. These marks are imperceptible in the final product. 6) Fine cuts are made. The piece takes the final shape. 7) Parts are prepared for future steps. Then, *Sewing*: It is made the union of the pieces in 2 or 3 dimensions that shapes the final product. At this stage, the technology has been limited by the complexity that implies the union of pieces in 3 dimensions. Finally, *Detailed*: Here you can apply the formulas (buttons, locks, etc.). This process then comprises a systematic and orderly succession of steps, the last cannot be performed without having done all the previous ones. Therefore, in processes as dependent as this is necessary to have a control of errors in each of the stages of the process, otherwise these errors would represent costs for manufacturing companies.

3.2 Analysis of a digital image

It is common to hear that the term quality is associated or used when talking about high definition images, however this is not totally true. While the current formats present large amounts of pixels, 980x720 pixels for the High Definition (HD), this does not guarantee the image quality. This is simply the size of the image. The true quality of the image is directly related to the colour depth of each pixel, namely its representation, the more bits to represent a pixel, the longer the colour palette. This way our images will have as much quality as we want. Also, the quality of the image will depend on external factors, such as lighting, it is not the same to take an image in an open place where the irregularities of light are constant, that a closed place with lighting controls. To solve these problems of quality and thus extract the information of the image with more facility exist methods as they are the smoothing and enhanced. By one hand, the purpose of smoothing is to eliminate noise, which is presented as colour irregularities between pixels. Some smoothing techniques are: Average neighbourhood environment and Medium, fashion, maximum and minimum filtering. By the other hand Enhanced methods for the study of this work, these methods of improvement are not necessary, since being a quality control the lighting conditions will be optimal. Furthermore, Histogram is a representation of an image in the form of discrete function, that is, with a finite number of values. This represents the number of pixels that are in the image according to its intensity level. For example, the histogram of a grayscale image whose intensity levels oscillate between 0 and 255, results in a bar graph from which we can obtain valuable information according to its distribution at the moment of analysing digital images [5]. Finally, for the analysis of the image it is necessary of the segmentation that is the process through which we will obtain the information of the edges and regions, being based in this way in the discontinuity and the similarity. An edge will be represented by a drastic change in intensity levels, i.e. there will be discontinuity in the image, in a region there will be similarity [6].

4. Main Algorithm

In the manufacture of garments, garments that go through the cutting process, come from a previous process of dyeing, so that the colour posed a challenge in obtaining the correct image for analysis and detect errors. So, it is necessary to contrast the background with the garment in our acquisition of the image. Due to these colour variations in the garments, it was necessary to add a dynamic background with RGB LEDs. Step 1; Acquisition of the image, the image of the garment is captured. Step 2; Average image, it averages each channel of the complete image, red, green and blue. That is, we will have an average of red, green and blue. Step 3; Change background colour, with these values the colour that is less present in the image is determined, this will be the new colour of the background. For electrical convenience (less current consumption) only combinations that have a single voltage level will be used in each RGB led. That is, the primary colours of the model, red, green, blue, Table 1. Finally, Step 4; Acquisition of the new image, With the background well contrasted with the garment a new image is taken to verify the correct contrast with the garment [7-8].

The number of RGB LED's in the matrix, 100 in our case, requires a certain current, each LED has a working current I_0 of 20mA, so our matrix needs to be fed with 2A of current to work properly, with a proper luminescence for our application. To achieve this luminescence, it is necessary to adequately saturate the circuit, for that we used TIP100 transistors with current handling up to 8A. Considering a future expansion of the LED array [9]. To make possible the colour change in the RGB matrix, it is necessary to have a control for each RGB primary colour (red, green, blue), using the output ports of the Raspberry Pi 3 that must be connected to each Transistor to exert the switch function and power the circuit with the 2A current. However, a problem arises here, because the base-emitter voltage (V_{BE}) that most transistors handle is 5v and the maximum output voltage of the Raspberry Pi 3 is 3.3v. This prevents the saturation of the transistor and therefore represents a malfunction.

Table 1. Colour combinations in RGB LEDs.

Red	Green	Blue	Combination
0	0	0	Without lighting
0	0	1	Blue
0	1	0	Green
0	1	1	Cyan
1	0	0	Red
1	0	1	Magenta
1	1	0	Yellow
1	1	1	White

The solution to this is the implementation of an photocoupler, which allows to use the 3.3v voltage level of the Raspberry Pi 3 as a switch to power the transistors with a voltage level of 5v. The next step is the comparison of histogram and binarization. Then, any histogram analyses pixel by pixel the intensity of the image, in our case, the acquired image will be transformed into grayscale, with intensity values from 0 to 255, i.e. with 8 bits of depth. The histogram shows us the concentration of pixels according to their tonalities, in our case the garment takes a dark tone due to the illumination, this helps us, since these dark colours will always be found at the beginning of the histogram, Figure 2(a). For each model, a mask is defined that represents the final pixel intensity value to be taken so that the garment is binarized and thus perfectly segmented from the colour background, Figure 2(b). The comparison will then be made with two images that in theory should be the same, one will be the original model to repeat in the cutting process and another will be the garment in production line, Figure 2(c). A great advantage of using this method for the analysis is that a prior alignment of the garment is not necessary, since no matter the location within the image, the garment contributes the same number of pixels. For this to be true care must be taken that the distance from the sensor to the garments is the same, otherwise, the objects in the image appear different sizes, changing the histogram [10].

Figure 2. (a)Histogram of the model. (b)Segmentation of the main image, and (c) Binarized image.

5. Experimental Results

Thus, the acquired image of the garment in production line is compared with the image of the model, stored in the system. The comparison will result in the following scenarios, depending on their similarity or difference expressed as a percentage:

1) Similarity between 100-95%; It is the ideal scenario and means that the garment in production line is the same as the desired model. The percentage difference is due to lighting defects that come to be presented. So, the process continues without interruption.

2) Similarity between 95-80%; In this scenario, there is a clear difference, which can mean 2 errors: a) Garments of different sizes have been mixed. b) The garment presents bad cuts or ruptures in the fabric of previous processes. In this case, a check indicator is illuminated for the operator. This person will determine, depending on the severity of the error, the use of the defective garment.

3) Similarity less than 80%; Here the difference is remarkable and can represent 2 errors: a) The garment in production line has been mixed from another process, which explains the little similarity. b) The change of the model analysis has undergone some error there is a problem in the prototype, having thus to enter in the equation the operator. In the 3 cases the proposed algorithm responds with a 98.07% of success. The errors were found when proposed algorithm processes a garment with similarity less than 80\%, i.e. the third case.

6. Conclusions

A broad investigation of the topics was developed within different areas that are related by this work, areas such as the textile, such as medicine and computational. Then, Raspberry Pi 3 embedded system was implemented along with the Ubuntu operating system and the OpenCV compiler for the development of computer vision codes. It was possible to make a perfect segmentation of garments regardless of the colour that was given them in previous processes within the textile manufacturing, this was possible thanks to the implementation of a dynamic opaque background composed of RGB LEDs that change colour depending on contrast with the garment. The error detection performed by this system improves performance by finding errors that are difficult to detect by employees, which undoubtedly results in an improvement in the quality control of any company.

7. Acknowledgments

This work is supported by National Polytechnic Institute of Mexico (Instituto Politécnico Nacional, México) by means of Project No. SIP-20171179, the Academic Secretary and the Committee of Operation and Promotion of Academic Activities (COFAA) and National Council of Science and Technology of Mexico (CONACyT). It is important to mention that this article is part of the degree thesis supported by Arturo Aguila Rodríguez and Eduardo Partida Robles.

8. References

[1] J. R. Zaruma, J. L. S. Santos, D. V. Escudero, and M. O. Vivanco, "Innovation and technology in the Ecuadorian textile industry," in CISTI 2017, pp. 1–7.

[2] M. Iwaniec, A. Holovatyy, "Development of vibration spectrum analyzer using the raspberry pi microcomputer and 3-axis digital mems accelerometer adxl345," in MEMSTECH 2017, pp. 25–29.

[3] A. M. L. López and J. E. A. Uribe, "Visual servo control law design using 2d vision approach, for a 3 dof robotic system built with lego ev3 and a raspberry pi," in STSIVA 2016, pp. 1–7.
[4] M. A. Yokus, R. Foote, and J. S. Jur, IEEE Sensors Journal 16, 7967–7976 Nov (2016).
[5] P. P. Kotgire, J. M. Mori, and A. B. Nahar, "Hardware co-simulation for chroma-keying in real time," in ICCCCA 2015, pp. 863–867.
[6] P. Upadhyay and J. K. Chhabra, "An un-supervised image segmentation technique based on multi-objective gravitational search algorithm (mogsa)," in IICIP 2016, pp. 1–4.
[7] Abeynanda, H. K. A study of factors affecting the exports of the garment industry in Sri Lanka 2017 6th National Conference on Technology and Management (NCTM), 2017, 69-74.
[8] Mongan, W. M.; Rasheed, I. and Ved, K, A. "Real-time detection of apnea via signal processing of time-series properties of RFID-based smart garments" in SPMB 2016, 1-6.
[9] Sayed, M. S. "Robust fabric defect detection algorithm using entropy filtering and minimum error thresholding" in MWSCAS 2016, 1-4.
[10] M. P. Pacaux-Lemoine, L. Habib, G. Tartare, L. Koehl and X. Zeng, "Human-robots cooperation through intelligent garment," in ISIE 2017, pp. 1996-2001.

IOP Conf. Series: Materials Science and Engineering **272** (2017) 012015 doi:10.1088/1757-899X/272/1/012015

Overall equipment efficiency of Flexographic Printing process: A case study

S Zahoor [1], A Shehzad[1], NA Mufti[1], Z Zahoor[1] and U Saeed[1]

[1] Department of Industrial & Manufacturing Engineering, University of Engineering & Technology Lahore, 54480 Pakistan

Abstract. This paper reports the efficiency improvement of a flexographic printing machine by reducing breakdown time with the help of a total productive maintenance measure called overall equipment efficiency (OEE). The methodology is comprised of calculating OEE of the machine before and after identifying the causes of the problems. Pareto diagram is used to prioritize main problem areas and 5-whys analysis approach is used to identify the root cause of these problems. OEE of the process is improved from 34% to 40.2% for a 30 days time period. It is concluded that OEE and 5-whys analysis techniques are useful in improving effectiveness of the equipment and for the continuous process improvement as well.

1. Introduction

To remain competitive in the global world, organizations seek to produce better quality and cost - effective products. This need forces the organizations to adopt several approaches related to production process and plant maintenance in order to run the plant at its full capacity by increasing productivity hence reducing the cost of manufactured parts. These approaches include Total Quality Management (TQM), Total Productive Maintenance (TPM) and Lean Manufacturing. Plant maintenance is playing an important role in the organization's profitability as 15-60% manufacturing cost is associated with the maintenance activities [1].

Overall equipment efficiency (OEE) is a quantitative TPM tool which is employed to investigate and analyze the production processes. OEE provides a metric to gauge the machine efficiency in order to improve the productivity through identification of potential areas for improvement. OEE evaluates three main factors related to equipment and process i.e. availability, performance and quality [2].

OEE mainly concerns with six main categories of losses which are. breakdown, reduced speed losses, minor stoppage losses, setup adjustment losses, rework and yield losses [3].

OEE can be defined as

$$\text{OEE} = \text{Availability} \times \text{Performance} \times \text{quality} \quad (1)$$

Whereas;

$$\text{Availability} = (\text{loading time-down time})/\text{loading time} \quad (2)$$

$$\text{Performance} = (\text{ideal cycle} \times \text{time output})/\text{actual operating time} \quad (3)$$

$$\text{Quality} = \text{input - volume of quality defects}/\text{input} \quad (4)$$

Today's manufacturing industry focuses on the importance of equipment efficiency and effectiveness in order to access organization's performance in terms of productivity. TPM plays a significant role in enhancing the productivity of an organization aiming at OEE [4]. Dogra et al give a study on implementation of TPM in rolling and HPH annealing machines. The study concludes that major downtime losses are caused by process adjustment. After one month TPM training to top management and the employees, the OEE of rolling and HPH annealing machine has been increased to 80% and 83%

respectively. It is also revealed from the study that maintenance cost has also reduced by the implementation of TPM [5]. Almeanazel et al present a study based on a steal manufacturing company. The main study is carried out by selecting three processes named ovens, dies, cutting and cooling bed. It is revealed through the research that OEE was 72% before the implementation of TPM. It is than decided by the company that TPM should be implemented in order to improve the company's performance. After the implementation of TPM, company becomes able to mark out potential areas for waste reduction and improvements [6]. Parihar et al calculate the OEE for assembly process of a tractor manufacturing company. The OEE of the process is found 62% which is very low as compared to the world class standard of 85%. It is recommended by the study that in order to improve the OEE, attention should pay to reduce unplanned maintenance, setup and change over losses through implementing new techniques, tools, employee training and inventory storage systems [7]. Tsarouhas et al conclude a study related to the implementation of TPM in food industry. The study is carried out on a pizza production line as this line requires continuous operation and transfer lines. Any interruption or stoppage in the production can make the pizzas present in the work stations useless. OEE is calculated on the basis of previous five years data after implementing the TPM and data shows an increase in OEE from 62% to 80% within five years. TPM implementation is not only influenced the reduction of downtime losses, also improves the quality of the product, safe workplace, reduced product lead time [8]. Sharma et al carry out a study on TPM in piston manufacturing industry. The study is carried out in two phases comprised of qualitative and quantitative analysis. Qualitative analysis reveals that availability of machines differs between 65-70% due to recurrent breakdowns and shortage of spare parts. Likewise, quantitative analysis also reveals that 70% downtime losses are unplanned and 15% are planned [9].

Mhetre et al report a study on TPM review for fabricated parts manufacturing company. The study aims at evaluating company's effectiveness through OEE evaluation. The main process is comprised of six operations and their computed OEE are 66.95, 52.79, 20.67, 15.12 and 47.97% respectively. It is obvious that process five which is bending needs improvement in order to improve OEE. It is also revealed from the study that workforce management is the reason behind reduced OEE for bending operations. Therefore, company advised to focus on preventive maintenance to train the workers [10].

The 5-why analysis is widely used in lean manufacturing for root cause analysis. Taiichi Ohno "the father of Toyota Production Systems" is considered one of the pioneers who used 5-why method as a root cause solving method during his days in Toyota. He noticed that people used to blame each other whenever problems occurred during production. Therefore, he figured out the best approach that is to find out the root cause of the respective problem and work on it rather to blame each other. Traditionally 5-whys analysis requires the "why" question to be asked five times in order to identify the root cause, however, the asking of the "Why" can be stopped if common sense tell us that no more "Why" questions are needed to solve the problem [11]. Five whys method is used for root cause analysis in order to investigate cause-failure relationship for a failure problem [12]. Banjamin et al use five why approach to investigate the root causes of wastage for a Toyota production system. The study suggests that use of 5whys in manufacturing industry gives a realistic approach to the problem from its identification to correction in order to reduce and eliminate the defects [11]. Breakdown time of equipment and quality losses are the two major concerns for any organization in order to achieve production targets and remain competitive.

This research is carried out in a renowned paper board and packaging company located in Lahore, Pakistan. The company is pioneer and the biggest supplier of packaging product to different industries all across Pakistan such as beverages, food and fast moving consumers goods etc. The present case study reports the OEE improvement for a flexographic printing machine aiming at reducing the breakdown time of the machine. Main problem areas are identified using Pareto diagram followed by 5-whys analysis approach used for root cause analysis.

2. Research Methodology

In the current study, a flexographic printing machine from Business Unit has been selected. The machine specification and production capacity are listed in table 1

Table 1. Specification of Flexographic Printing Machine.

Machine Name	Key Raw Material	Machine Speed (m/min)	Type of printing Process	Actual Production (meters/ year)
Cerruti Green (CG)	PET, Polyester, PVC and BOPP	220	In-Line Flexographic Printing	80 million

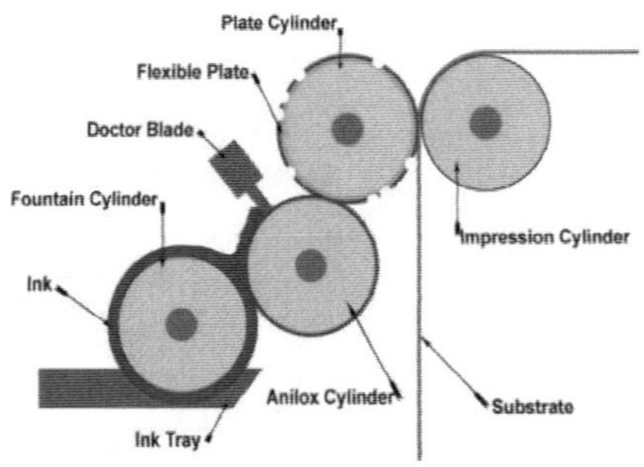

Figure 1. Process Schematic.

There are five basic steps involved in flexographic printing. A fountain roller is dipped in ink tray that is transferred to next anilox roller. Anilox roller transfers the ink to the flexible plates. In the next step a doctor blade removes an excessive amount of ink pasted in the previous step. In the next step uniform distribution of ink is achieved using a soft rubber like printing plate. The process completes when a pressure is applied by the last cylinder called print anvil. The schematic of process under study is shown in the figure 1.

For OEE calculation, data for last one year breakdown time for each machine was selected. Calculated OEE based on the one year breakdown time data was analysed with the help of 5-whys analysis approach and Pareto charts were used to prioritize the failures in order to rectify these failures in future to improve the OEE. After the analysis, the data was collected on the same machines for 30 days.

3. Experimental results and analysis

3.1 *OEE calculation*

Initial OEE was calculated on the basis of one year downtime data which was maintained by the production department of the organization. OEE was calculated by using equation 1-4 mentioned above which was found 34%. To improve OEE, four main problem areas were identified for the flexographic printing process by using Pareto diagram as shown in table 2.

Table 2. Problem areas identified by Pareto diagram.

Sr #	Problem Area	Major Effects	Total Breakdown Time/Year
1	Matching with Lacquer	Duplicate Printing	15065 min
2	Shade Card Preparation	Fade prints on poly sheet	13474 min
3	Doctor Blade Problem	Grainy dots on poly sheet	14407 min
4	Ink Mottling Problem	Missing of even and smooth prints	20895 min

IOP Conf. Series: Materials Science and Engineering **272** (2017) 012015 doi:10.1088/1757-899X/272/1/012015

To avoid failures occurred due to the above-mentioned problems, root cause analysis was carried out using 5-why analysis technique. During the present study, the "why" question was asked only three times because the root cause was clearly identified. Figure 2 illustrates the root cause analysis for problem 1 i.e. matching with Lacquer using 5-why technique.

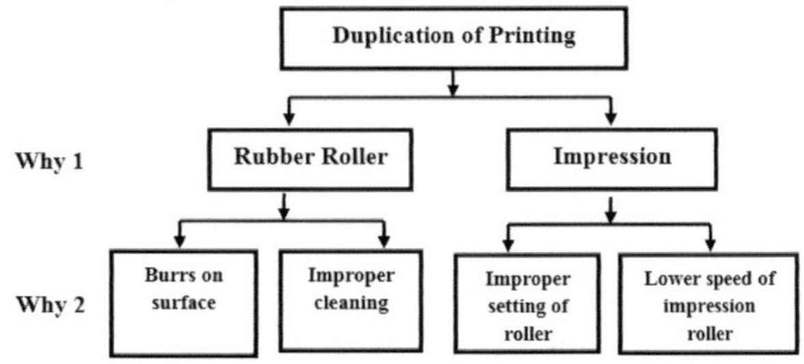

Figure 2. 5-why analysis for matching with lacquer.

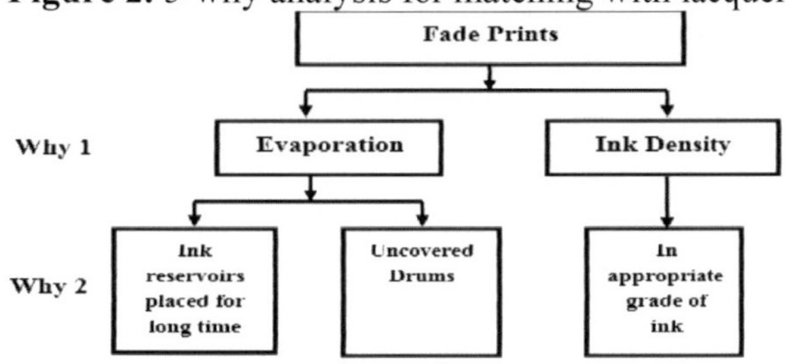

Figure 3. 5-why analysis for shade card problem.

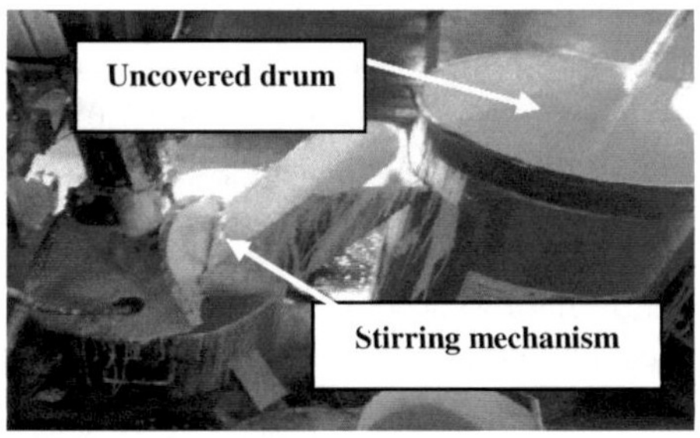

Figure 4. Uncovered drum.

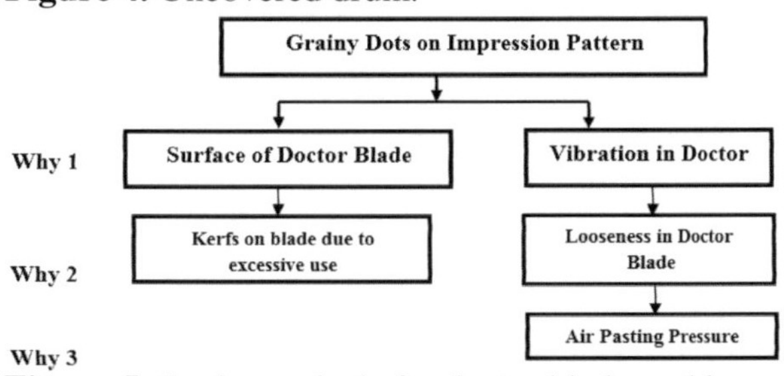

Figure 5. 5-why analysis for doctor blade problem.

IOP Conf. Series: Materials Science and Engineering **272** (2017) 012015 doi:10.1088/1757-899X/272/1/012015

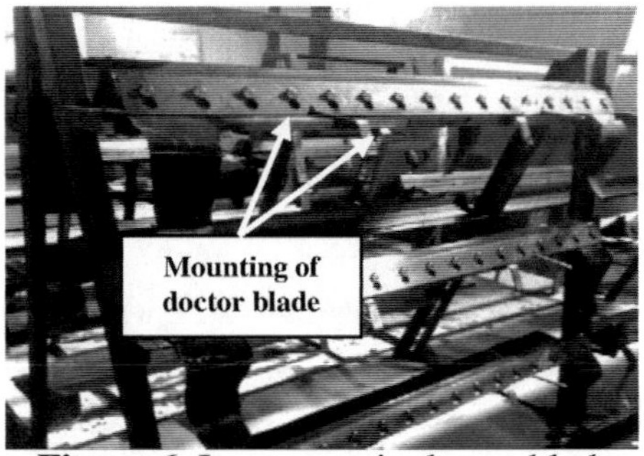

Figure 6. Looseness in doctor blade.

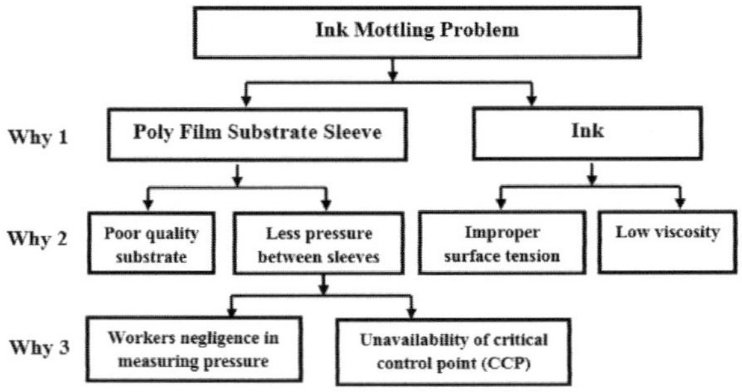

Figure 7. 5-why analysis for Ink Mottling problem.

Lacquer performs solvent evaporation from poly film to give the best and durable printing finish. The malfunctioning of this step may results in duplicate printing on the product. It is evident from the figure 2 that the cause of the problem is primarily due to improper alignment of the impression roller caused by worker's negligence

Ink and solvent are mixed together in a drum to prepare shade card before printing. As drums are exposed to air during mixing causing the evaporation of solvent, consequently reducing sticking effect of ink that results into fade printing on the poly sheet. It can be seen from the figure 3 that evaporation is the main cause for the problem due to uncovered drums and improper stirring of ink and solvent as shown in figure 4. Looseness in doctor blade causing vibration due to interaction with the Anilox roller resulting in grainy dots on the impression patterns. Figure 5 reveals that air pasting pressure is the possible reason for the looseness of doctor blade as shown in figure 6. Ink mottling problem results in to poor printing and figure 7 shows that lower pressure between the sleeves is the main cause of this problem.

3.2. Remedial actions

Lean manufacturing tool, in general requires small quantitative analysis than six sigma analysis tools. 5-whys is widely used analysis tool for root cause analysis without performing statistical analysis. During this 5-why technique, a question is asked five times to identify the main cause of the failure. During the present study, root cause analysis is carried out to three whys as compared to the traditional 5why technique. The reason behind this is the clear identification of root cause for above cited problem which requires no further why question to be asked.

Corrective measure implemented after 5-why analysis to reduce the problems:

- Proper inspection via Preventive Planned Maintenance after every shift so to have a regular check for not going towards the breakdown while running.
- Drums were covered.
- A proper channel flow rotation should be made to keep in movement so that the solvent would not be evaporated, otherwise keeping it idle, will lessen its concentration.
- Proper procedures for clamping and waxing the doctor blade on the jaw type die to the workers

- Proper planned maintenance schedule and formation of critical control points (CCP) for the future hazard concerns.
- Inspection of required pressure to be maintained during the job setting.
- Critical control points should be displayed above the machine on a readable location.

3.3. Post implementation results

These proposed solutions were implemented and final OEE was calculated for a time period of 30 days. The improved results are given in the table 3.

Table 3. Breakdown data for final OEE calculations.

Sr.#	Problem Areas	Current Breakdown Time (min)	Percentage Improved Breakdown Time	Time Saved (min)
1	Matching with Lacquer	15065	53 %	7985
2	Shade Card Preparation	13474	31 %	4177
3	Doctor Blade Problem	14407	42 %	6051
4	Ink Mottling	20895	80 %	16716

By using the equation 1-4 and data available in table 3, the final OEE was found 40.2%. A positive trend has been observed in OEE and if overall equipment efficiency is considered as a continuous process improvement tool than further improvement in OEE can be made by rectifying the problems and their root cause analysis.

4. Conclusions

Following conclusions are drawn from the study:

1. OEE behaves as the quantitative measurement for total productive maintenance which is used to identify major areas of improvements for a process.
2. It is also concluded that 5-why approach is a useful technique for root cause analysis which ultimately improved the OEE from 34% to 40.2% within time period of 30 days.
3. Overall equipment efficiency can be used as measure of continuous process improvement.

5. References

[1] Mobley R K 2002 *An Introduction to Predictive Maintenance, A Volume in Plant Engineering* (Elsevier Science, New York) 2nd ed.

[2] Nayak D M, Naidu G S and V. Shankar 2013 Evaluation of OEE in continous processs industry on an insulation line in a cable manufacturing unit *International Journal of Innovative Research in Science* **2** 1629-1634

[3] Besterfield D H, Michna C B, Besterfield G H and Sacre M B 2004 *Total Quality Management* (Pearson education India) 3rd ed p 431-442

[4] Saleem F, Nisar S, Khan M A, Khan S Z and Sheikh M A 2017 Overall equipement effectiveness of tyre curing press: a case study *Journal of Quality in Maintenance Engineering* **23** 39-56

[5] Dogra M, Sharma V S and Sachdeva A, 2011 TPM – a key strategy for productivity improvement in process industry *Journal of Engineering Science and Technology* **6** 1-16

[6] Almeanazel O T R 2010 Total productive maintenance review and overall equipment effectiveness measurement *Jordan Journal of Mechanical and Industrial Engineering* **4(4)** 517-522

[7] Parihar S, Jain S and L. Bajpai 2012 Calculation of OEE for an assembly process *International Journal of Research in Mechanical Engineering & Technology* **2(2)** 25-29

[8] Tsarouhas P, Implementation of total productive maintenance in food industry: a case study *Journal of Quality in Maintenance Engineering,* **13(1)** 5-18.

[9] Rohit S and Trikha V 2011 TPM, implementation in piston manufacturing industry for OEE *International Journal of Current Engineering and Technology* **1(1)** 118-125

[10] Mhetre R S and Dhake R J 2012 TPM review & OEE measurement in a fabricated parts manufacturing company *International Journal of Research in Engineering & Applied Sciences* **2(2)** 1084-1094
[11] Alukal G 2007 *Lean kaizen in the 21st century* Quality Progress **40(8)** 69-70
[12] Benjamin S and Muthaiyah S 2010 Scrap loss reduction using 5-whys analysis *International Journal of Quality & Relaiability Management* **27** 527-540

Research on width control of Metal Fused-coating Additive Manufacturing based on active control

Ren Chuan qi, Wei Zheng ying, Wang Xin, Du Jun, Zhang Shan, Zhang Zhitong ,Bai Hao

State key Laboratory of Manufacturing System Engineering, Xi'an Jiaotong University, Xi'an 710049, China

Corresponding author: legend7718@163.com

Abstract. Given the stability of the shape of the forming layer is one of the key problems that affect the final quality of the sample morphology, taking a study on the forming process and the control method of morphology make a significant difference to metal fused-coating additive manufacturing (MFCAM) in achieving the efficient and stable forming. To improve the quality and precision of the samples of single-layer single pass, a control method of morphology based on active control was established by this paper. The real-time acquisition of image was realized by CCD and the characteristics of morphology of the forming process were simultaneously extracted. Making analysis of the characteristics of the width during the process, the relationship between the relative difference of different frames and moving speed was given. A large number of experiments are used to verify the response speed and accuracy of the system. The results show that the active system can improve the morphology of the sample and the smoothness of the width of the single channel, and increase the uniformity of width by 55.16%.

1. Introduction

Additive manufacturing (AM) technology can break through the limits of parts in complexity of shape and form three-dimensional solid parts directly, which has several inimitable advantages over other traditional manufacturing technology [1]. Metal Fused-coating Additive Manufacturing (MFCAM) has been proposed as a commercial manufacturing that is a prospective additive manufacturing (AM) technology for fabricating 3D metal functional parts. Making a comparison with other metal additive manufacturing technologies, it possesses the obvious advantage of the lower cost of equipment.

However, the poor shaping precision and surface quality have seriously restricted the application of MFCAM. In order to overcome these shortcomings and expand the application prospects of technology, monitoring the forming process is of great significance. As far as we know, process monitoring is widely used in the field of welding rather than MFCAM in quality control. Pinto-Lopera et al. presented a novel technique for real-time measuring of the width and height of weld beads in gas metal arc welding (GMAW) using a single high-speed camera and a long-pass optical filter in a passive vision system [2]. Cruz J G et al. presented a methodology to perform the modeling, optimization and control of the weld bead width, enabling the adjustment of process parameters in real time. [3]. A common method to monitor geometric variables in GMAW processes is the indirect mode, by using a mathematical model to relate the geometric parameters to another process variable that is easier to measure, such as current or voltage [4-5]. Hu Xiao dong from Xi'an Jiao tong University, who used CCD to monitor the metal forming process and designed a fuzzy PID controller that was adjusted by capturing the image information of the weld pool, which made welding samples obtain better processing accuracy [6]. Heralic et al. designed a system for directly observing the height of the molten pool and structured photo-detection based on CCD and established a dual-input-output closed-loop control system adjustment system through adjusting the wire feed speed to control the width of the weld and the laser power ,which ensured a high degree of stability [7-8]. Liu Xiao Yan et al from Hunan University, proposed the control algorithm of material state by using the principle of image feedback and designed the real-time control system of the motion state of the rotating drum, which can realize real-time monitoring of the movement state of the rotating cylinder realize the real-time monitoring of the rotary cylinder motion state [9]. Huang Jin et al. from Northwestern Polytechnic University, proposed an automatic target recognition algorithm based on sequence images, and the results of the simulation for various posture aircraft images show that the algorithm can effectively and accurately identify the aircraft types [10].

2. Experimental procedure

Molten metal is transported from the channel of the fused-coating nozzle to the region between the horizontal moving substrate and the thermal capillary flow. The melt encounters the relatively cold substrate or pre-solidified layers, cooling and solidification begin at the interface of the melt and substrate or pre-solidified layers. The solidified layers will be subjected to the same forming motion along with the substrate. The strong shear stress generated at the solidification front will be consequent ancillary products as the rapidly moving and advancing solidification front encounter cooling conditions. What's more, the newly formed dendrites might shed to a certain extent due to the interfacial shear stress. What is important is that the new forming process can significantly enhance metal forming efficiency. The conclusion can be drew that if the forming accuracy and surface quality got be further improved, MFCAM will have a great prospect in the field of national defense industry, automobile industry and so on.

2.1. Experimental System

The schematic diagram of MFCAM and forming platform are illustrated in detail in Fig. 1 (a). The main subsystems of experimental system consist of a pressure control system, an inert gas (argon) protection system, and a machine control system based on a three-dimensional motion platform. In most actual conditions, the specimens were successfully manufactured through the effective combination and mutual cooperation of each module according to model's information. Under the combined action of gas pressure, hydrostatic pressure, and surface tension, the molten metal flows through the channel in the fused-coating head, as shown detailed in Fig. 1 (b).

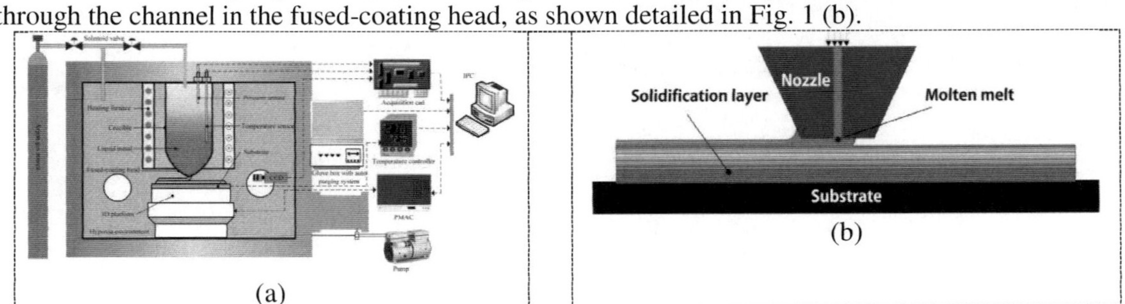

Figure 1. (a) Molten metal micro-coating equipment schematic diagram.
(b) Forming principle diagram

2.2. Arrangement of the Experiments

The raw material of Single-layer single pass specimens in the experiment is the long block of Sn63Pb37 alloy. Taking the thermal properties of material into comprehensive consideration, the crucible temperature was heated to 270 °C, the initial distance between the micro-coating nozzle and

IOP Conf. Series: Materials Science and Engineering **272** (2017) 012016 doi:10.1088/1757-899X/272/1/012016

the substrate was set to 1.6 mm, the substrate temperature was set to 90 °C, and the argon mass flowmeter pressure was set to 100KPa. Single-layer single pass (SLSP) experimental conditions are shown in Table 1.

TABLE 1. parameters of metal fused-coating experiments.

Parameter	Value
Coating Head Temperature	270 ℃
Argon Mass Flow Rate	10~80 mm^3/s
Deposition Velocity	9~27 mm/s
Initial Distance	1.6 mm
Coating Nozzle	0.3 mm
Pressure	100 KPa
Glove Box	Ar (99.999%) （20 ppm）
Size of Copper-clad Substrate	300*200*10 mm
Single-track Deposited Length	130 mm

3. Problems and solutions

It is found that there is an inconsistency in the width of the single-layer single pass alloy specimens under normal forming process, through visual direct observation. The width of the middle section is in the smaller fluctuations with a high consistency, however, there are more fluctuations in the start and stop segments. Given the above phenomena, the first width sampling point is set at 5mm from the start position, and the length of the model is more than 100 mm. To reflect the variation of the width of the sample, the single-layer single pass sample is segmented by the sampling as Fig. 2 (b). Where L represents the total length of the single-layer single pass sample, ΔL (ΔL for different samples is different) represents the sampling segment spacing, the number 1-20 represents the number of sampling points, and the length related in Fig. 2 (b) satisfies L = (ΔL + 20 * 5+10) mm. With reference to the sampling model, the width of the sample point is measured using the solftware of Halcon after the necessary calibration and the value of width retains four decimal places as shown in Fig. 2(a). The statistical curves of 6 groups of samples' width have been drew according to the measured results, which could be found in Fig. 2(c).

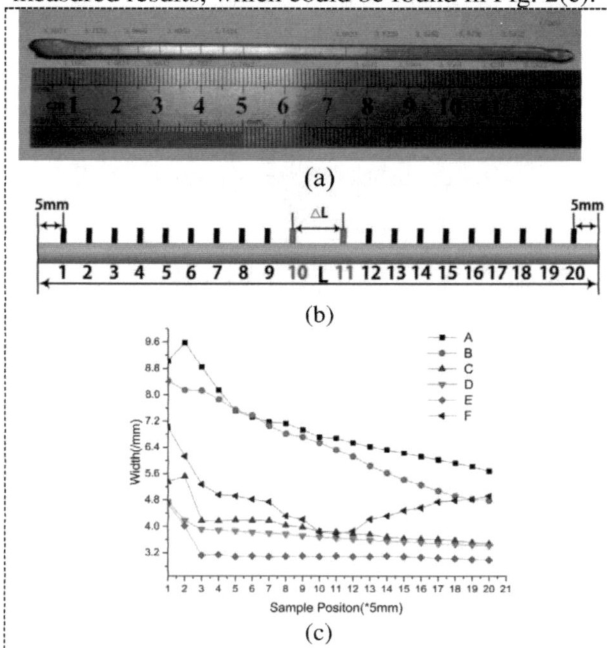

Figure 2. (a) Width measurement practicality picture. (b) Width sampling model.
(c) Width variation curves of single-layer single pass in 6 group

In order to quantify the changes of the width of the single-layer single pass alloy samples and obtain the necessary information which comprises maximum width, minimum width, mean width and width standard deviation , the measured width of the above six groups was analyzed statistically. More details could be found in Table 2. The standard deviations of the B and C samples in Table 2 are 1.30929 and 1.23926, respectively, which are shown in Fig. 2 (c) with large fluctuation. The standard deviations of E and F samples in Table 2 are 0.41369 and 0.36477, respectively, reflected in Fig. 2 (c) shows a small range of width fluctuation. Analysis found that the standard deviation can better reflect the width of single-layer single pass specimen changes. Taking the above into consideration, the paper uses the standard deviation to measure the uniformity of the sample width.

TABLE 2. statistics of width information of single-layer single-pass samples.

Sample	Mean (mm)	Minimum (mm)	Maximum (mm)	Standard Deviation
A	3.88	3.32	5.54	0.57109
B	6.54	5.02	9.6	1.30929
C	6.65	4.44	8.42	1.23936
D	4.24	3.56	7.02	0.86077
E	3.08	2.88	4.68	0.41369
F	3.56	3.18	4.72	0.36447

There are many factors affect the sample width, such as the physical properties of the raw material, the system pressure and substrate speed. The phenomenon found by experiment was that among these factors, the substrate velocity had a great influence on the sample width and it had faster response than other factors. The paper proposes the solution that the speed of the substrate can be adjusted automatically by the width change, and thus the control of sample width can be realized to a certain extent. Therefore, the sample shape and the width consistency can be improved greatly.

4. Image capture system

In order to verify the proposed ideas, the vision sensor of CCD was added to current MFCAM system. The camera used in system has a USB2.0 data transfer interface and the camera can be used for a maximum resolution of 2048 * 1536 pixels. It is worth emphasizing that the camera has a frame rate range of 12-114 PFS related to the resolutions. The relative position of the camera, nozzle and substrate is shown in Fig. 3 (a)

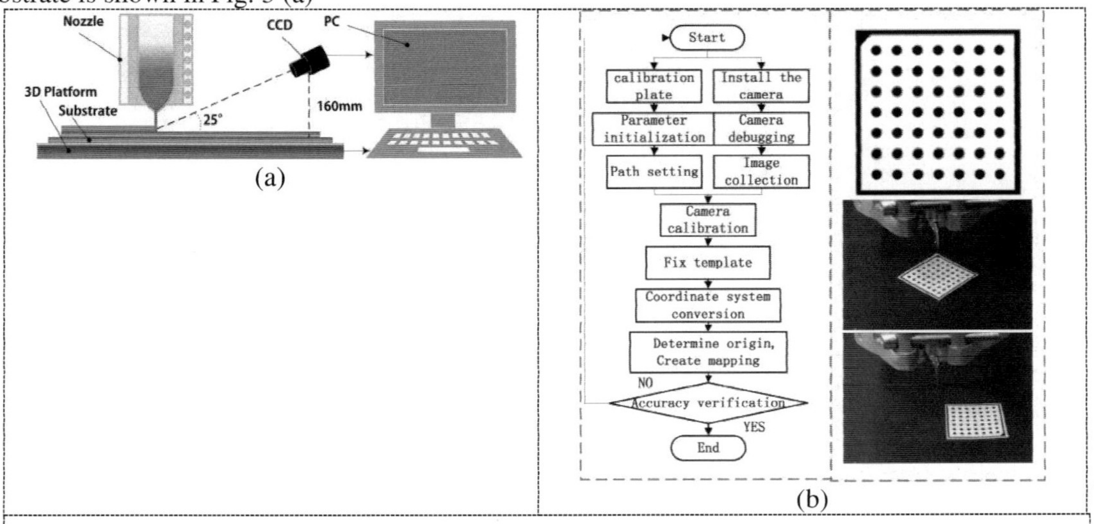

Figure 3. (a) Relative position of the camera, nozzle and substrate.

(b) Calibration process and related images

The CCD in the equipment was calibrated to minimize the effect of optical distortion to the maximum extent. Camera calibration is the process of solving the camera internal and external parameters through a certain number of different spatial position images. In the practical application, we need to recalibrate the CCD camera, if we need to adjust the camera position. The process of camera calibration is completed in the Halcon software platform by capturing the images of the calibration plate at different positions. The calibration process and related images are shown in Fig. 3 (b), which is characterized by the follows: the number of marked points and rows of 7*7, the distance of the neighbor marker points is 3mm, the diameter of the mark point is 1.5mm, and the width of the black frame is 0.75 mm. 20 calibration images were used in Polynomial model of traditional calibration method. The calibration process is completed in the Halcon calibration assistant, and the camera internal and external parameters are obtained as follow:

SatrtCamPar:=[f, k, S_x, S_y, C_x, C_y, Width, Height] 1)

{Where f is the focal length, k is the distortion coefficient, initialized to 0, Sx is the horizontal distance of the adjacent pixels of the, Sy is the vertical distance of two adjacent pixels, Width is the image width, Height is the image height. Cx , Cy is the center of projection in the vertical projection imaging plane.

5. Automatic control algorithms

Dynamic monitoring the width of single-layer single pass is a prerequisite for automatic control. In the forming process, the crucible nozzle is relatively stationary and 3D platform follow the predetermined trajectory. It is possible to obtain the increment of single channel width and the width of the last node, as for the length of samples on different time nodes is different. Considering that the system depends on the motion of the substrate, the image captured by CCD is the relative displacement at the actual condition. The automatic control algorithm of width based on active control is proposed for the MFCAM process, which mainly includes the method of automatic detection and control of the motion state of the substrate based on the image width. The block diagram of single-layer single specimen's width control system can be found in Figure 4 in detail.

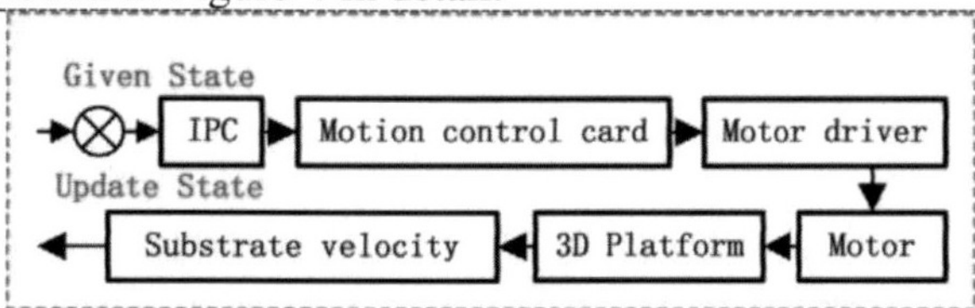

Figure 4. Block diagram of single-layer single specimen's width control system.

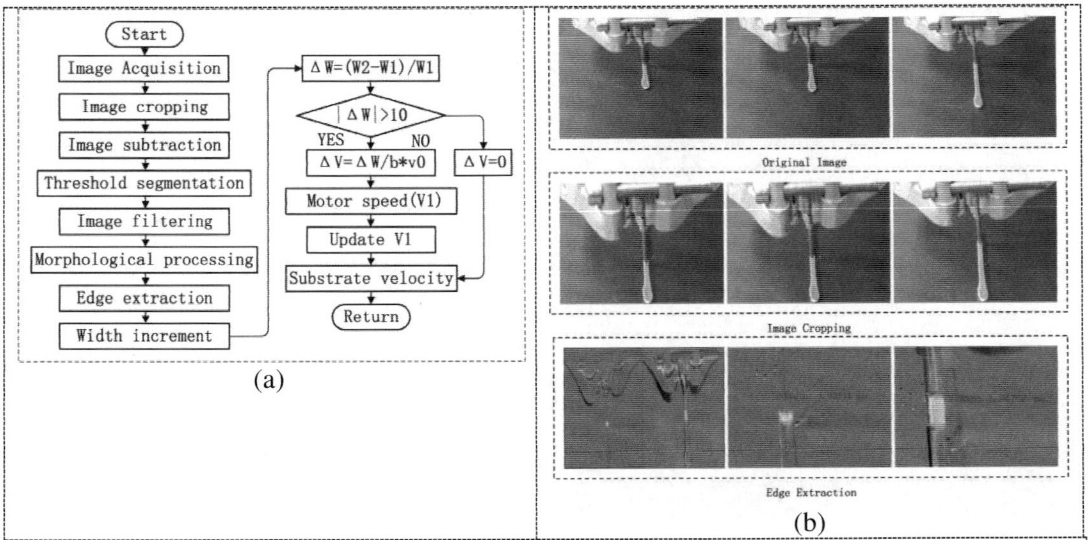

Figure 5. (a) Automatic control algorithm. (b) Pictures of measurement of incremental width
To format a paragraph, use the Microsoft Word template style: *Paragraph.*

The relative difference between the two adjacent widths ΔW could be calculated on the basis of the obtained width of single-layer single pass, and the update velocity of substrate is determined by the dynamic change of ΔW. Many experiments show that when $|\Delta W| <0.05$, the current substrate speed shouldn't be changed, and when $|\Delta W| > 0.05$, the substrate speed needs to be changed according to empirical Formula 2.

$$\Delta V = \begin{cases} -\frac{\Delta W}{a} * V0, & (\Delta W \geq 0.05) \\ 0, & (|\Delta W| \leq 0.05) \\ +\frac{\Delta W}{a} * V0, & (\Delta W \geq 0.05) \end{cases} \quad 2)$$

The ΔV that can be converted into the velocity of motor could be calculated quantitatively by ΔW.IPC uses the pulse increment which is transformed from ΔV to achieve the control of substrate velocity. As far as we know that the pressure at the beginning is unstable, so the automatic control program of width delays 20ms after the implementation. In the actual forming process, the CCD acquisition speed is 8 frames per second, and the width increment is dynamic refreshed to actualize real-time control.

6. Experiment and result analysis

The hardware and software of system combined effectively to take the forming process in single direction for verification. The width data of single-layer single pass specimens were made into statistics in Fig. 6 (b) according to the Fig. 2 (b) sampling model. The trend of width change is weakened than before and the subsequent forming process is stable respectively like Fig. 6 (a). The standard deviation of the sample width (Table 3) is found to be reduced, and the sample has better consistency than ever. The experimental results show that the control system based on image feedback can meet the requirements of real-time response, and improve the forming accuracy of the sample to some extent.

TABLE 3. statistics of width information of single-layer single-pass samples based on active control.

Sample	Mean (mm)	Minimum (mm)	Maximum (mm)	Standard Deviation
A	4.212	3.92	5.68	0.44568
B	3.39	5.28	3.98	0.15252
C	4.926	4.62	6.56	0.52114
D	5.808	5.52	7.22	0.37721
E	2.588	2.46	3.78	0.28153

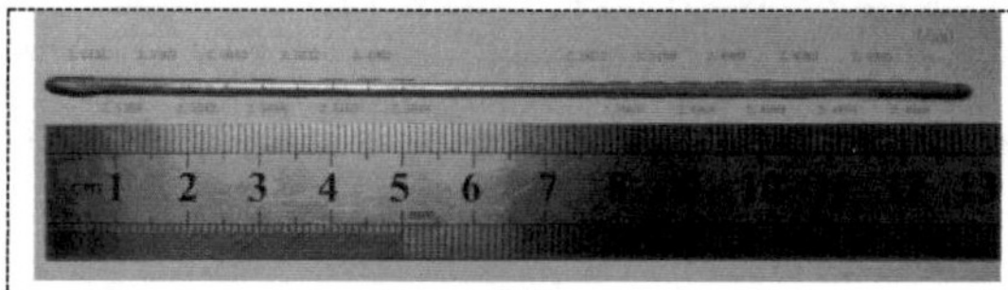
(a)

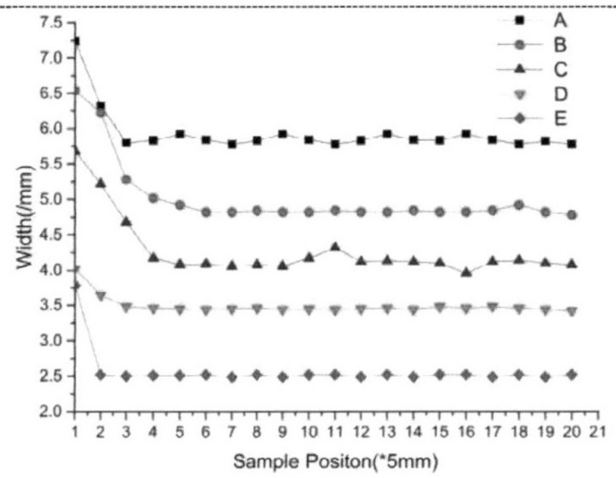

(b)

Figure 6. (a). A test sample of control in track width;
(b). Statistical curves of 5 different samples with control in track width)

7. Conclusions

To conclude, the method based on active control was carried out in micro-coating metal additive manufacturing (MCMAM). The forming morphologies of the width of single layer single pass specimens were analyzed in detail, and the main findings are summarized as follows:

After automatic width automatic control algorithm is used, the accuracy has been greatly improved, and the average increase of sample width uniformity reaches 55.16% than before .

In Metal Fused-coating Additive Manufacturing (MFCAM), the monitoring data and process parameters are closely combined to obtain the better specimens. The developed algorithm based on active control has great universality and versatility to improve the quality of specimens in MFCAM.

Although the method could improves the surface quality and forming accuracy of the forming parts, to a certain extent, there is still a long way to go from the precise industrialization application. In the further work, we will pay attention to further improve the response speed of the method to improve its efficiency.

8. References

[1] Han Man-man, Wurikaixi Aiyiiti ,Teng Yong, Zhang Wen-ju. Application of 3D printing technology in the foot fracture-assisted surgery [J].Machinery Design & ManuFacture, 2016,10 (10):36-40.

[2] Pinto-Lopera J E, Motta J M S T, Alfaro S C A. Real-Time Measurement of Width and Height of Weld Beads in GMAW Processes[J]. Sensors, 2016, 16(9):1500.

[3] Cruz J G, Torres E M, Alfaro S C A. A methodology for modeling and control of weld bead width in the GMAW process[J]. Journal of the Brazilian Society of Mechanical Sciences & Engineering, 2015, 37(5):1529-1541.
[4] Wang Z, Zhang Y M, Wu L. Adaptive interval model control of weld pool surface in pulsed gas metal arc welding ☆[J]. Automatica, 2012, 48(1):233-238.
[5] Li X R, Shao Z, Zhang Y M, et al. Monitoring and control of penetration in GTAW and pipe welding[J]. Welding Journal, 2013, 19(6):S21-S22.
[6] Hu Xiao-dong, Zhao Wan-hua, Xu Chao, Lu Bing-heng. Monitoring system of metal direct forming process based on CCD [J]. Acta Photonica Sinica, 2003,32 (3):375-377.
[7] Heralic A, Christiansson AK, Lennartson B. Height Control of Laser Metal-wire Deposition based on Iterative Learning Controland 3D Scanning [J]. Optics and Lasers in Engineering, 2012,50(9),1230-1241.
[8] Heralic A, Christiansson AK, Ottosson M, et al. Increased Stability in Laser Metal Wire Deposition through Feedback from Optical Measurements [J]. Opticsand Lasersin Engineering, 2010,48:478-485.
[9] Liu Xiao-yan, Lu Xin-yue,Zhang Yao-yao. Image-Feedback Control System for the Solid Motion in Rotating Cylinders [J].Control Engineering of China,2016,23(3):313-317.
[10] Huang Jin, Liang Yan,Cheng Yong-mei,Pan Quan,Hu Jin-wen. Automatic target recognition method based on sequential images [J]. Acta Aeronautica Et Astronautica Sinica,2006,27(1):87-93

Acknowledgments
Project supported by the State Key Development Program Research of China (2017YFB1103201), the State Key Development Program Research of China (2016YFB1100400), the Fundamental Research Funds for the Central Universities (xjj2016124) and the Equipment pre-research project from Ministry of Education (6141A02022109).

Morphology Analysis and Process Research on Novel Metal Fused-coating Additive Manufacturing

Wang Xin, Wei Zheng ying, Du Jun, Ren Chuan qi, Zhang Shan, Zhang Zhitong ,Bai Hao

State key Laboratory of Manufacturing System Engineering, Xi'an Jiaotong University, Xi'an 710049, China

Corresponding author: wx624753588@gmail.com

Abstract. Existing metal additive manufacturing equipment has high capital costs and slow throughput printing. In this paper, a new metal fused-coating additive manufacturing (MFCAM) was proposed. Experiments of single-track formation were conducted using MFCAM to validate the feasibility. The low melting alloy was selected as the forming material. Then, the effect of process parameters such as the flow rate, deposition velocity and initial distance on the forming morphology. There is a strong coupling effect between the single track forming morphology. Through the analysis of influencing factors to improve the forming quality of specimens. The experimental results show that the twice as forming efficiency as the metal droplet deposition. Additionally, the forming morphology and quality were analyzed by confocal laser scanning microscope and X-ray. The results show that the metal fused-coating process can achieve good surface morphology and without internal tissue defect.

1. Introduction

High performance metal components are made of alloy powder or metal wire as raw material, through high power laser or electron beam to layer by layer rapid melting [1]. However, this equipment and raw materials are usually very expensive, and not conducive to the promotion of application[2-3]. Fabrication of metal fused-coating additives manufacturing (MFCAM) is a new method for fabricating three-dimensional metal functional components [4].

To overcome the shortages of traditional AM, metal fused-coating metal additive manufacturing (MCMAM) has been proposed as a commercial manufacturing technology. Compared with traditional metal AM technologies, MCMAM has shown several advantages. At first, it provides a higher material utilization than selective laser melting (SLM) with a high deposition rate. Second, it produces less dust pollution than powder-based equipment when the powder material was recycled. Third, it has a lower equipment cost than SLM and electron beam machining (EBM) [5].

Xiong et al. investigated the forming characteristics of a multilayer single pass with the application of GMAW-based additive manufacturing [6]. Jorge et al. developed a fused deposition modeling (FDM) system for metals that can deposit electronic structures directly [7]. Yao et al. adopted a metal droplet deposition manufacturing process to reduce product development time as well as the cost of manufacturing [8]. However, deposition accuracy was difficult to control. In addition, the novel metal additive manufacturing process proposed in this paper was analyzed using a numerical simulation method. Nevertheless, the influences of process parameters on the forming morphology have never been mentioned [9].

Therefore, this paper is aimed at revealing the key factors that affect the morphology of MFCAM, such as flow rate, deposition velocity and the initial distance. The statistical analysis of the forming morphology has obtained abundant information. "Experimental details" section describes the experimental details, followed by "Results and discussion" section, which present the effect of process parameters on forming appearance. The last section ends the paper with some central conclusions.

2. Experimental procedure

2.1. Experimental System

In order to increase productivity and reduce cost, a novel metal fused-coating technology called MFCAM was proposed. A special fused-coating nozzle was designed to assist in fabricating dense metal parts. The way to build metal parts can be defined as continuous deposition. Molten metal is transported from the fused-coating nozzle to the substrate and thermal capillary zone is formed between the nozzle end and the directional moving substrate. Controlled by our 3D manufacturing software (software copyright: 2016SR070404), melt solidifies and forms layer by layer. The technology integrated the parts of the design, forming and processing, which reduces energy consumption and has the sophisticated processing application potentials. Therefore, they are well suited for aerospace, automotive, national defense transportation and other fields, etc.

Fig. 1 provided a conceptual view of the overall process. The equipment of MFCAM includes a pressure control system, an inert environment control system and a machine control system with a movable platform. The pressure control system was used to produce molten metal on demand. It consisted of a pressure control device, a solenoid valve, a crucible, a heating furnace, a thermocouple, a temperature control device and argon gas source. The inert environment control system consists of the glove box and gas circulation device, which are used to protect molten metal from oxidation. The 3D platform system has a PMAC (programmable multi-axis controller), which is consists of a deposition substrate, a substrate heating device, 3D movement platform, a motion control device and a temperature control device. The form parts according to data information through controlling the motion of the 3D platform, and the whole process were coordinately controlled to complete the fabrication of prototype parts by industrial computer. Crucible and nozzle inner wall wipe with hydrochloric acid, Sn63-Pb37 alloy in argon environment refining again after being washed with alcohol. Finally, the raw materials into the crucible within the 220 mesh stainless steel mesh for metal smelting.

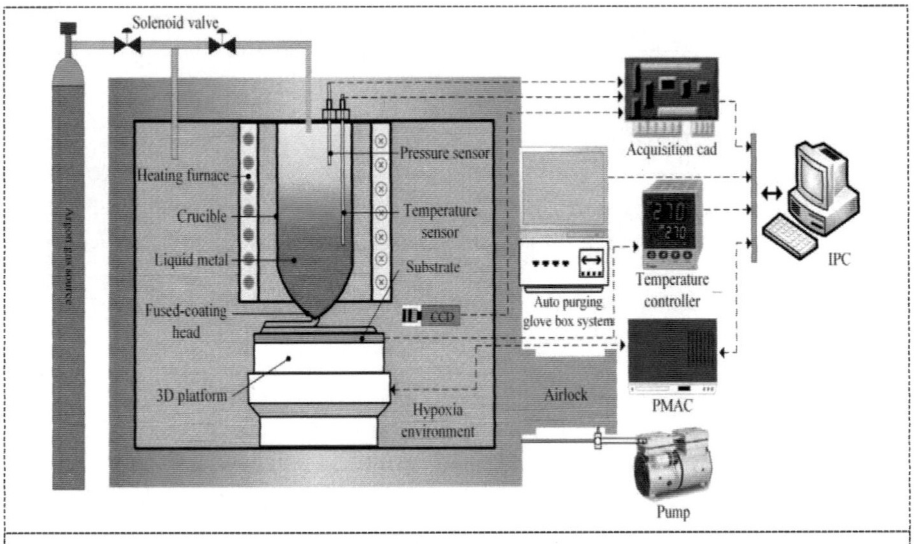

FIGURE 1. Schematic of the metal fused-coating additive manufacturing.

2.2. Arrangement of the Experiments

In the course of the experiment, the temperature of the molten metal is too low to be squeezed out of the nozzle, and the temperature is too high, which leads to the phenomenon of the collapse. Therefore, the choice of crucible temperature and substrate temperature is very important. This paper is based on the simulation results of the research group combined with the actual conditions preferred temperature value. The nozzle pressure corresponds to the flow rate, and its selection is based on the fluid Bernoulli equation, taking into account the loss along the way, local loss and hydrostatic pressure. The data on pressure-flow are obtained after analytical calculations. In order to study the influence of flow rate, deposition velocity and the initial distance. According to previous experience we will be set the crucible temperature was 270 ℃, the initial distance between the nozzle and the substrate is set to 1.6 mm, the substrate temperature was set as 90 ℃, argon mass flowmeter pressure was set as 150 KPa.

TABLE 1. parameters of metal fused-coating experiments

Parameter	Value
Coating Head Temperature	270 ℃
Argon Mass Flow Rate	10~80 mm^3/s
Deposition Velocity	9~27 mm/s
Initial Distance	1.6 mm
Coating Nozzle	0.3 mm
Pressure	100 KPa
Glove Box	Ar (99.999%)（20 ppm）
Size of Copper-clad Substrate	300*200*10 mm
Single-track Deposited Length	130 mm

3. Results and discussion

Metal fused-coating process evaluation standard includes surface morphology and internal tissue, and the geometry size has directly affected the quality of the single-track. The single-track height has an important influence on the bonding strength between layers. The single-track width has an important influence on the lap ratio. As a basic unit, the forming process of a single-track specimen should be primarily investigated. There are many process parameters in the MCMAM forming process, including the deposition velocity, the flow rate, the distance from the nozzle to the workpiece, the heating temperature, the nozzle size, and the heat dissipation conditions, having great impacts on the final forming quality of specimens. The present paper focuses on the deposition velocity, the flow rate, and the initial distance; thus, other factors are kept constant.

3.1. Influence of Deposit Velocity on Forming Appearance in Surface Morphology

The deposition velocity is a very important process parameter. According to the preliminary experimental data, the coating head temperature is set at 270 ℃, nozzle diameter 0.3 mm, flow rate of 50 mm3/s, the initial distance between nozzle and substrate is 1.6 mm, the deposition velocity were 9 mm/s, 12 mm/s, 15 mm/s, 18 mm/s, 21 mm/s, 24 mm/s, 27 mm/s were deposited, as shown in Fig. 2(a) – (g), there are seven different shapes of the surface morphology with single factor experiment on deposition velocity.

From Fig. 2 a, b we can see that the surface quality is not satisfactory, which indicates that the flow rate is too large and the deposition velocity is small. The single-track width increased significantly, and the end of shape had piled up. From Fig.2 f, g we can see that the deposition velocity too fast to produce fewer overlaps by surface tension. And this phenomenon has led to necking into droplets, the necking in place will lead to declining mechanical properties. From Fig.2 c, d, e we can see that the deposition velocity and flow rate matching are good with 15 mm/s ~ 21 mm/s. The process parameter matching test can be carried out in this range.

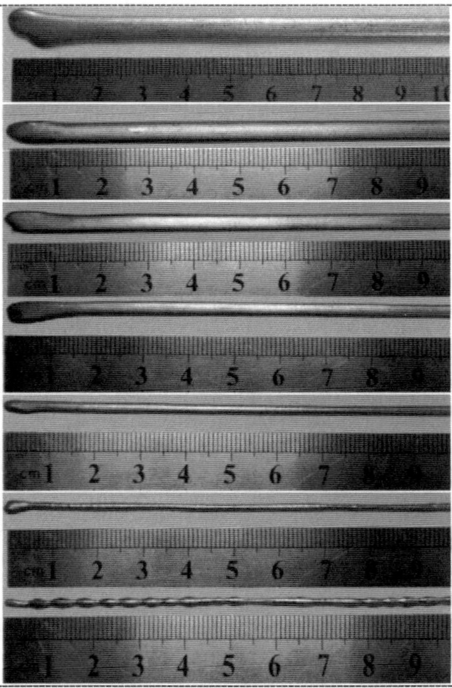

FIGURE 2. different deposition velocity from 9~27mm/s.

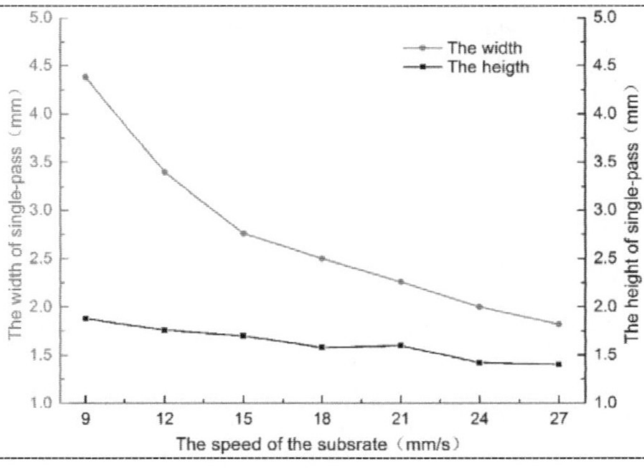

FIGURE 3. The forming width and height with different deposition velocity.

In order to study the process parameters more directly, we have made the curve of width and height with different deposition velocity. The depositing dimension includes the height and the width of the single-track part. In this work, the width and the height after layer finished were measured and recorded by the vernier caliper after deposition of each layer. The average value was calculated according to the measurements. as shown in Fig.3. The width and height decrease gradually with the increase of deposition velocity. However, the effect of deposition velocity on the width is more significant, and the trend is slowing down. In order to quickly find the test parameters, we choose medium deposition velocity 18 mm/s medium for subsequent experiments.

3.2. Influence of Flow Rate on Forming Appearance in Surface Morphology

The effects of different flow rates on the width and height as shown in Fig. 4. The deposition velocity is 18 mm/s and the initial distance is 1.6 mm. From the graph, we can see that the width and height increases with the increase of flow rate, but the impact of flow rates on the width is more significant. When the flow is 50 mm3/s, the inflection point appears, and the influence of gravity and surface tension reaches a balance value.

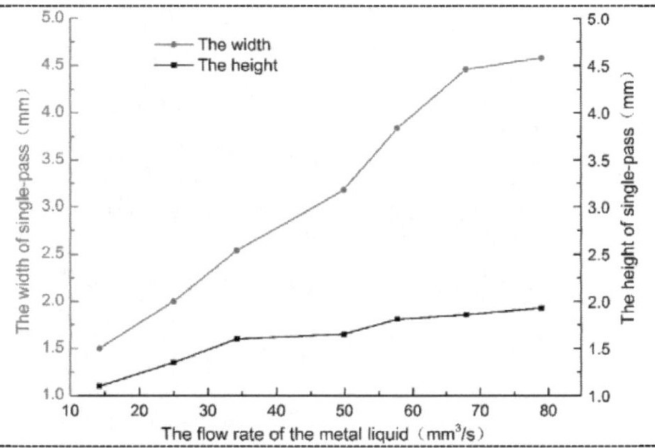

FIGURE 4. The forming width and height with different flow rate.

3.3. Influence of Initial Distance on Forming Appearance in Surface Morphology

The distance from the nozzle to the substrate is defined as the initial distance; the effect of different initial distances on the height is shown in Fig. 5. The deposition velocity is 18 mm/s, and the flow rate is about 30 mm3/s, 50 mm3/s, 70 mm3/s. From the graph, we can see that the height increase with the increase of the initial distance. The surface tension of the metal liquid reaches a relatively stable state at the initial distance. When the flow rate is different, the initial distance of the steady state is different. The larger the flow rate, the greater the initial distance to the steady state. The flow rate is 50 mm3/s, the height change trend is relatively stable.

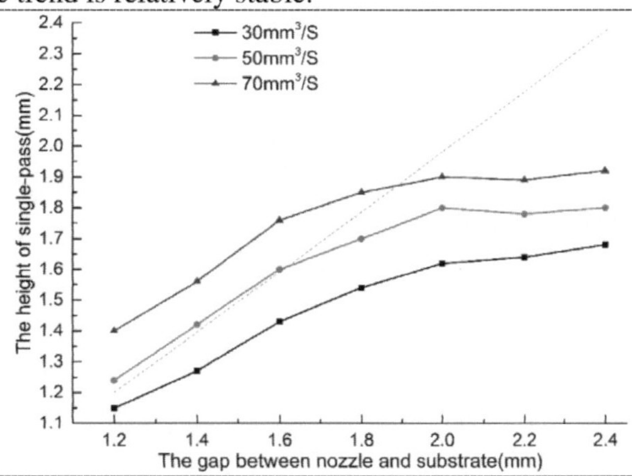

FIGURE 5. Forming height with different initial distance.

The effect of different initial distances on the width is shown in Fig. 6. From the graph, we can see that the width decrease with the increase of the initial distance, and finally tends to be stable, and when the flow rate is 50 mm3/s, the width change trend is relatively stable. The change tendency of width is opposite to that of height.

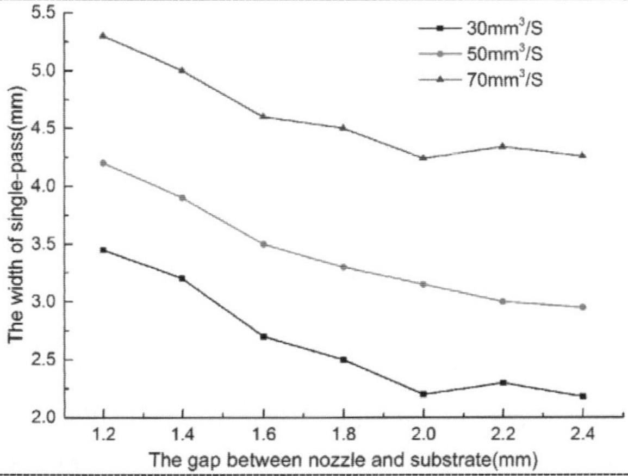

FIGURE 6. Forming width with different initial distance.

3.4. Surface Morphology Analysis

As a basic unit, the forming process of a single-track specimen should be primarily investigated. In single forming the high-temperature melt being extruded from the nozzle once contacts with the pool, as the melt flow is very complex and the melt shape fluctuates frequently, they would influence the quality and performance of the formed parts at the initial stage of fused-coating process. Therefore, it is necessary in order to explore the single-track forming process.

In this section, A micro-focus X-ray multi-functional three-dimensional imaging detection system with 1.5 um dimensional accuracy to observe the internal microstructure, as shown in Fig. 7 (a), it can be seen in single-track internal microstructure uniform and no cracks, and has good spread ability, similar to the shape of oval section. A Confocal Laser Scanning Microscope (CLSM) with 0.12 um dimensional accuracy to observe the surface morphology, as shown in Fig. 7 (b), It can be seen that it has a good surface morphology, roughness average of about Ra3.2, with a good subsequent processing potential.

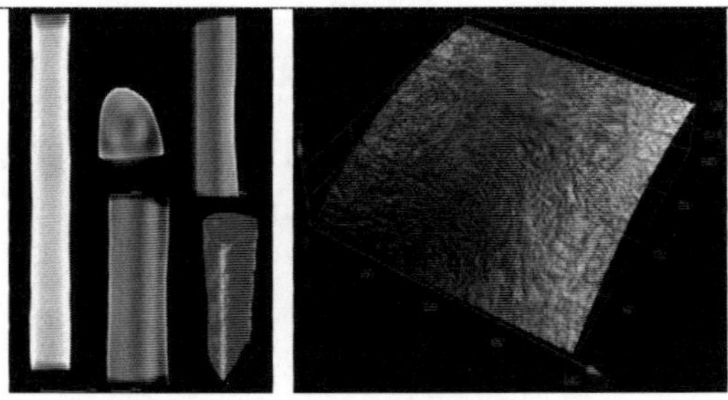

FIGURE 7. single-track X ray three-dimensional imaging and CLSM imaging.

4. Conclusions

In this study, based on the accumulation of experimental data, optimal process parameters of MFCAM were analyzed to achieve good appearance and internal morphology. The main results can be summarized as follows:

- The MFCAM technology is considered an economical and effective forming process. Forming efficiency of the MCMAM up to 50 mm3/s (viz. 1490g/h) is doubled, compared to the metal droplet deposition.

- There is a strong coupling effect between the single track forming morphology, the deposition velocity, the flow rate and the flow initial distance. The experimental results show that the optimal process parameter for the initial distance is 1.6 mm, the flow rate is 50 mm3/s and the velocity is 18 mm /s.
- Single-track specimen's surface morphology was observed by confocal laser scanning microscopy and internal microstructure was observed by X - ray diffraction. Results show that the metal fused-coating process can achieve good surface morphology and without internal tissue defect. The research foundation is established for the next print of the whole parts.

5. References

[1] Frazier W E. Metal Additive Manufacturing: A Review[J]. Journal of Materials Engineering & Performance, 2014, 23(6):1917-1928.

[2]. Everton S K, Hirsch M, Stravroulakis P, et al. Review of in-situ process monitoring and in-situ metrology for metal additive manufacturing[J]. Materials & Design, 2016, 95:431-445.

[3]. Xiong J, Zhang G, Zhang W. Forming appearance analysis in multi-layer single-pass GMAW-based additive manufacturing[J]. International Journal of Advanced Manufacturing Technology, 2015, 80.

[4]. DU Jun, Zhengying WEI. Numerical analysis of pileup process in metal microdroplet deposition manufacture [J]. International Journal of Thermal Sciences, 96(2015), 35-44.

[5]. Heralic A, Christiansson AK, Ottosson M, Lennartson B (2010)Increased stability in laser metal wire deposition through feedbackfrom optical measurements. Opt Lasers Eng 48:478-85.

[6]. Xiong J, Zhang G. Adaptive control of deposited height in GMAW-based layer additive manufacturing[J]. Journal of Materials Processing Technology, 2014, 214(4):962-968.

[7]. Mireles, J.; David Espalin.; David A Roberson.; B. Zinniel.; Francisco Medina.; R. Wicker. Fused Deposition Modeling of Metals. In Proceedings of the Solid Freeform Fabrication Symposium, Austin, Texas, USA, 6–8 August 2012; pp. 836–845.

[8]. Yao Y, Gao S, Cui C. Rapid prototyping based on uniform droplet spraying[J]. Journal of Materials Processing Technology, 2004, 146(3):389-395.

[9]. Du, J.; Wei, Z.; Wang, X.; Fang, X.; Zhao, G. A novel high-efficiency methodology for metal additive manufacturing. Appl. Phys. A[J]. 2016, 122, 945.

Acknowledgments

Project supported by the State Key Development Program Research of China (2017YFB1103201), State Key Development Program Research of China (2016YFB1100400), the Fundamental Research Funds for the Central Universities (xjj2016124) and the Equipment pre-research project from Ministry of Education (6141A02022109).

IOP Conf. Series: Materials Science and Engineering **272** (2017) 012018 doi:10.1088/1757-899X/272/1/012018

Effect of suspension kinematic on 14 DOF vehicle model

T Wongpattananukul and C Chantharasenawong

Department of Mechanical Engineering, Faculty of Engineering, King Mongkut's University of Technology Thonburi, 126 Pracha Uthit Rd., Bang Mod, Thung Khru, Bangkok 10140, Thailand

Abstract. Computer simulations play a major role in shaping modern science and engineering. They reduce time and resource consumption in new studies and designs. Vehicle simulations have been studied extensively to achieve a vehicle model used in minimum lap time solution. Simulation result accuracy depends on the abilities of these models to represent real phenomenon. Vehicles models with 7 degrees of freedom (DOF), 10 DOF and 14 DOF are normally used in optimal control to solve for minimum lap time. However, suspension kinematics are always neglected on these models. Suspension kinematics are defined as wheel movements with respect to the vehicle body. Tire forces are expressed as a function of wheel slip and wheel position. Therefore, the suspension kinematic relation is appended to the 14 DOF vehicle model to investigate its effects on the accuracy of simulate trajectory. Classical 14 DOF vehicle model is chosen as baseline model. Experiment data is collected from formula student style car test runs as baseline data for simulation and comparison between baseline model and model with suspension kinematic. Results show that in a single long turn there is an accumulated trajectory error in baseline model compared to model with suspension kinematic. While in short alternate turns, the trajectory error is much smaller. These results show that suspension kinematic had an effect on the trajectory simulation of vehicle. Which optimal control that use baseline model will result in inaccuracy control scheme.

1. Introduction

Computer simulations have made a significant impact in modern science and engineering. They are used to reduce time and computation resources on study and design. Accuracy of simulation depends on mathematical model that is implemented. Highly sophisticated model is able to represent natural phenomena but the simulation requires longer time to complete. Therefore, models with sufficient accuracy for a specific task is a better choice in simulation. Road vehicle simulations have been extensively studied with mathematical models representing dynamics of vehicles ranging from single track bicycle models to multibody models. These models are developed for their specific tasks that their requirements are different. For race car applications, optimal control is often coupled with the vehicle dynamic model to determine its control scheme to yield the minimum lap. Standard four-wheel vehicle models that are often used in this task are 7 DOF in [1] and [2], 10 DOF in [3], and 14 DOF in [3] and [4]. The standard 14 DOF model incorporate 6 DOF in body motion depicting surge, sway, heave, roll, pitch and yaw, with additional 8 DOF from the wheels. There are 2 DOF per wheel, which represent wheel-ground distance and wheel spin. The 10 DOF model neglects the wheel ground distance while the 7 DOF further neglects the following motions: body roll, body pitch and body. Even though the 14 DOF model is one of the most accurate models for minimum lap time optimization purpose, it still lacks one significant aspect of the vehicle that is suspension kinematic.

Suspension kinematic refers to the relative motion between the wheels and body whose movement is defined by linkage in between. The wheel angle with respect to body such as toe angle, camber angle can be expressed as functions of the suspension travel. These relative motions are required in the simulation because tire forces predicted by Pajaceka's magic formula tire model [5] uses the slip angle and vertical force as independent variables for tire force function.

The aim of this work is to identify the significance of suspension kinematic on the accuracy of trajectory in simulation results. The classical 14 DOF without suspension kinematic model is chosen as baseline model for comparison. The suspension kinematic will be added to the baseline vehicle model to study its heading and trajectory deviations from the experimental data.

2. Vehicle modeling

The baseline vehicle model is developed base on [3] and [4]. These models incorporate 6 DOF of body motion including surge, sway, heave, roll, pitch and yaw. Additional 8 DOF represent the four wheels with 2 DOF from each wheel denoting wheel height from ground and wheel spin. The wheel height from ground is represented by suspension spring compression displacement. This change in the generalized coordinates is made to simplify the incorporation of suspension kinematic into the baseline model. The tire model used in this study is taken from Pajaceka's magic formula [5] which is widely used in literature [1], [6] and [7].

2.1. Suspension kinematic

Suspension kinematic is a geometrical problem of linkages between the car body and the wheels. The double wishbone suspension type with pushrod actuated shock absorber is used in the derivation of these relations. Geometric problems are solved using numerical methods described in [8]. Solution for the roots of the equations encounter some instability, therefore, numerical relation result is fitted with polynomial equations to ensure there always exists a solution within the boundaries.

In front suspensions, there are two independent variables including steering rack displacement (component of control vector, $\boldsymbol{u}$) and suspension travel (component of state vector, $\boldsymbol{x}$). These 2 variables define the position and angle of the wheel on front suspension (front suspension kinematic relation vector, $\boldsymbol{S}_\mathbf{f}$). While the rear suspension has only one independent variable which is the suspension travel. These are expressed as equation (1) and equation (2) for front suspension and rear suspension respectively. Suspension kinematic relation vector are a combination of functions where each function represents the position or the angle of wheel with respect to body.

$$\boldsymbol{S}_\mathbf{f} = f(\boldsymbol{x}, \boldsymbol{u}) \tag{1}$$

$$\boldsymbol{S}_\mathbf{r} = f(\boldsymbol{x}) \tag{2}$$

2.2. Integration of suspension kinematic in 14 DOF model

Energy equations in Lagrangian system of vehicle model are modified for suspension kinematic function. These equations are used in the Euler-Lagrange equation to derive the left-hand side of the equation of motion. Equation (3) shows the original front left tire spring potential energy equation where k denotes the tire vertical stiffness, $x_{\mathrm{ss,fl}}$ denotes front left spring deflection, $\mathrm{r_t}$ denotes the tire radius. While equation (4) shows a modified tire spring potential energy equation where $S_{\mathrm{f,mo}}$ denotes distance of front tire to ground which is component of equation (1). Deflection in the original equation changes to suspension kinematic relation of motion ratio between spring deflection and vertical movement of wheel. Another important modification is in the tire force generation function where camber and slip angles now depend on spring compression length and steering rack travel.

$$V_{\mathrm{st,fl}} = \frac{1}{2}\mathrm{k}(x_{\mathrm{ss,fl}} - \mathrm{r_t})^2 \tag{3}$$

$$V_{\mathrm{st,fl}} = \frac{1}{2}\mathrm{k}(S_{\mathrm{f,mo}})^2 \tag{4}$$

3. Verification and simulation

An experiment is carried out with a formula student style car to verify the vehicle model. The car weighs 280 kg with driver. It has a double wishbone suspension with front track width of 1.260 m, rear track width of 1.190 m, wheelbase of 1.550 m and a maximum suspension travel length of 0.05 m. The front spring stiffness is 43,781.7 N/m and rear spring stiffness is 52,538 N/m. Ten inch wheels and tires are selected.

Testing is performed in two cases. First, a 180-degree constant radius turn with a track inner radius of 5 m and outer radius of 9 m. The second test is a single lane change at constant. These two methods are chosen because they generate lateral acceleration during turning so that differences in left and right suspension travel can be detected. Longitudinal acceleration affects left and right suspension travel equally, hence it is not of our interest. These tests are carried out at constant speed of 40 kph. The sensors collect data from driver input i.e. steering rack travel and vehicle heading, position, speed, suspension travel as experiment data.

Both baseline and suspension kinematic model vehicle simulation use the steering rack travel as driver input. The simulation results will show comparison of suspension travel, trajectory and heading differences between the baseline model, suspension kinematic model and experiment data. The total trajectory and total heading errors are calculated using equations (5) and (6), respectively. Where $x_{e,i}$ is the trajectory error in x direction, $y_{e,i}$ is the trajectory error in y direction $\phi_{e,i}$ is yaw deviation ω_i is the yaw rate. The suspension travel error is calculated using the root mean square method.

$$Err_{\text{traj}} = \left\{\sum_{i=1}^{n}\left[\left({x_{e,i}}^2 + {y_{e,i}}^2\right)|\omega_i|\right]\left(\sum_{i=1}^{n}|\omega_i|\right)^{-1}\right\}^{1/2} \quad (5)$$

$$Err_{\text{head}} = \sum_{i=1}^{n}\left[\left(\phi_{e,i}\right)|\omega_i|\right]\left(\sum_{i=1}^{n}|\omega_i|\right)^{-1} \quad (6)$$

4. Result and Discussion

Experiment data is collected from actual tests on a formula student car on a dry track with 35°C ambient temperature. Driver's steering controls are recorded and used as simulation inputs for both vehicle models. Two experiments as described in Section 3 are conducted. The results of the experiments and simulation are shown, compared and discussed in the following subsections.

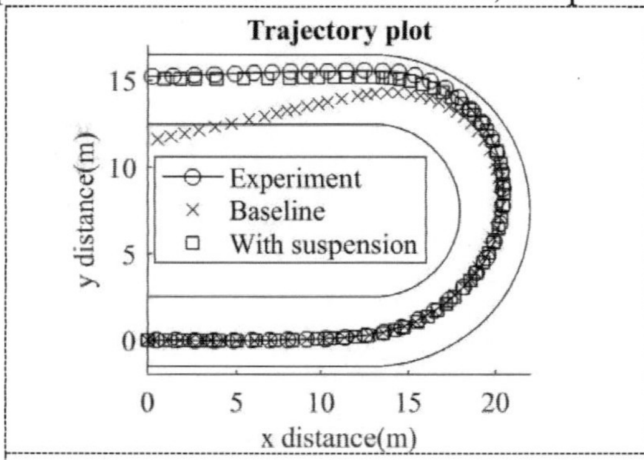

Figure 1. Trajectory plot of simulated results and experiment data on constant radius and constant speed of 40 kph.

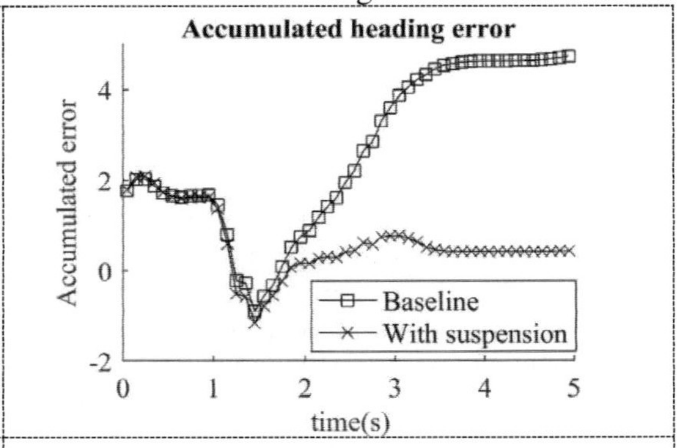

Figure 2. Accumulated of trajectory deviation plot on constant radius and constant speed of 40 kph.

4.1. *Turning at constant radius and constant speed*

The car makes a constant radius turn at a constant speed of 40 kph. The car trajectory from experiment data and simulation results are illustrated on figure 1 to show a deviation of baseline path and 14DOF with suspension kinematic path in comparison to experiment data. The large deviation at the end of the turn of baseline trajectory is a result of accumulated heading error over the turn. The 14 DOF with suspension kinematic trajectory displays a small deviation at the end of the turn but its heading is in the same direction as the experiment data, therefore the trajectory error is significantly smaller compared to baseline simulation. The accumulated error of heading over the track is shown in figure 2.

4.2. Single lane change at constant speed

In the single lane change at constant speed of 40 kph test, the simulation results show that the deviations of trajectory of baseline model and model with suspension kinematic are minimal as shown in figure 3. The heading accumulated error is investigated further to identify the difference. The heading error as shown in figure 4 begins to accumulate after 0.8 sec when the driver begins to steer left, but the accumulated error decreases after 1.45 sec when the steering is reversed to turn into the new lane. The suspension kinematic effect seems to be reduced because the error cancel itself out due to the opposite steering maneuvering.

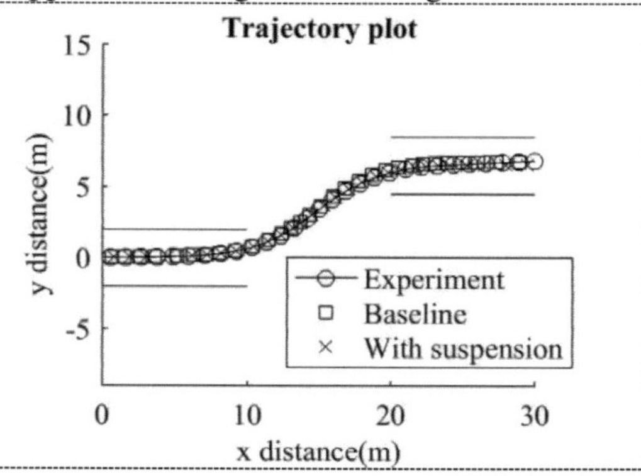

Figure 3. Trajectory plot of simulated results and experiment data on Single lane change at constant speed of 40 kph.

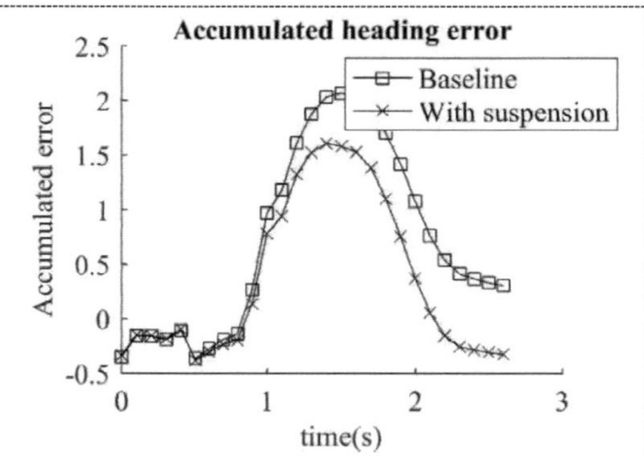

Figure 4. Accumulated of heading deviation plot which shown that the deviation of baseline model and model with suspension kinematic had increase and decrease in tandem with the alternate turn.

4.3. Discussion

Table 1. An error of simulation result compare to experiment data on constant radius track and single lane change. Suspension travel error of each wheel, heading error and trajectory deviation of the car was shown.

	Constant radius		Single lane change	
	Baseline	With suspension kinematic	Baseline	With suspension kinematic
Front left suspension travel (m)	0.0014	0.0012	0.0021	0.0009
Front right suspension travel (m)	0.0032	0.0021	0.0021	0.0009
Rear left suspension travel (m)	0.0012	0.0011	0.0021	0.0010
Rear right suspension travel (m)	0.0041	0.0029	0.0020	0.0013
Heading (degree)	4.7356	0.4432	0.3119	-0.3221
Trajectory (m)	0.8491	0.1892	0.2609	0.1878

These two experiments and simulation results show that deviation of trajectory also depend on the track. Suspension kinematic effects accumulate over the track especially on tracks with long turn. Tracks with many short alternate turns will have minimum effect on trajectory as described in the single lane change simulation. The heading errors are caused by the compression steer which further affects the wheel angles. In baseline model, these dynamic changes due to spring compression are neglected. Hence, these effects become more pronounced long turns but in short alternate turns they quickly cancel themselves out. Another indicator is the suspension travel data in Table 1, which also shows that the errors in baseline model are higher in long turn than in short alternate turn.

Although, the suspension kinematic is added to the 14 DOF baseline model there is no additional computational cost. This is because the suspension kinematic relation is integrated into baseline model. However, the major disadvantage of model with suspension kinematic is acquiring accurate vehicle data for the suspension kinematic simulation.

IOP Conf. Series: Materials Science and Engineering **272** (2017) 012018 doi:10.1088/1757-899X/272/1/012018

5. Conclusion

In this work the 14 DOF vehicle model with suspension kinematic has been successfully developed and investigated its accuracy in comparison to a baseline 14 DOF vehicle model without suspension kinematic and experiment data on formula student style car. The model with suspension kinematic exhibits higher accuracy compared to baseline model in case of constant radius turn with constant speed. However, under single lane change with constant speed situation, these two models show comparable accuracy. To complete a lap around a closed circuit race track, the car usually has to turn one way more than the other, therefore this can potentially result in significant accumulated heading errors when predicted with the baseline model. The error is caused by the neglect of suspension kinematic which affects the wheel position, wheel angle and tire force. Optimal controls with the use of suspension kinematic model yield high accuracy of control scheme. This is useful in minimum lap time optimization problem.
An important aspect of this work depends on the range of suspension travel and suspension kinematic of such car. This aspect should be further investigated to define level of impact on accuracy of vehicle model. Thus, define the level of impact which depend on type of car or range of suspension travel that need vehicle model with suspension kinematic.

6. Reference

[1] Cassanova D 2000 *On minimum time vehicle maneuvering: the theoretical optimal lap* (Bedford: Cranfield University)

[2] Koutrik S V 2015 *Optimal control for race car minimum time maneuvering* (Delft: Delft University of Technology)

[3] Lot R and Bianco N D 2016 *The Dynamics of Vehicles on Roads and Tracks* ed M Rosenberger (London: Taylor & Francis Group) chapter 58 pp 553–62

[4] Setiawan J D, Safarudin M and Singh A 2009 Modeling simulation and validation of 14 dof full vehicle model *Proc. Int. Conf. Instrument Communication Information Technology Biomedical Engineering* (Bandung: IEEE) pp. 1-6

[5] Pacejka H B 2002 *Tire and Vehicle Dynamics* (Cornwall: Butterworth-Heinemann)

[6] Rucco A, Notarstefano G and Hauser J 2010 *IEEE Conf. on Decision and Control* (Atlanta) pp.4934-39

[7] Siegler B 2002 *Lap time simulation for racing car design* (Leeds: University of Leeds)

[8] Dixon J C 2009 *Suspension geometry and computation* (Wiltshire: John Wiley & Sons Ltd)

IOP Conf. Series: Materials Science and Engineering **272** (2017) 012019 doi:10.1088/1757-899X/272/1/012019

Water Adsorption on Various Metal Organic Framework

H W B Teo[1] and A Chakraborty[1]

[1] School of Mechanical and Aerospace Engineering, Nanyang Technological University, 50 Nanyang Avenue, Singapore 639798

Abstract. In this paper, Metal Organic Framework (MOF) undergoes N_2 and water adsorption experiment to observe how the material properties affects the water sorption performance. The achieved N_2 isotherms is used to estimate the BET surface area, pore volume and, most importantly, the pore size distribution of the adsorbent material. It is noted that Aluminium Fumarate and CAU-10 has pore distribution of about 6Å while MIL-101(Cr) has 16 Å. The water adsorption isotherms at 25°C shows MIL-101(Cr) has a long hydrophobic length from relative pressure of $0 \leq P/P_s \leq 0.4$ with a maximum water uptake of 1kg/kg sorbent. Alkali metal ions doped MIL-101(Cr) reduced the hydrophobic length and maximum water uptake of original MIL-101(Cr). Aluminium Fumarate and CAU-10 has lower water uptake, but the hydrophobic length of both materials is within relative pressure of $P/P_s \leq 0.2$. The kinetic behaviour of doped MIL-101(Cr), Aluminium Fumarate and CAU-10 are faster than MIL-101(Cr).

1. Introduction

The specialty of adsorption cooling system has caught many attentions in recent years. To provide necessary cooling, solar heat or waste heat is required to supply to the system. This cooling application can reduce the harmful effects relating to global warming and ozone depletion. To operate, the system requires an adsorbate and an adsorbent for the adsorption-desorption process. Water is usually preferred to be the adsorbate as it is non-toxic, easily available and has a high evaporation of enthalpy [1]. The adsorbent material is usually a high porosity and high heterogeneous material. In present, the existing adsorbents that are used for adsorption system are silica gels and zeolites [2-5]. These adsorbent materials assisted in thermal compression in the adsorption bed of the cooling application, thus, replacing the compressor unit of the conventional vapour compression system.

However, to adopt the existing adsorbents for cooling application, large amount of space is required to install such system for cooling application purpose. The reason for the bulky size is mainly due to low uptake difference of the current adsorbent materials between adsorption and desorption process [5]. Thus, low working capacity is available over the cooling cycle from a small part of the total adsorption capacity of the adsorbent. Hence, large amount of adsorbent is required to be installed in the bed for sorption purposes.

Currently, a new class of microporous material namely Metal Organic Frameworks (MOFs) has been drawing many attentions due to their designable and high micro-porosity. These materials are made by the self-assembly of organic ligands and metal-containing nodes. Many researches have been ongoing for MOF + water system [6-9] with high potential of replacing the conventional material due to higher water uptake – offtake difference based on its 'S' – shaped isotherm. These material properties are able to reduce the size of the adsorption chiller, it may improve the performance of the cooling application. Thus, these materials show great potential in replacing the existing adsorbents for adsorption cooling system.

To understand the behaviour of MOF + water system, this paper targets potential MOFs to undergo surface characterisation and water adsorption experiment. The purpose of these experiments is to understand how material properties like pore size distribution and BET surface area affects the water adsorption behaviour of each individual material.

2. Experiments

The adsorbent for this paper includes MIL-101(Cr), alkali metal (Li^+, Na^+) doped MIL-101(Cr), Aluminium Fumarate and CAU-10. These materials will be synthesized and characterized using N_2 isotherm fitted with computed BET isotherm model to achieve the surface area, pore volume and pore size distribution. Then, a gravimetric water adsorption experiment is conducted to achieve the isotherm and kinetic results for all materials.

2.1. Synthesis

MIL-101 was performed by hydrothermal reaction method. 4g of chromium (III) nitrate nonahydrate and 1.66g of terephthalic acid were added into the reactor with 1ml of concentrated hydrochloric acid and 50 ml of deionized water before heating at 200 °C for 10 hours. The mixture was separated using a centrifugal spinning machine. After separation, the synthesized material was mixed with N, N-dimethylformamide (DMF) and heated at 80°C for purification. The mixture was then separated by the centrifugal spinning machine. Last, the synthesized material was dried at 80 °C.

Doped MIL-101 was a modification of MIL-101 structure. The synthesized MIL-101 were activated via vacuum oven at 150°C for 3 hours. The samples were mixed with hexane and stirred for 30 mins. Upon stirring, alkali metal solution (5% wt LiCl or NaCl with 1ml of deionized water) was added slowly into the suspension. The mixture was separated and the modified samples were dried at 80°C.

Aluminium Fumarate was synthesized slightly different from BASF patent [10]. 2.45g of aluminium chloride hexahydrate and 1.4g of fumaric acid were added into the suspension of DMF (100ml) in a beaker. The suspension was stirred at 130°C for 4 days. The mixture was separated by a centrifugal spinning machine. The white, synthesized material was rinsed with acetone and methanol twice. Last, the synthesized material was dried at 80°C.

CAU-10 was prepared by hydrothermal reaction method. 1g of isophthalic acid (1,3-H_2BDC) were dissolved in 5ml of DMF and 4g of aluminium sulphate hydrate were dissolved in 20ml of water. The solutions were mixed into an autoclave reactor and it was heated at 135°C for 12 hours. The mixture was separated by using a centrifugal spinning machine. The synthesized material was rinse with water via sonication for three times. Then, the material was dried at 80°C.

2.2. N2 Adsorption

The specific surface area, pore volume and pore size of the samples can be achieved by N_2 sorption isotherms at 77K, measured by Autosorb iQ (Quantachrome). The samples are degassed at 120°C to 160°C for 3 hours before N_2 adsorption. The measured isotherms data were computed with BET (Brunuer-Emmet-Teller) equation to calculate BET surface area. The isotherms data also estimate the pore size and the pore volume of the material.

2.3. Water Adsorption

The water adsorption isotherm and kinetics results of the samples are measured by Aquadyn DVS (Quantachrome) at 298K to 333K. The samples are heated at about 120°C to 160°C for 1.5 hours to remove any moisture adsorbed from the atmosphere. The samples are held by a microbalance in the chamber feel with mixture of dry N_2 and water vapour. The relative pressure in the chamber is controlled by the ratio of N_2 and water vapour measured. The mass of the samples increased as water molecules are adsorbed. Thus, the water adsorption isotherm is measured based on the weight differences from the sample's dry weight at different relative pressure. Figure 1 shows the schematic of this analyser.

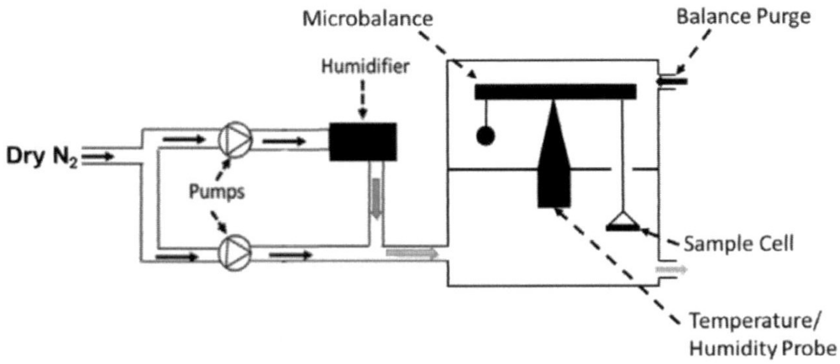

Figure 1. Water Adsorption Analyzer.

3. Results and Discussion

Figure 2 shows the N_2 isotherms of the samples. Based on the results shown, MIL-101 has the highest uptake among the samples. As the structure is modified, the N_2 uptake decreases. This may be due to the formation of micro pores after the modification, as MIL-101 are mainly meso pores. Aluminium Fumarate has a low N_2 uptake as compared to MIL-101. However, the uptake increases tremendously at the saturated pressure region. This is due to the large existence of macro pores in Aluminium Fumarate. Higher N_2 uptakes have the tendency of higher surface area and pore volumes. Table 1 shows the values of the BET surface area, pore volume and pore sizes of the samples. It is noted that MIL-101 has higher BET surface area and pore volume than the other samples. Although the BET surface area and pore volume decreases, the pore sizes of the modified MIL-101 are similar. On the other hand, Aluminium Fumarate has large pore volume with a small BET surface area due to the large existence of marco pores. CAU-10 has the smallest values for all three categories since the N_2 uptake is the lowest.

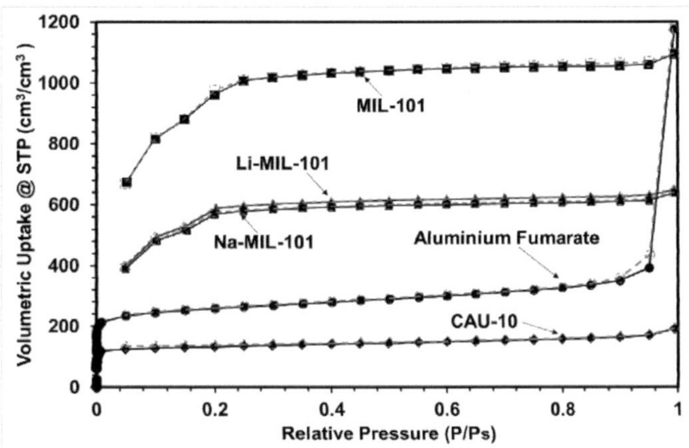

Figure 2. N_2 Adsorption Isotherms.

Table 1. Properties of samples

	BET Surface Area (m^2/g)	Pore Volume (cm^3/g)	Pore Size (Å)
MIL-101	3402.69	1.59	18.469, 23.129
Li-MIL-101	2054.03	0.94	18.469, 23.129
Na-MIL-101	2000.66	0.91	18.469, 23.129
Aluminium Fumarate	792.26	0.926	5.996, 7.509, 11.258
CAU-10	553.32	0.25	5.996, 11.258

IOP Conf. Series: Materials Science and Engineering **272** (2017) 012019 doi:10.1088/1757-899X/272/1/012019

Figure 3 shows the water adsorption uptake at 25°C. It is shown that MIL-101 has the highest water uptake at saturated pressure region. This is common since MIL-101 has higher surface area and larger pores. The hydrophobic length and water uptake are reduced after the modification of MIL-101. It is also noted that both CAU-10 and Aluminium Fumarate have much shorter hydrophobic length and lower water uptake than MIL-101. By comparing the results with Table 1, the size of micro pores for the adsorbent is rather ideal for short hydrophobic length. Further observation shows that CAU-10 has shortest hydrophobic length since CAU-10 has the smallest pores distribution. Another interesting cross reference between Figure 3 and Table 1 is all MOFs have at least 2 major pore sizes in the structure. It is observed that the higher water uptake is related to the size of larger pores. Hence, this shows that the location of the second peak of the pore size distribution is closely related to the maximum water uptake.

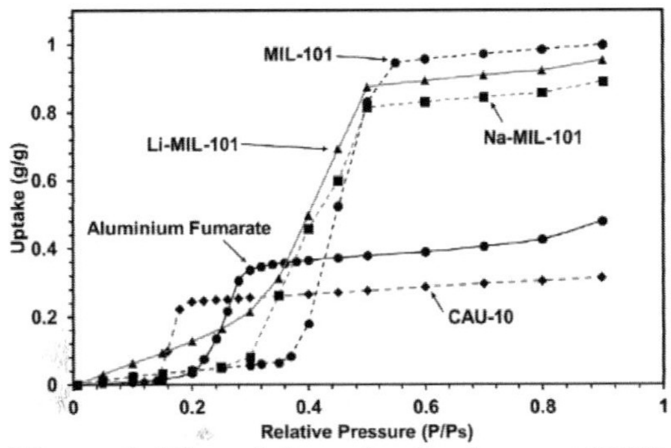

Figure 3. Water Adsorption Isotherm at 25°C

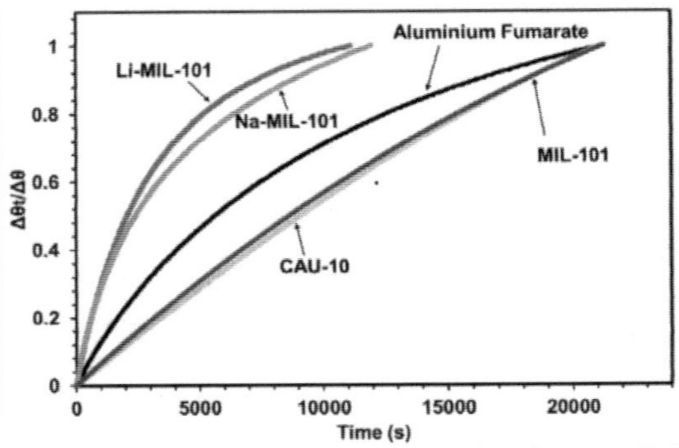

Figure 4. Water Adsorption Kinetics at 25°C

Figure 4 shows the adsorption kinetics of the MOFs at a temperature of 25°C. The original MIL-101 and CAU-10 have a slow kinetics to reach its equilibrium state. Based on the Figure, both MIL-101 and CAU-10 might require a longer time to reach its steady state region. Aluminium Fumarate also displays a slow kinetic behaviour. However, the trend indicates Aluminium Fumarate has a faster kinetic than MIL-101 and CAU-10. The curve also suggests the time taken for Aluminium Fumarate to reach its steady state region. On the other hand, modified MIL-101 samples show a much faster kinetic trend from its original sample, reaching steady state around 10000s to 12000s. This shows the modification via doping of alkali metal ions increase the interaction of water molecules for the adsorbent material. Hence, the time taken for the adsorption process of water molecules is shorten.

4. Conclusions

MOF adsorbents were experimented through N_2 adsorption and water adsorption test. The N_2 isotherms showed the large number of macro pores can be found in Aluminium Fumarate surface. It is also estimated that MIL-101 samples have better surface properties than CAU-10 and Aluminium Fumarate. A pore size distribution graph showed that the MOF materials have at least two distinct distribution peaks. Further comparison with water adsorption isotherms showed that smaller distribution of pores can be a factor to reduce the hydrophobic length of the adsorbent. On the other hand, the location of the second distribution peak can determine the maximum water uptake of the adsorbent. Hence, the combination of these two factors may be the key to achieve a water adsorbent with an isotherm with a high uptake difference. The kinetics data has shown that modified MIL-101 has better capability in reaching steady state at a shorter time. Thus, in terms of higher uptake difference and faster kinetic performance, these comparisons of the MOF material would provide a good direction on new MOF adsorbent for water adsorption applications.

Acknowledgements

The authors acknowledge the financing support from Ministry of Education, Singapore (grant no. MOE 2014-T2-2-061)

5. References

[1] Jeremias, F., Fröhlich, D., Janiak, C., and Henninger, S. K., Water and methanol adsorption on MOFs for cycling heat transformation processes, *New Journal of Chemistry,* vol. 38, p. 1846, 2014.

[2] Anyanwu, E. E. and Ogueke, N. V., Thermodynamic design procedure for solid adsorption solar refrigerator, *Renewable Energy,* vol. 30, pp. 81-96, 2005.

[3] Saha, B. B., Akisawa, A., and Kashiwagi, T., Silica gel water advanced adsorption refrigeration cycle, *Energy,* vol. 22, pp. 437-447, 1997.

[4] Thu, K., Chakraborty, A., Saha, B. B., and Ng, K. C., Thermo-physical properties of silica gel for adsorption desalination cycle, *Applied Thermal Engineering,* vol. 50, pp. 1596-1602, 2013.

[5] Myat, A., Kim Choon, N., Thu, K., and Kim, Y.-D., Experimental investigation on the optimal performance of Zeolite–water adsorption chiller, *Applied Energy,* vol. 102, pp. 582-590, 2013.

[6] Canivet, J., Fateeva, A., Guo, Y., Coasne, B., and Farrusseng, D., Water adsorption in MOFs: fundamentals and applications, *Chem Soc Rev,* vol. 43, pp. 5594-617, Aug 21 2014.

[7] Ehrenmann, J., Henninger, S. K., and Janiak, C., Water Adsorption Characteristics of MIL-101 for Heat-Transformation Applications of MOFs, *European Journal of Inorganic Chemistry,* vol. 2011, pp. 471-474, 2011.

[8] Frohlich, D., Henninger, S., and Janiak, C., Multicycle water vapour stability of microporous breathing MOF aluminium isophthalate CAU-10-H, *DALTON TRANSACTIONS,* vol. 43, pp. 15300-15304, // 2014.

[9] Henninger, S. K., Schmidt, F. P., and Henning, H. M., Water adsorption characteristics of novel materials for heat transformation applications, *Applied Thermal Engineering,* vol. 30, pp. 1692-1702, 2010.

[10] Kiener, C., Muller, U., and Schubert, M., Organometallic Aluminum Fumarate Backbone Material, USA Patent US20090092818A1, 2009.

IOP Conf. Series: Materials Science and Engineering **272** (2017) 012020 doi:10.1088/1757-899X/272/1/012020

Effect of fiber content on tensile retention properties of Cellulose Microfiber Reinforced Polymer Composites for Automobile Application

J R Aseer[1*] and K Sankaranarayanasamy[2]

[1]School of Mechanical Engineering, Galgotias University, Greater Noida, India

[2]Department of Mechanical Engineering, NIT, Tiruchirappalli, Tamil Nadu, India

E-mail: j.ronaldaseer@galgotiasuniversity.edu.in; ksnsamy@nitt.edu

Abstract. Today, the utilization of biodegradable materials has been hogging much attention throughout the world. Due to the disposal issues of petroleum based products, there is a focus towards developing biocomposites with superior mechanical properties and degradation rate. In this research work, Hibiscus Sabdariffa (HS) fibers were used as the reinforcement for making biocomposites. The HS fibers were reinforced in the polyester resin by compression moulding method. Water absorption studies of the composite at room temperature are carried out as per ASTM D 570. Also, degradation behavior of HS/ Polyester was done by soil burial method. The HS/ polyester biocomposites containing 7.5 wt% of HS fiber has shown higher value of tensile strength. The tensile strength retention of the HS/ Polyester composites are higher than the neat polyester composites. This value increases with increase of HS fiber loading in the composites. The results indicated that HS/ polyester biocomposites can be used for making automobile components such as bumper guards etc.

1. Introduction

Recently, natural fibers like jute, hemp, banana, flax and sisal have been a part of high-tech development and began to remove their negligence. Automobile industries have got interested in new biomaterials, which can be partially decomposable or recyclable for the current global trends of the environmental protection and development of sustainable technology. The application of natural fiber composites has increased and is gaining preference over glass fiber and carbon fiber due to their low-cost and low weight characteristics [1]. Continuous efforts are on developing newer or alternative materials to achieve cost effectiveness, fuel efficiency, reduced emissions, increased safety, and always with a target on future ability to recycle or biodegrade [2]. Wotzel et al. [3] studied that the natural fiber composites were more environmentally friendly than glass fiber composite as a candidate for automobile applications. Singha et al. [4] studied the mechanical and thermal properties of Hibiscus Sabdariffa (HS) fiber reinforced phenolic resin matrix based polymer composites. Different fiber loading (in terms of weight) such as 10, 20, 30 and 40% was investigated. The result showed a general trend of increase in properties with fiber loading up to 30% and beyond this loading the properties decreases. Sonia et al. [5] investigated the extraction of cellulose microfibers from HS fibers by steam explosion technique. Structural and surface analysis of the microfibers showed a reduction in diameter and changes in surface morphology from that of the raw fibers. The chemical composition of fibers showed increase in α-cellulose content and decrease in lignin and hemicelluloses for the microfibers. The effect of laccase and Xylanase enzyme treatment on chemical and mechanical properties of bananafiber was studied. The results revealed that removal of lignin content is high in laccase treatment [6]. The present study deals with the effect of fiber content on tensile retention properties of cellulose microfiber composites. Also, biodegradation behaviour of HS/ Polyester was investigated by soil burial method.

2. Materials and methods

2.1 Raw materials

Hibiscus sabdariffa var. altissima is a commercial and cheaper fiber crop known by its common name Mesta. Especially, these fibers were collected from the local cultivators of West Bengal for this study. The raw fibers were dried in an oven and they were chopped by the length of 10cm. All the chemicals such as sodium hydroxide (NaOH), sodium chlorite ($NaClO_2$), oxalic acid ($C_2H_2O_4$), epoxy resin (LY556), and hardener (HY951) (99.5% purity) used for this work were purchased from fisher scientific and HI-media, Delhi, India. The chemical treatment of raw fibres is explained elsewhere [5]

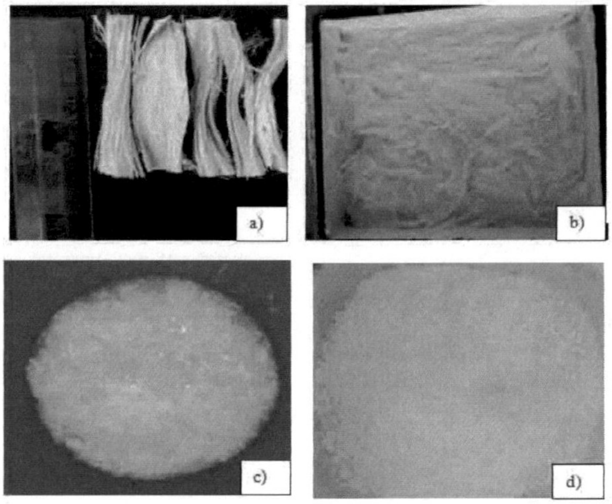

Figure 1. Various stages of microfiber extraction a) raw fiber b) bleached fiber c) oxalic treated fiber d) dried cellulose microfiber.

2.2 Extraction of cellulose microfibers from hibiscus sabdariffa var. altissima

Cellulose microfibers are extracted from HS fibers by steam explosion technique [5]. The raw fibers are collected, cleaned, washed and dried before subjected to a various chemical treatments. The figure 1 shows various stages of extraction of cellulose microfiber from HS fiber.

2.3 Composite fabrication

The composite materials are fabricated by hand layup process. The cellulose microfibers are used in the preparation of the specimen. The processed composite is pressed by Hydraulic press under 550 kPa of pressure at 100 ° C for 2 hours and dried for 24 hours. The composites are prepared by varying the fiber weight ratio such as 1, 2.5, 5, 7.5, 10 wt. % respectively. The epoxy resin hardener ratio is maintained at 10:1 in all formulations of composites.

2.4 Testing

2.4.1 Water absorption testing

Water absorption studies of composites such as molar percentage of water uptake, diffusion, sorption coefficient and permeability are carried out at room temperature as per ASTM D 570. Square specimen of 30×30 mm is cut from the composite panel. In each case, five specimens were tested to obtain average value.

2.4.2 Soil burial testing

Composites are kept in a series of perforated boxes containing moisturized soil. The specimens (30×10 mm) are buried 100 mm under the surface of soil. The samples are removed at predetermined time interval and washed with distilled water. Then, it is dried at room temperature [7]. The weight loss of composite is measured using the following equation (1).

$$\text{Weight loss} = \frac{w_0 - w_t}{w_0} \quad (1)$$

where w_0 - initial mass and w_t - remaining mass at any time

2.4.3 Tensile strength testing

The tensile test of composites is carried out on Instron tensile testing machine 8801 with the cross head speed of 5 mm/min as per ASTM D638-03. Samples after water absorption (2.4.1) or biodegradation (2.4.2) are also tested to determine the tensile strength. Tensile strength retention is also determined using the following equation (2)

$$\text{Tensile strength retention} = \frac{\tau_t}{\tau_0}\times100 \quad (2)$$

where τ_t - tensile strength after water absorption/biodegradation and τ_0 -initial tensile strength

3. Results and discussion

Tensile strength and water absorption properties of the Hibiscus sabdariffa (HS)/ polyester composites with different fibre loadings are given in Table 1. The tensile strength values of neat resin are also given in this table. The tensile strength and Young's modulus are higher for the fiber composites compared to those of the neat resin. There is an increasing trend in the value of tensile strength and modulus with fibre content can be identified from Table 1. Optimum mechanical property is obtained at 7.5% HS fiber loading after which it decreases. The improvement in tensile strength with HS fiber loading is due to the micro-level interactions. So, more effective stress transfer takes place between the components. The decrease in tensile strength at higher loading may be due to the imperfection in filler dispersion at higher loading [7]. The water uptake, diffusion and permeability values are increased with the addition of fiber due to the hydrophilic character of the natural fiber (Table 1). The neat resin has the lowest sorption value. The permeability coefficient increases linearly as the fiber loading increases from 1% to 10 wt%. Similarly, Sreekala et al. observed increased uptake of water with increasing fiber loading when cellulose fiber loading was less than 3 wt.% [8].

Table 1. Tensile strength and water absorption properties of composites.

Hibiscus sabdariffa fiber content (wt.%)	Tensile strength (MPa)	Tensile modulus (MPa)	Elongation at break (%)	Mole percentage water uptake (mole/g)%	Diffusion coefficient D ($\times10^{-6}$) (cm^2/sec)	Sorption coefficient S	Permeability coefficient P($\times10^{-6}$) (cm^2/sec)
Neat	17	70	475	0.0938	0.0225	0.016	0.38
1	19	153	525	1.96	3.3	0.362	1.27
2.5	22	228	550	2.26	4.85	0.408	1.98

5	24	256	575	2.31	6.96	0.453	2.75
7.5	26	301	750	2.88	7.06	0.519	3.66
10	21	248	625	3.11	7.16	0.566	3.86

The tensile strength values of neat resin and HS fiber composites are given in the Table 2. It is understood that water absorption affects the polymer composites. The tensile strength of the neat resin and HS fiber composite are decreased after moisture absorption. The neat resin shows reduction of tensile strength values 10% and 30% respectively, at 24 and 182 h. However, fiber reinforced composites values are lower in comparison to the neat resin. This is due to the less sensitive of composites to moisture. The different retentions of tensile strength values between the composites and neat resin indicate that incorporation of HS fibers leads to an improvement in strength retention properties.

Table 2. Tensile strength retentions after water absorption and soil burial.

Hibiscus Sabdariffa Fiber loading (Wt %)	Tensile strength retention after water absorption (%)			Tensile strength retention after soil burial (%)		
	0 hr	24 Hr	182 hr	0 hr	120 hr	480 hr
Neat resin	100	90	70	100	79	65
1	100	92	75	100	81	67
2.5	100	94	82	100	83	70
5	100	96	85	100	85	73
7.5	100	98	90	100	87	75

The weight loss with degradation time is shown in figure 2. The weight loss shows an approximately linear relation with degradation time for both neat and 7.5 wt% HS fiber composites. It is observed that the weight loss of the 7.5 wt% composite is lower than that of the neat resin. The tensile strength of the neat resin and the HS fiber composite are decreased by 21% and 13%, respectively, after incubation in soil (Table 2). There is a reduction in mechanical properties of composites due to the weight loss. The composites lose their structural integrity when microorganisms consume the resin surrounding the fiber. As reported earlier [9], the attack on the HS fiber by microorganisms causes deterioration of the mechanical properties.

4. Conclusion

The mechanical properties of the Hibiscus sabdariffa /polyester biocomposites are increased as the HS fiber loading is increased up to 7.5%. The HS/polyester biocomposites have higher tensile strength and modulus than the neat polyester resin. The tensile strength of the HS /polyester composites is deteriorated after water absorption. The HS/polyester biocomposites has shown higher tensile strength retention when compared to the neat due to the higher resistance to microrganism attacks on the HS fiber. The experimental results indicated that the developed HS/ polyester biocomposites has shown better environmental protection and can be used for making automobile components.

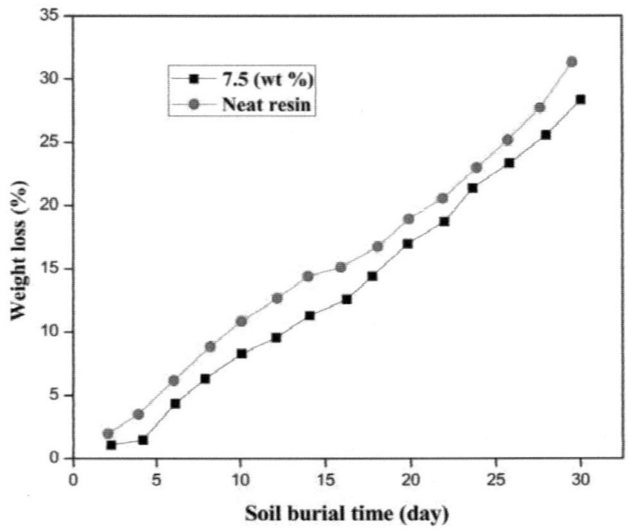

Figure 2. Weight loss as a function of time.

5. References

[1] Ahmad F, Choi HS and Park M K 2015 *Macromol. Mater. Eng.* **300** 10-24
[2] Elaya Perumal A and Venkateshwaran N 2008 *Int J Adv. Eng. Appl.* **6** 68-74.
[3] Wotzel K, Wirth R and Flake M 1999 *Die Angew. Makromol. Chem.* **272** 121-7
[4] Singha AS and Thakur VK 2008 *Int J Polym. Mats.* **31** 791
[5] Sonia A and Priya DK 2013 Carbohydr. Polymer. **92** 668-74.
[6] Vardhini KJV and Murugan R 2017 *J Nat. Fiber,* **14** 217-227
[7] Wan YZ, Luo H, He F, Liang H, Huang Y and Li XL 2009 *Comp. Sci Technol.* **69** 1212-17
[8] Sreekala MS, Goda K and Devi PV 2008 *Compos. Interface.* **15** 281-99.
[9] Zuchowska D, Steller R, Meissner W 1998 *Polym. Degrad. Stabil.* **60** 471-80.

IOP Conf. Series: Materials Science and Engineering **272** (2017) 012021 doi:10.1088/1757-899X/272/1/012021

Development of a tension system and influence of ultimate tension on the properties of composites

Yuqing Cui and Zhongwei Yin*

School of Mechanical Engineering, Shanghai Jiao Tong University, Shanghai 200240, China

*Corresponding author e-mail: yinzw1972@163.com

Abstract. The demand for high tension winding has increased in recently years. However, in-depth studies on high–tension winding have not been conducted. In the current study, a device was developed to apply tension to the winding process, and several fibers were adopted. A new method of tension controlling tension was applied in the experiments, allowing accurate and digital control of tension in the winding. Ultimate tension (pretension) in filament winding process was evaluated. Samples with varying tension were selected for the experiment and their physical and mechanical properties were measured. Experimental results indicated that tension in winding significantly affect composites. The function of the tension was subsequently analyzed. This study concluded that the ultimate pretension in winding was far less than the fracture tension, ultimately influencing the physical and mechanical properties of composites.

1. Introduction

High-speed composite rotors, composite tubes and pressure vessels are typically manufactured by filament winding [1–3]. In filament winding, winding tension (pretension) is considered a vital parameter to measure the performance of a product [4, 5]. A numerical control method was applied in a winding tension system to achieve accurate and real-time tension adjustment [6–11]. Currently, tension on the filament can be controlled using two methods. One method is with the use of a tensioner connected to the fiber package on a card or metal tube. The tensioner in a tension control system could be a magnetic brake, torque motor, or servo-motor [6–8]. The other method is with the use of a drag wheel, which is connected to the magnetic brake or torque motor, when the filament winds through it [9]. High pretension can be required in a composite product, such as, a high-speed PM rotor composite sleeve and a composite flywheel [1]. However, high-tension winding for the current winding machine is difficult to obtain. In summary, studies are rarely been reported on high-tension winding and the development of a pretension control system that could supply sufficient tension for filament winding. In addition, studies on the maximum allowable tension in filament winding are rarely conducted.

In the present study, a new tension control system for filament winding was developed. This new device does not only apply sufficient pretension on the filament; it is also low-cost and has a less occupied area. In the test, the device used in filament winding and carbon fiber (CF), Kevlar fiber, and polybenzoxazole (PBO) fiber were adopted. Meanwhile, the maximum allowable tension of these fibers in winding was tested. The properties of the composites with varying tension were then characterized. Research shows that the developed tension control system can effectively apply high tension on filaments, and a significant difference in the properties of the composites with varying tension is indicated.

2. Experimental

2.1 Method and Device

The method used to apply tension is similar to the second method [9] mentioned earlier, which produces static friction resistance through the tensioner during filament delivery. This numerical tension control system mainly consists of a tension meter, a tension actuator and a controller. The tension actuator is mainly composed of two magnetic powder brakes, two roll shafts, and a shell frame. The controller can automatically adjust the excitation current of the magnetic powder brake according to the set target tension and the measured tension after PID (proportional-integral-derivative) operation. The two roller shafts are connected to a magnetic powder brake. The axes of the two roller shafts form an angle between them. The schematic of the tension control system is presented in Figure 1, and an image of the system is shown in Figure 2. The first roller shaft is connected to the second magnetic powder brake, which slants the second roller shaft. The axis of the second roller shaft is perpendicular to the line of the filament on the surface of the shaft. The second roller shaft is connected to the first magnetic powder brake. After the filament is dispensed from the fiber package, the filament should be wound from the large opening end formed by two shafts. The principle of tension application is that the tension is obtained by static friction generated from filament winding on the surface of two roller shafts. The schematic of the control principle of the tension system is shown in Figure 3. This device can produce greater tension than the single drag wheel device. The filament was then delivered to the tension meter. The tension detection mechanism consists of six guide wheels and a torque arm.

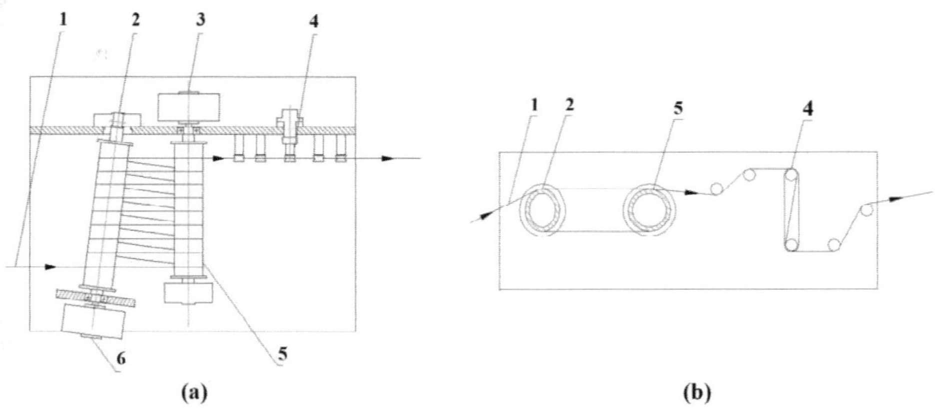

Figure 1. Schematic of the tension control system:(a) top view;(b) side view, (1) filament, (2) first roller shaft, (3) first magnetic powder brake, (4) tensionmeter, (5) the second roller shaft, (6) second magnetic powder brake.

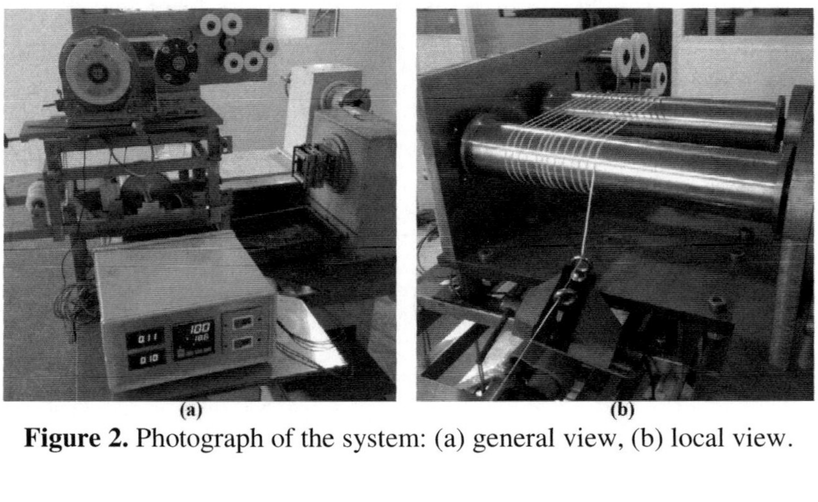

Figure 2. Photograph of the system: (a) general view, (b) local view.

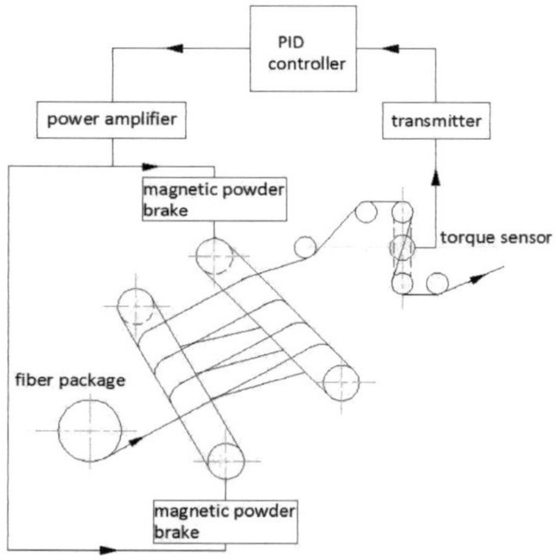

Figure 3. Control principle schematic of the system.

The torque sensor passes the measured value to the controller after the transmitter conversion. The digital controller acts a closed-loop feedback control. The PID control algorithm was adopted in the experiment. The operator can digitally input tension and select the tension control mode. Winding is completed automatically, and the parameters can be digitally displayed and controlled. Therefore, the tension can be controlled accurately and digitally in tape winding.

2.2 Materials

High-strength fibers, such as carbon, Kevlar, and S-glass are usually chosen as materials for winding composites, and wet winding is the commonly used and suitable method. Compared with Kevlar fiber, PBO has a higher strength and a higher modulus. In the current study, CF, Kevlar and PBO were used. Kevlar was provided by DuPont Co. and of type K29 with a linear density 1670 dtex. The PBO (Zylon HM) was supplied by Toyobo Co., with a linear density is 1670 dtex. The properties of the fibers and the normal winding tension are listed in Table 1. A heat-resistant epoxy resin was selected as the matrix of the composites. The properties of the resin are listed in Table 2.

Table 1. Properties of the fibers and winding tensions.

Fiber type	Number of filament	Tensile strength (MPa)	Tensile modulus (GPa)	Tow cross-sectional area (mm^2)	Winding tension (N)
T700S	12K	4900	230	0.444	32→30→28→26
T800H	12K	5490	294	0.245	22→20→18→16
T1000G	12K	6370	294	0.269	26→24→22→20
M40J	6K	4410	377	0.127	15→14→13→12
M50J	6K	4120	475	0.116	15→14→13→12
T700S towpreg	12K	4900	230	0.444	35→33→31→29
Kevlar	–	3600	83	0.116	15→14→13→12
PBO	–	5800	270	0.105	15→14→13→12

IOP Conf. Series: Materials Science and Engineering **272** (2017) 012021 doi:10.1088/1757-899X/272/1/012021

Table 2. Properties of the resin.

Property	Density (g/cm^3)	Tg (°C)	Tensile strength (MPa)	Tensile modulus (GPa)	Elongation at break (%)	Flexural strength (MPa)	Flexural modulus (GPa)
Value	1.21	205	70	2.3	3.6	120	2.5

2.3 Ultimate Tension of Filament Testing

Breakage and damage in the fiber bundles were inevitable in high-tension winding. Larger damage and breakage in fiber bundles indicated difficulty in winding and a decrease in product strength. Therefore, the maximum pretension of the filament in winding process should be evaluated. In the tension control system, the applied tension increased gradually until the filament was destroyed. Ultimate tension was measured, and ultimate tensile stress was calculated based on the basic parameters of the fiber.

2.4 Winding with Different Pretensions

2.4.1 Physical Property Testing

The density of the composites with different winding tensions was measured with a density meter. The fiber volume fraction and porosity of the composite specimen were measured with the microscope method based on the Chinese standard GB/T3366.

2.4.2 Mechanical Property Testing

In accordance with the ASTM: D2290-12, the hoop tensile strength of the winding composite was measured. These tests were carried out on a mechanical test machine (CTM Co., 10T) using a cross-head displacement rate of 1.3 mm/min. The procedure of inter-laminar shear strength (ILSS) is outlined amply in ASTM: D2344-16. The lengths of the samples cut from rings were six times the thickness, and the widths of arcs measured 10mm.The test were carried out on the testing machine (CTM Co., 500Kg) with a crosshead displacement rate of 1mm/min.

3. Results and Discussion

3.1 Ultimate Tension in Winding Process

Figure 4(a) shows that CFs in wet winding have lower ultimate tension compared with towpreg, Kevlar, and PBO. In these CFs, the T700 12K type exhibits the highest ultimate tension, whereas the M50 6K type exhibits the lowest tension. The fibers are likely to break when the filament passes through the guide pulley in winding because CFs are more brittle than organic fiber. The towpreg has a high ultimate tension because of the resin. The resin protects the fiber from damage during fiber delivery and transfers stress between the fibers while filaments in a bundle of fibers cannot be uniformly stressed at a loading state. Ultimate tensile stress is the ratio of ultimate tension to the cross section of the fiber bundle. As presented in Figure 4(b), ultimate tensile stress is higher in Kevlar and PBO than in CFs. The ultimate tensile stress of T700 towpreg was not higher than that of CFs in wet winding.

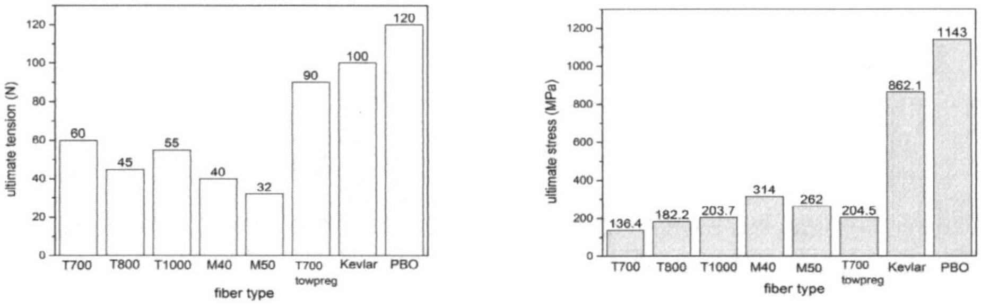

(a) (b)

Figure 4. Ultimate tensile property of the filament: (a) ultimate tension of fibers, (b) ultimate tensile stress of fibers.

3.2 Physical Property

The fiber volume fraction, density, and porosity of composites with ultimate tension (UT) and normal tension (NT) are listed in Table 3. The table shows a clear trend of increasing fiber content and density with ultimate tension. Delamination and lack of resin could be observed in the Kevlar and PBO composites with ultimate tension. Thus, density, fiber content, and porosity were difficult to measure accurately. The porosity of CF composite decreased when ultimate tension was applied.

Table 3. Physical property of composites.

Fiber type		T700	T800	T1000	M40	M50	T700 towpreg	Kevlar	PBO
Density (g/cm^3)	NT	1.561	1.602	1.553	1.608	1.627	1.5	1.356	1.433
	UT	1.593	1.613	1.601	1.628	1.647	1.515	-	-
Fiber content	NT	64.6	62.8	63.1	63.2	63.7	61.8	65.3	64.9
(%)	UT	69.7	66.4	70.3	68.4	69.5	64.5	-	-
Porosity	NT	2.8	2.4	2.2	2.3	2.0	3.7	-	-
(%)	UT	2.3	2.1	2.0	2.1	1.9	3.2	-	-

3.3 Mechanical Property

As shown in Figure 5(a) (b), the tensile strength and inter-laminar shear strength (ILSS) of the CF composite in wet winding decreased with ultimate tension. The range of decline in the hoop tensile strength of the samples was lower than that of the ILSS. In Figure 5(b), the ILSS of values of the high-modulus fibers M40 and M50 are higher than those of high-strength fibers. For the T700 towpreg, the ILSS of the samples with ultimate tension was higher than that with normal tension, which varied from the others because the viscosity of resin in towpreg winding was higher than that in wet winding. Thus, excess resin cannot be easily extruded for normal tension. Under the influence of ultimate tension, the resin content can reach the optimal proportion. In addition, the porosity of the towpreg composite decreases with an increase in tension, which significantly affects ILSS. As shown in Figure 5(c), the tensile strengths of the Kevlar and PBO composites increase rather than decrease with ultimate tension. For CF, the damage of the fibers and the inadequacy of the resin content lead to a low mechanical strength in the composite when ultimate tension was applied. Although the ILSS of Kevlar and PBO composite was close to zero, reduced lower resin content did not decrease the tensile strength of the composite. Owing to the superior toughness of organic fibers, considerably few fibers were damaged and broken during the filament delivery.

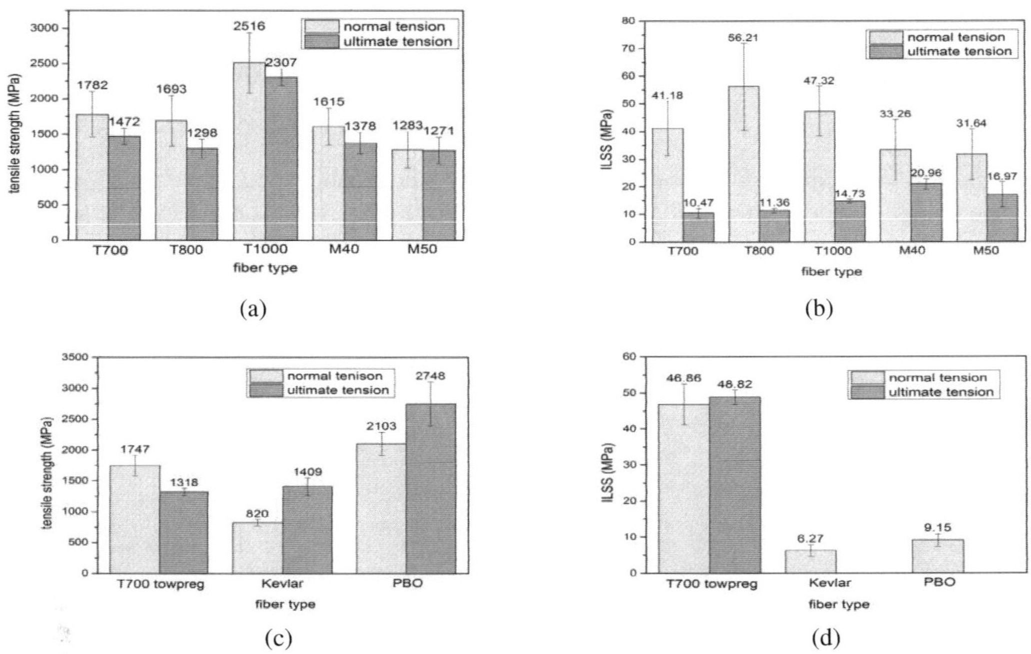

Figure 5. Mechanical strength of composites: (a) tensile strength of CF composites , (b) ILSS of CF composites, (c) tensile strength of towpreg and organic fiber composites, (d) ILSS of towpreg and organic fiber composites.

4. Conclusions

In this study, a new method of tension control was adopted, and a related device was developed. The developed device was proven capable of applying specific winding tension. Both method and device were used in filament winding, and ultimate tension testing of filament was conducted using the developed device. The results indicate that the ultimate tensile stress of fiber is far less than the strength of the fiber. Kevlar, PBO, and T700 towpreg can exert high ultimate tension. The Kevlar fiber and PBO can achieve a high ultimate tensile stress. M40 and M50 have higher ultimate tensile stress than do other CFs. The ultimate tensile force can reduce the porosity, as well as increase the density and fiber content of the material. The ultimate tension can significantly affect the mechanical properties of wound composite. The tensile strength of the CF composite is decreased with ultimate tension, whereas the tensile strengths of Kevlar and PBO are increased. The shear strengths of most of composites decreases, except that for the T700 towpreg.

References

[1] Wang B, Bi L and Chen L 2013 Journal of Zhejiang University (Engineering Science) **47** 2101–10.(in Chinese)
[2] E. Zīle, M. Daugevičius and V. Tamužs 2009 *Mech. Compos. Mater* **45** 457–66
[3] Tamuzs V, Tepfers R and You C S 2006 *Mech. Compos.* Mater **42** 13–32
[4] Mertiny P and Ellyin F 2002 *Compos. Part A-Appl S* **33** 1615–22
[5] Cohen D 1997 *Compos. Part A-Appl S* **28** 1035–47
[6] Imamura T, Kuroiwa T and Terashima K 1999 *Proc. of the IEEE Int. Conf. on Systems, Man and Cybernetics* (Tokyo) vol 2 (New York: IEEE) pp 660–70.
[7] Knittel D, Laroche E and Gigan D 2003 *IEEE Trans. Ind. Appl.* **39** 113–20
[8] Wang C, Yang R and Wang Y 2005 *Int J Adv Manuf Tech* **27** 7–13
[9] Akkus N, Genc G and Girgin C 2008 *Acta Mechanica Et Automatica* **2**.

[10] Ren S L, Lai Y N and Wang Y Z 2010 *Int. Conf. on Material and Manufacturing Technology*, (Chongqing) vol 129–131 (Clausthal-Zellerfeld: Trans Tech Publications) pp 370–75.
[11] Ming M and Cao S Z 2013 *3rd In. Conf. on Computer Science and Network Technology* (Dalian) (New York: IEEE) pp1290–93.

Acknowledgments
This research was supported by National Natural Science Foundation of China 51705310, China Postdoctoral Science Foundation of 2017M611555, and Marine Low Speed Engine Project–phase 1 of CDGC01-KT11.

IOP Conf. Series: Materials Science and Engineering **272** (2017) 012022 doi:10.1088/1757-899X/272/1/012022

Tribological behaviors of UHMWPE composites with different counter surface morphologies

Yanzhen Wang, Zhongwei Yin*, Hulin Li and Gengyuan Gao
State Key Laboratory of Mechanical System and Vibration, Shanghai Jiao Tong University, Shanghai, China
* Corresponding author. e-mail address: yinzw1972@163.com

Abstract. The influence of counter surface morphologies on hybrid glass fiber (GF) and carbon fiber (CF) filled ultrahigh molecular weight polyethylene (UHMWPE) were studied under various contact pressure and sliding speed against GCr15 steel in dry condition. The goals were to investigate the tribological behavior of GF/CF/UHMWPE composite as a kind of water lubricated journal bearing material. The friction and wear behavior of composites were examined using a pin-on-disc tribometer. The morphologies of the worn surface were examined by scanning electron microscopy (SEM) and laser 3D micro-imaging and profile measurement. Generally, the wear rate and friction coefficient of composites increase as the increment of counter surface roughness. The friction coefficient increases firstly and then decrease with an increase in sliding speed and contact pressure for counterface with Ra=0.2 and 3.5 μm, while the friction coefficient decreased for counterface with Ra=0.6 μm.

1. Introduction

Water-lubricated journal bearings are drawing increasing attention due to its environmental friendliness, low cost, and low friction coefficient. They have been widely used in shipbuilding and industrial equipment [1, 2]. UHMWPE is an advanced engineering plastic with ultra-low water absorption, excellent chemical stability, and high impact strength, these advantages making it potentially suitable for application in water-lubricated journal bearings [3, 4]. The study of tribological behavior of bearing materials in different counter surface morphology is crucial for the simulation of bearing operation. At present, it is lack of relevant research on how the counter surface morphology influence the performance of UHMWPE composites. Therefore, it is necessary to study the performance of UHMWPE composites in different counter surface morphology and operation condition.

Research have suggested that counter surface morphology plays a critical role in the occurrence of different wear mechanisms, which are intrinsically connected to the friction coefficient values [5-9].

El-Domiaty et al. [10] investigated the wear characteristics of UHMWPE, the results showed that the wear rate of polyethylene (PE) increased with surface roughness. Blanchet et al. studied the coupled effects of nano-filler content and counter surface roughness on polytetrafluoroethylene (PTFE). The results showed that the wear rate at different counter roughness are affected by nano-fillers contents. Wang et al. [11] studied the tribological behaviors of five composites in seawater, and the results showed that the wear rate of composites in seawater increased sharply for the surface roughness of the counterface increases largely owing to the corrosion of seawater.

Operate conditions also have significant influence on the friction and wear performance of UHMWPE composites. Sumer et al. [12] studied the friction and wear performance of pure polyetheretherketone (PEEK) and 30 wt% glass fiber reinforced PEEK in different contact pressure and sliding speed. The results showed that the friction coefficient and wear rate of PEEK and PEEK composites increase slightly with the increment of pressure. Besides, the friction coefficient decrease while the wear rate increase with an increase in sliding speed. Srinath et al. [13] studied the sliding wear behavior of polyamide 6-clay nanocomposites at different contact pressure. The results showed that the specific wear rate increase as the increase of normal pressure, while the change of friction coefficient with normal pressure varies with filler content.

UHMWPE is a potential material for the application in water lubrication. Therefore, it is necessary to fully study the performance of UHMWPE composites with various counter surface roughness and working condition to provide basis for the application in bearing. Thus, this study is proposed to study the performance of 12.5 wt.% GF and 12.5 wt.% CF filled UHMWPE sliding against GCr15 steel in dry condition with various counter surface roughness, sliding speed and contact pressure.

2. Material and methods

2.1 Materials

12.5 wt.% E-glass fiber (average diameter of 10 μm, the length to radius ratio 4, density 2.25 g/cm^3) and 12.5 wt.% PAN-based carbon fiber (average diameter of 7 μm, the length to radius ratio 4, >95% carbon content, density of 1.75 g/cm^3) were filled in UHMWPE powders (average particle size 40 μm, molecular weight 3 000 000, density 0.94 g/cm^3). The molding method of composites was first cold pressing and sintering, then hot-pressing. These manufactured composites were machined to pins with 8 mm diameter and 20 mm long. The disc material was GCr15 (AISI 52100 steel) with the hardness of HRC62. The dimension of the disc was 50 mm diameter and 8 mm long.

Three types of surface morphology were processed to investigate its influence on tribological behavior. They are grinding-finishing (Ra=0.2 ± 0.05), grinding (Ra=0.6 ± 0.1), and grinding -sandblasting-finishing (Ra=3.5±0.1). The measured roughness characteristics of metal disc were shown in Table 1, where R_p is the maximum height of the profile above the mean line within the assessment length, R_v the maximum depth of the profile below the mean line within the assessment length, R_z the average maximum height, R_q root mean square roughness, R_c the average height of profile, R_a the arithmetic mean roughness of the profile [5, 14]. The values in Table 1 were an average of three measurements. The arithmetic mean roughness R_a was used to describe roughness of counterface. All the samples were cleaned with mixed solution (acetone and ethanol 1:1) before tests.

Table 1. The roughness characteristics of the metal disc before tests (units: μm; I, II, III: group number of surface roughness).

Number	R_p	R_v	R_z	R_c	R_a	R_q
I	0.37	0.82	1.19	0.62	0.14	0.20
II	1.38	1.88	3.26	2.03	0.51	0.64
III	6.73	9.59	16.32	12.98	3.43	4.14

2.2 Tribological test

A pin-on-disc tribometer (MFT-5000, RTEC, San Jose, CA) was used for friction and wear tests. The schematic diagram of the test configuration was shown in Fig.1. The tribological test was conducted according to standard ASTM G99 [15]. The experiments were performed under the temperature of 22 ^{o}C, and relative humidity 50±5%. The surface of the GF/CF/UHMWPE composite pins was examined by scanning electron microscope (SEM) after tests. The worn surfaces of steel disc were tested by the laser microscopic measurement apparatus with a repeated accuracy of 0.012 μm (Keyence, Osaka, Japan). The test conditions of friction coefficient are shown in Table 2, and the specific wear rates are tested using sliding speed of 0.2 m/s and pressure of 5.0 MPa.

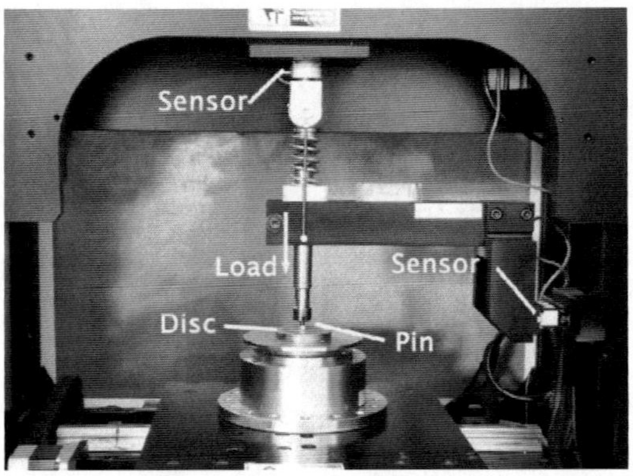

Figure 1. The schematics of the test configuration.

The wear of the polymer pins was determined by measuring their mass loss, which was subsequently converted to their volume loss using the densities of the composites. The specific wear rate of composite was determined using the following equation:

$$W=\frac{\Delta V}{F_N \times L} \quad [mm^3/(Nm)]$$

where W is the specific wear rate, ΔV the volume loss, F_N the normal load, and L the sliding distance.

Table 2. Experimental conditions.

Parameters	Value
Specific pressure (MPa)	1.0, 2.0, 3.0, 4.0, 5.0
Speed (m/s)	0.2, 0.4, 0.6, 0.8, 1.0
Pin material	12.5 wt%GF+12.5wt%CF+UHMWPE

Disc material	GCr15
Operating temperature (°C)	22
Relative humidity	50 ± 5 %

3. Results and discussion

Fig. 2a shows the relation between the friction coefficient and sliding speed using a pressure of 2.5 MPa in dry condition for GF/CF/UHMWPE composites with a counterface roughness R_a of 0.2, 0.6 and 3.5 μm. For counterface roughness R_a of 0.2 μm and 3.5 μm, the friction coefficients first increased and then decreased with an increase in sliding speed. While for a counterface roughness R_a of 0.6 μm, the friction coefficients decreased with an increase in sliding speed. With an increase in counterface roughness from R_a=0.2 to R_a=0.35 μm, the friction coefficient increases slightly for rough counterface conducive to the transfer of GF/CF/UHMWPE to disc surface and the transfer film has excellent antifriction properties.

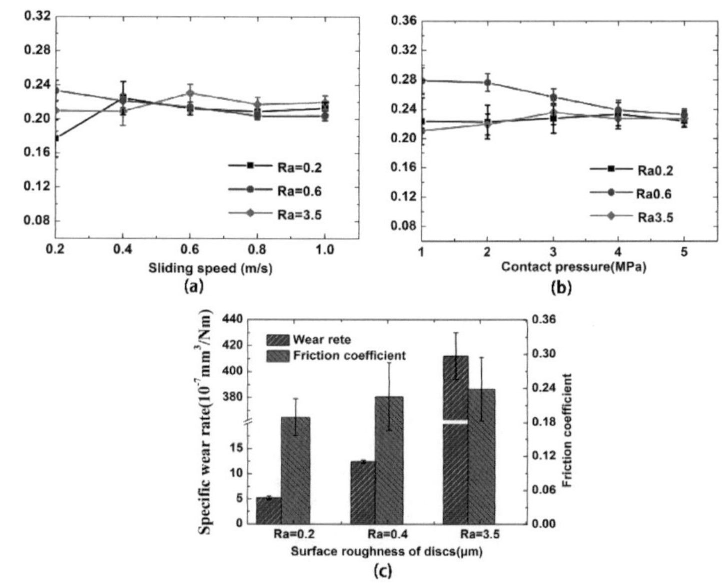

Figure 2. Friction coefficients and specific wear rate of GF/CF/UHMWPE composites.

Fig. 2b shows the relation between the friction coefficient and contact pressure using a sliding speed of 0.2 m/s in dry condition with counterface roughness R_a of 0.2, 0.6 and 3.5 μm. For counterface roughness R_a of 0.2 μm and 3.5 μm, the friction coefficients first increased and then decreased with an increase in contact pressure. While for counterface roughness R_a of 0.6 μm, the friction coefficient decreased with contact pressure increasing from 1.0 MPa to 5.0 MPa.

Fig. 2c shows the specific wear rates and friction coefficient of GF/CF/UHMWPE composites using contact pressure 5.0 MPa and sliding speed of 0.2 m/s. It demonstrates that the wear rate of composites with counterface roughness R_a=0.2 μm is the lowest with the value of 5.23×10^{-7} mm^3/Nm, followed by that of counterface roughness R_a=0.6 μm with a value of 12.4×10^{-7} mm^3/Nm and that of counterface roughness R_a=3.5 μm with a value of 412.1×10^{-7} mm^3/Nm. Thus it can be concluded that the specific wear rate increase with an increase in counterface roughness. The friction coefficient increases only slightly as counterface roughness increase from R_a=0.2 to R_a=3.5 μm. This means that the rough surface accelerates the transfer of composites, which decrease the friction coefficient but increase the wear rate of composites.

Fig. 3 shows the SEM micrographs of the worn surfaces of GF/CF/UHMWPE composites after specific wear rate experiments. Each composite pin was sprayed with a gold coating to obtain electrical conductivity. As shown in Fig.3 (a) and (A), the worn surface of GF/CF/UHMWPE after sliding against counterface with R_a of 0.2 exhibited many scratches which indicated the remarkable abrasive wear. The worn surface of GF/CF/UHMWPE after sliding against counterface with R_a of 0.6 (Fig.3 (b) and (B)) was smooth with some spalling fiber on the surface. Besides, the worn surface also showed abundant conjunct fiber in the matrix, resulting in the anti-wear properties of GF/CF/UHMWPE composites. The worn surface of GF/CF/UHMWPE after sliding against counterface with R_a of 3.5 (Fig.3 (c) and (C)) exhibited serious exfoliation of matrix, which resulted in the sharp increase in wear rate. For GF/CF/UHMWPE composites sliding against different counterface morphologies (R_a=0.2, 0.6, and 3.5 μm), all of the worn surface showed bulges of glass fibers and carbon fibers. The abundant fiber bulges reduced the contact area, which resulted in the anti-friction properties of GF/CF/UHMWPE composites.

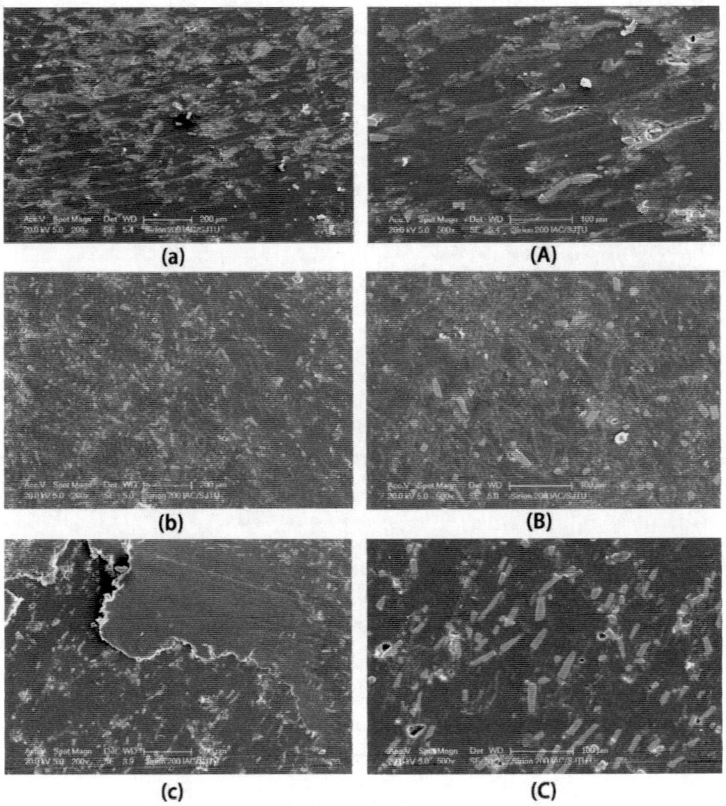

Figure 3. Scanning electron microscopy (SEM) images of worn surfaces of GF/CFUHMWPE composites (a, surface of GF/CF/UHMWPE after test using R_a=0.2 μm; A, enlarged view of a; b, surface of GF/CF/UHMWPE after test using R_a=0.6 μm; B, enlarged view of b; c, surface of

GF/CF/UHMWPE after test using R_a=3.5 μm; C, enlarged view of c).

Fig.4 shows the digital camera photo of disc after tests and the 3D and profile views. As shown in Fig.4 a-c, the worn surfaces of discs with R_a=0.2 μm had little transferred GF/CF/UHMWPE composites, and worn surfaces of discs with R_a=0.6 μm had a certain amount of transferred GF/CF/UHMWPE composites. While the worn surface of discs with R_a=3.5 μm had a thick layer of GF/CF/UHMWPE transfer film. It indicated that the rougher the disc surface, the more conducive to the transfer of GF/CF/UHMWPE composites. And the formation of transfer films reduced the friction coefficient further.

As shown in Fig.4 A-D, the morphology of worn surface of discs with R_a=0.2 μm and R_a=0.6 μm had little difference before and after tests, while the morphology of worn surface of discs with R_a=3.5 μm had obvious characteristic of smear. It suggested that the formation of transfer film for the rough counterface (R_a=3.5 μm) decreased the surface roughness values and friction coefficient.

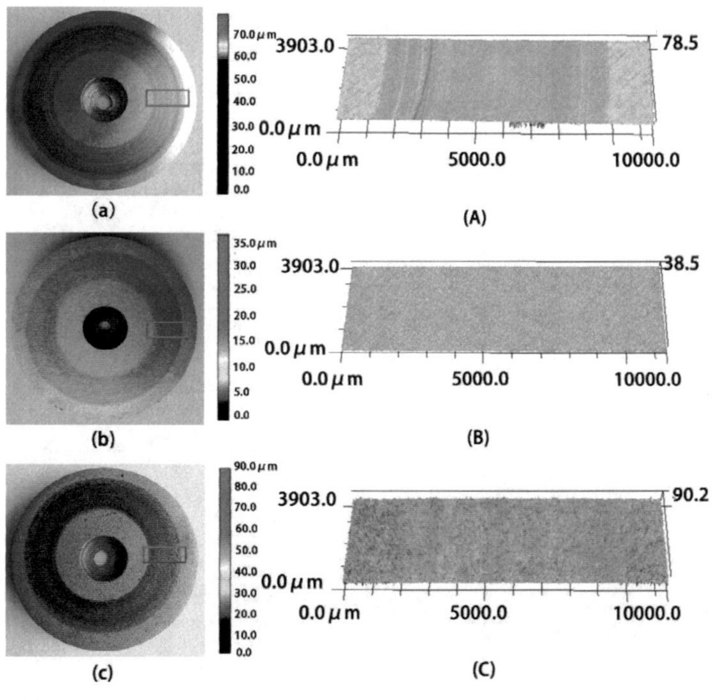

Figure 4. Surface of discs after tests (a, digital camera photo of disc with R_a=0.2 μm; A, 3D and profile views of a; b, digital camera photo of disc with R_a=0.6 μm; B, 3D and profile views of b; c, digital camera photo of disc with R_a=3.5 μm; C, 3D and profile views of c;).

4. Conclusion

In this work, the tribology behaviors of 12.5wt.% glass fiber and 12.5 wt.% carbon fiber filled UHMWPE sliding against GCr15 steel under dry friction at different counter surface morphology were studied. The following conclusions can be drawn:

a. It was found that the tribological behaviors of 12.5 wt.% GF and 12.5 wt.% CF filled UHMWPE were significantly influenced by counter surface morphology. The wear rate of GF/CF/UHMWPE composites increases with an increase in counter surface roughness, while the increase in friction coefficient is not obvious for the formation of transfer film.

b. Sliding speed has an important influence on the tribological behavior of GF/CF/UHMWPE composites. With an increase in sliding speed, the friction coefficient first increased and then decreases for Ra=0.2 and 3.5 μm, while the friction coefficient decreases for Ra=0.6 μm.

c. Also, contact pressure has an important influence on the tribological behavior of GF/CF/UHMWPE composites. With an increase in contact pressure, the friction coefficient first increased and then decreases for R_a=0.2 and 3.5 μm, while the friction coefficient decreases for R_a=0.6 μm.

d. It indicated that the rougher the disc surface, the more conducive for the transfer of GF/CF/UHMWPE composites. And the formation of transfer films reduced the friction coefficient but increased the wear rate of composites.

Acknowledgements

The authors thank the editor and anonymous reviewers for their helpful comments and suggestions. The work was financially supported by Marine Low Speed Engine Project-Phase I (Grant No. CDGC01-KT11).

Reference

[1] Santos E. N., Blanco C. J. C., Macêdo E. N., Maneschy C. E. A. and Quaresma J. N. N. 2012 *Tribol Int* **52**, 161.
[2] Pai R. and Hargreaves D. J. 2012, 347.
[3] Wang J., Yan F. and Xue Q. 2009 *Tribol Lett* **35**, 85.
[4] Xiong D. and Ge S. 2001 *Wear* **2001**, 242.
[5] Golchin A., Friedrich K., Noll A. and Prakash B. 2015 *Tribol Int* **88**, 209.
[6] Golchin A., Nguyen T. D., De Baets P., Glavatskih S. and Prakash B. 2013 *Proc IMechE Part J: J Engineering Tribology* **228**, 371.
[7] Quaglini V., Dubini P., Ferroni D. and Poggi C. 2009 *Mater Design* **30**, 1650.
[8] Barrett T S S. G. W., Batchelor A W. 1992 *Wear* **153**, 331.
[9] Stupak P R K. J. H., Donovan J A 1990 *Wear* **141**, 73.
[10] El-Domiaty A., El-Fadaly M. and Nassef A. E. 2002 *J Mater Eng Perform* **2002**, 577.
[11] Wang J., Yan F. and Xue Q. 2009 *Science Bulletin* **54**, 4541.
[12] Sumer M., Unal H. and Mimaroglu A. 2008 *Wear* **265**, 1061.
[13] Srinath G. and Gnanamoorthy R. 2007 *Compos Sci Technol* **67**, 399.
[14] Gadelmawla E S K. M. M., Maksoud T M A, et al. 2002 *J Mater Process Tech* **123**, 133.
[15] ASTM 2010 *Annual Book of ASTM Standards* **3**, 1.

An integrated study on the optimal shape design of cruciform specimen used in equibiaxial tensile test of a hyperelastic material

J A Pramudita[1*], H Kubota[2], Y Tanabe[1], M Ito[3] and R Watanabe[3]

[1]Mechanical Engineering Program, Faculty of Engineering, Niigata University, Niigata 950-2181, Japan
[2]Graduate School of Science and Technology, Niigata University, Niigata 950-2181, Japan
[3]Product Analysis Center, Panasonic Corporation, Osaka 571-8686, Japan

*E-mail: jpramudita@eng.niigata-u.ac.jp

Abstract. To conduct a finite element analysis of a material exhibiting large deformations (hyperelastic material), hyperelastic Ogden model is commonly used. However, hyperelastic Ogden model requires several tensile tests data including equibiaxial tensile test during the parameter identification. In equibiaxial tensile test, although cruciform-shaped specimen is often adopted as the test specimen, high stress concentration at the corner area causes an early rupture in the specimen. Therefore, it is difficult to obtain the mechanical response of the material under large equibiaxial strain condition. In present study, an experimental study using equibiaxial tensile test device developed previously by authors and a numerical study using finite element analysis were conducted to investigate the stress distribution and the stress-strain response of cruciform specimen under an equibiaxial tensile loading. The optimal shape design of the corner area of the cruciform specimen was then discussed according to the results.

1. Introduction

Hyperelastic Ogden model [1] is commonly selected to model the mechanical behaviour of material exhibiting large deformation (hyperelastic material), e.g. rubber, gel and human soft tissue. In order to identify the parameter value of hyperelastic Ogden model, the use of several tests data obtained from uniaxial, planar and equibiaxial tensile tests during the curve fitting process was recommended [2]. Although uniaxial and planar tensile tests can be performed easily using a universal testing machine equipped with proper chucks, equibiaxial tensile test required a special testing machine equipped with multiple actuators [3]. Therefore, several studies in the development of a simple low-cost test device for biaxial tensile test [4, 5] have recently been actively pursued. Authors have also developed an equibiaxial tensile test device targeting on hyperelastic material [6]. A low-cost equibiaxial tensile test could be delivered by attaching the test device to a single column universal testing machine.

An equibiaxial tensile test using this test device requires cruciform specimen. Although a cruciform specimen with specific dimensions has been proposed for the test, it was confirmed that high stress concentration occurred at the corner area of the cruciform specimen [6]. The stress concentration caused an early rupture of the corner, led to a difficulty in acquiring the mechanical behaviour of test material at high strain region. In order to obtain the mechanical behaviour of the test material, it is

IOP Conf. Series: Materials Science and Engineering **272** (2017) 012023 doi:10.1088/1757-899X/272/1/012023

necessary to reduce the stress concentration occurred at the corner area of the cruciform specimen. Modification of the shape and dimensions of the cruciform specimen, especially at the corner area is considered to give an advantageous effect on the mitigation of the stress concentration. In previous studies, shape optimization of cruciform specimen and investigation of the effect of shape on the mechanical parameters were frequently performed on metallic or composite materials [7-10]. Although the results of the studies were also considered as valuable references for designing cruciform specimen of hyperelastic materials, the shape and dimensions of the cruciform specimens varied widely and were different from the cruciform specimen used in the study conducted by authors [6]. Therefore, an independent study should be conduct in order to confirm the optimal shape design of the cruciform specimen designated for the equibiaxial tensile test device.

In this study, several shape designs of the corner area of cruciform specimen were proposed. The effect of the corner shapes on the stress distribution of a hyperelastic material specimen under an equibiaxial tensile loading was then investigated using finite element analysis. Furthermore, equibiaxial tensile tests on the target specimens were also conducted in order to validate the analysis results. The optimal corner shape design was then discussed based on the results of the finite element analysis and the tensile test.

2. Analysis of stress distribution on the cruciform specimens

Finite element analysis was considered to be an effective tool for investigating the stress distribution of the cruciform specimen. Therefore, finite element models of the target specimens were constructed and were then subjected to an equibiaxial tensile loading in simulations. Distributions of von Mises stress in the specimen model were obtained from the analysis results. The details of the finite element analysis are described below.

2.1. Finite element models of cruciform specimens

Finite element models of cruciform specimens and their dimensions are shown in figure 1. In this study, three types of corner shapes (type R, type RD and type RR) as shown in figure 2 were devised. The specimen models were constructed using hexahedral solid elements. The numbers of elements and nodes of each model were listed in table 1. In order to simulate the large deformation behaviour of hyperelastic material, Ogden rubber model was adopted as the material model. Since the base material of the specimen was assumed to be a RTV silicone rubber (KE-12, Shin-Etsu Chemical Co. Ltd., Tokyo, Japan) , the parameter values of Ogden rubber model used in this study were same to those in the previous study [6].

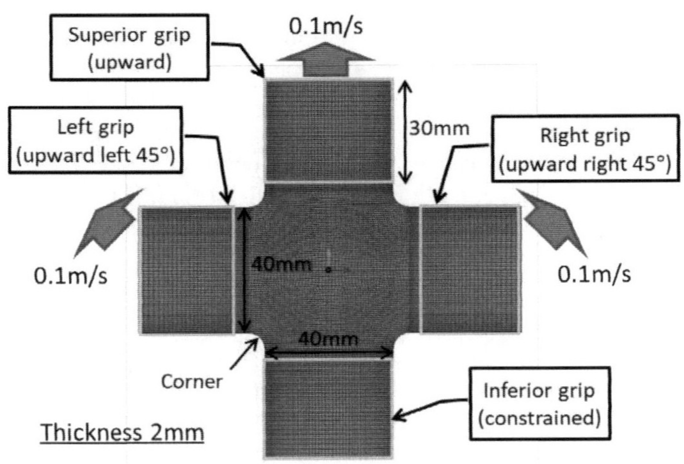

Figure 1. Finite element model of the cruciform specimen and the simulation conditions.

IOP Conf. Series: Materials Science and Engineering **272** (2017) 012023 doi:10.1088/1757-899X/272/1/012023

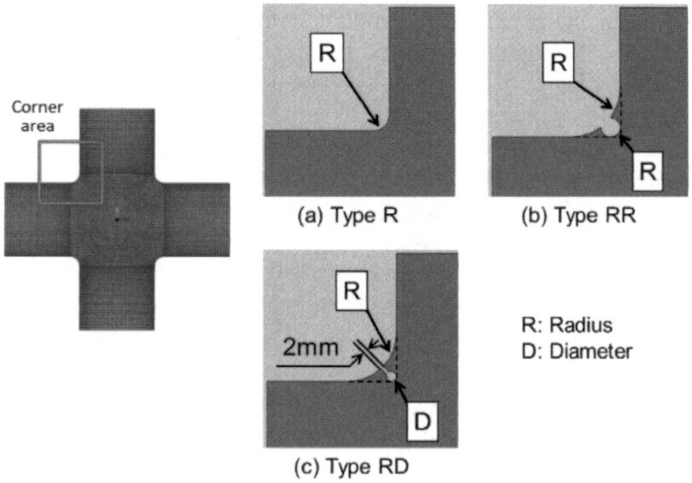

Figure 2. Shapes of the corner area of the cruciform specimen. (a) Type R; (b) Type RR; (c) Type RD.

Table 1. Numbers of nodes and elements of the models and the maximum von Mises stress at the corner area.

Specimen		Nodes	Elements	Maximum von Mises stress [MPa]
Type R	**R5**	63,965	50,080	0.793
	R10	43,594	34,144	0.696
Type RR	**R10R3**	44,245	34,464	0.908
	R10R4	47,705	37,264	0.641
	R10R5	42,405	32,944	0.754
Type RD	**R10D2**	44,805	34,848	1.233
	R10D3	46,207	36,016	1.224

2.2. Simulation of equibiaxial tensile test
As shown in figure 1, grip areas were defined as the 30 mm length areas from the edges of the four arms to the center of the specimen model. The inferior grip area was completely fixed during the analysis. In order to simulate an equibiaxial tensile test, an upward displacement was imposed on the superior grip area, while 45 degree obliquely upward displacements were also imposed on the right and left grip areas respectively. The displacements were imposed at a constant velocity of 0.1 m/s until the upward displacement of the superior grip area reached 10 mm. The finite element analysis was performed using a commercial software PAM-CRASH v2012 (ESI Group, Paris, France).

2.3. Maximum von Mises stress at the corner area
Distributions of von Mises stress were determined from the analysis results. Comparison of maximum values of von Mises stress at the corner area was shown in table 1. The R10R4 model showed the lowest value, while the R10D2 model showed the highest value of von Mises stress. Overall, the RD-typed models indicated a tendency to have higher von Mises stress rather than the R-typed models and the RR-typed models. An addition of a R4 notch on a R10 corner (R10R4) caused a decrease of maximum von Mises stress at the corner area. However, the decrease was considered to be relatively small.

3. Equibiaxial tensile test for validating the analysis result

In order to validate the result of the finite element analysis, equibiaxial tensile test on silicone rubber specimens which have similar shapes to the specimen finite element models was carried out. The equibiaxial tensile test was conducted considering also the variations that may occur due to the different defects in each specimen. Corner shape design that could provide more data at larger strain region was evaluated from the stress-strain responses. The details of the equibiaxial tensile test are described below.

3.1. Cruciform specimens

Some specimens that showed lower maximum von Mises stress were selected among the specimens used in the finite element analysis. R-typed specimens (R5 and R10) as well as R10D3 specimen and R10R4 specimen were finally selected and tested in order to investigate the differences in the stress-strain response and rupture point under an equibiaxial tensile loading. The test specimens were manufactured from RTV silicone rubber (KE-12, Shin-Etsu Chemical Co. Ltd., Tokyo, Japan) by vacuum casting method. The test specimens with different corner shapes are shown in figure 3.

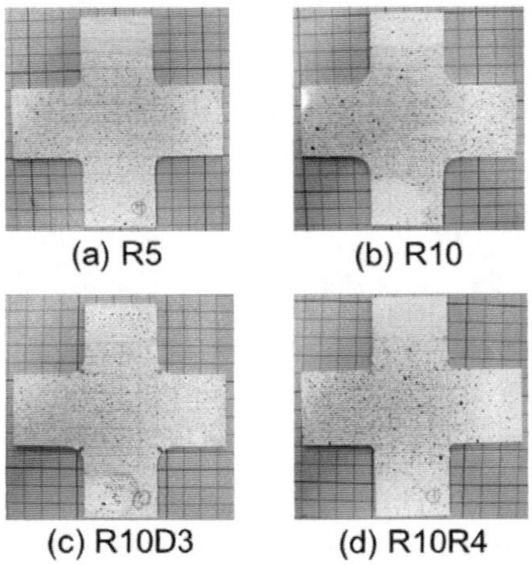

Figure 3. Test specimens with different corner shapes. (a) R5; (b) R10; (c) R10D3; (d) R10R4.

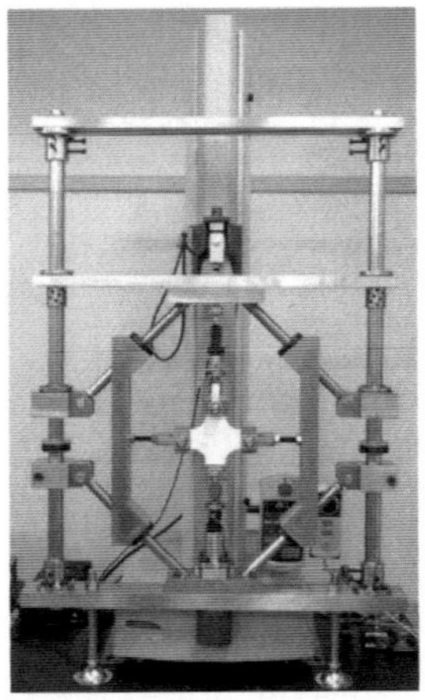

Figure 4. The equibiaxial tensile test device attached to a single column universal testing machine.

3.2. Equibiaxial tensile test

Equibiaxial tensile tests were conducted using a compact universal testing machine (EZ-LX, Shimadzu Corp., Kyoto, Japan) and an equibiaxial tensile test device developed by authors [6]. Loading rate was 100 mm/min. Number of specimens was six specimens for each specimen type. In total, 24 tests were carried out. A preloading of 1 N was applied to the specimens prior to the tests. Load was measured using a load cell (LUR-A-SA1, Kyowa Electronic Instruments Co. Ltd., Tokyo, Japan) attached to the test device. Elongation was determined from the video images taken using a video camera (HDR-S11, Sony Corp., Tokyo, Japan).

3.3. Comparison of the mechanical responses

Figure 5 shows the comparison of stress-strain curves obtained from the equibiaxial tensile tests. Although specimen variations could be seen from the result, the repeatability of the result was considered to be sufficiently high. The R10D3 specimen showed the lowest rupture strain, while the R10R4 specimen showed the highest rupture strain. This finding was well accorded with the finding obtained from the finite element analysis. In contrast, the R10 specimen and the R5 specimen indicated an opposite result to the result obtained from the finite element analysis. However, the difference was confirmed to be statistically insignificant. Furthermore, according to the analysis and test results, the R10 specimen demonstrated higher stress response compared to other specimen types. The constraint conditions might have a contribution on this behaviour.

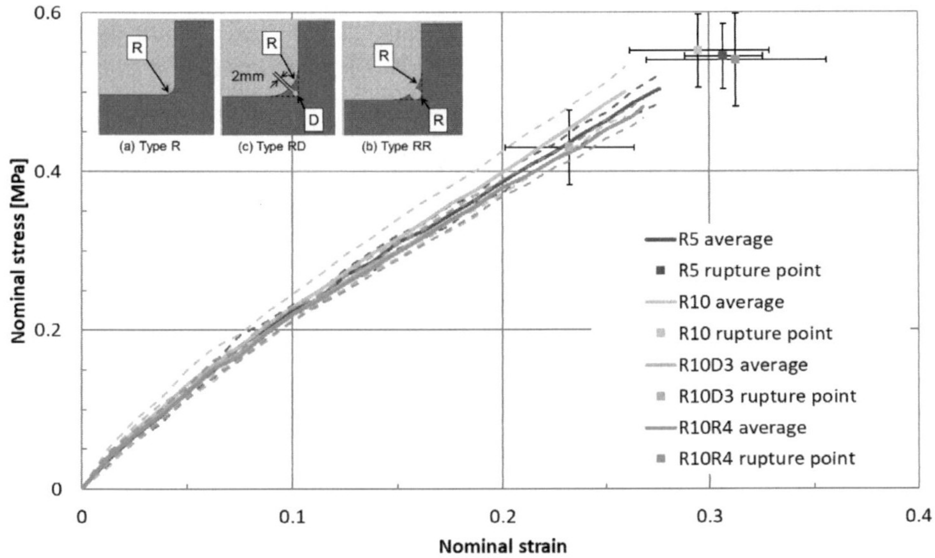

Figure 5. Comparison of stress-strain curves among four types of specimen. Solid lines indicate the average values. Dash lines indicate the average values ± standard deviation.

4. Conclusion

An integrated study combining finite element analysis and mechanical test was proposed to investigate the optimal shape design of the corner area of a cruciform specimen used in the equibiaxial tensile test. As a result, an addition of a small circle notch on the round corner area of the cruciform specimen could reduce the maximum von Mises stress at the area, indicating a possibility to obtain the mechanical behaviour of a material at larger strain region. In addition, the results of the equibiaxial tensile tests also supported the finding obtained from the finite element analysis.

The integrated method used in this study can save time and cost by reducing the number of tests that should be conducted. However, several limitations such as maximum stroke of the test device and defect in the specimens may provide different result to that obtained from the finite element analysis. Furthermore, introduction of an optimization method into the analysis should be done in the future.

5. References

[1] Ogden R W 1972 *Proc. R. Soc. Lond.* A *Math. Phys. Sci.* **326** 565-84
[2] Ogden R W, Saccomandi G and Sgura I 2004 *Comput. Mech.* **34** 484-502
[3] Fujikawa M, Maeda N, Yamabe J, Kodama Y and Koishi M 2014 *Exp. Mech.* **54** 1639-49
[4] Brieu M, Diani J and Bhatnagar N 2007 *J. Test. Eval.* **35** 1-9
[5] Shahzad M, Kamran A, Siddiqui M Z and Farhan M 2015 *Mat. Res.* **18** 918-24
[6] Pramudita J A, Kubota H, Tanabe Y, Ito M and Watanabe R 2017 *J. Test. Eval.* (in press)
[7] Lebedev A A and Muzyka N R 1998 *Strength Mater.* **30** 243-54
[8] Yu Y, Wan M, Wu X-D and Zhou X-B 2002 *J. Mater. Process. Technol.* **123** 67-70
[9] Smits A, Van Hemelrijck D, Philippidis T P and Cardon A 2006 *Compos. Sci. Technol.* **66** 964-75
[10] Hanabusa Y, Takizawa H and Kuwabara T 2013 *J. Mater. Process. Technol.* **213** 961-70

IOP Conf. Series: Materials Science and Engineering **272** (2017) 012024 doi:10.1088/1757-899X/272/1/012024

Influence of small particles inclusion on selective laser melting of Ti-6Al-4V powder

Haijun Gong[1, a)], J J S Dilip[2], Li Yang[2], Chong Teng[3] and Brent Stucker[3]

[1]Georgia Southern University, Statesboro, GA 30460, United States
[2]University of Louisville, Louisville, KY 40292, United States
[3]3DSIM LLC, Park City, UT 84098, United States
[a)]Corresponding author: hgong@georgiasouthern.edu

Abstract. The particle size distribution and powder morphology of metallic powders have an important effect on powder bed fusion based additive manufacturing processes, such as selective laser melting (SLM). The process development and parameter optimization require a fundamental understanding of the influence of powder on SLM. This study introduces a pre-alloyed titanium alloy Ti-6Al-4V powder, which has a certain amount of small particles, for SLM. The influence of small particle inclusion is investigated through microscopy of surface topography, elemental and microstructural analysis, and mechanical testing, compared to the Ti-6Al-4V powder provided by SLM machine vendor. It is found that the small particles inclusion in Ti-6Al-4V powder has a noticeable effect on extra laser energy absorption, which may develop imperfections and deteriorate the SLM fatigue performance.

1. Introduction

Powder-bed-fusion based Additive Manufacturing (AM) processes, such as selective laser melting (SLM), are increasingly employed for fabricating metal parts. In particular, the SLM titanium alloy parts (e.g. Ti-6Al-4V) are of great interest for aerospace, biomedical and industrial applications due to its geometry complexity, fracture resistance, fatigue behavior, corrosion resistance and biocompatibility [1]. A number of studies have been conducted to characterize the on the mechanical properties of SLM Ti-6Al-4V alloy. For example, Facchini et al. [2] and Ramosoeu et al. [3] characterized the microstructure and then tested tensile properties of SLM Ti-6Al-4V parts. Wycisk et al. [4] and Rafi et al. [5] carried out fatigue tests to evaluate the performance of as-built SLM Ti-6Al-4V parts under a high frequency periodical load cycles. Moreover, studies were also carried out on the applications of SLM Ti-6Al-4V parts in industry, aerospace, and medical implants. Caiazzo et al. [6] experimentally studies the use of SLM for manufacturing aircraft components. Cardaropoli et al. [7] discussed the feasibility of producing dental implants through SLM using Ti-6Al-4V pre-alloy powder.

SLM process selectively melts metallic powder layer by layer based upon a sliced CAD file to build a part on a base plate. Inside the SLM building chamber, a recoating unit is used to feed new powder over the build platform, as shown in figure 1. The widely usage of SLM parts attracts more metallic powder manufacturers to set foot into AM market. Many powder vendors provide metallic powders to SLM equipment users for research and process development. Whether a powder is applicable for SLM process to produce comparable parts is uncertain to the SLM user community. In order to clarify these uncertainties, this study attempts to employ a new Ti-6Al-4V powder (AP&C Inc, Canada) into a commercialized SLM equipment (EOS M270). The powder morphology and particle size distribution were analyzed and compared to a mature Ti-6Al-4V powder (machine vendor provided). Experiments were carried out to develop the optimum process parameters for fully dense parts. Metallography and elemental composition of as-built part was studied. Tensile and fatigue tests were also conducted.

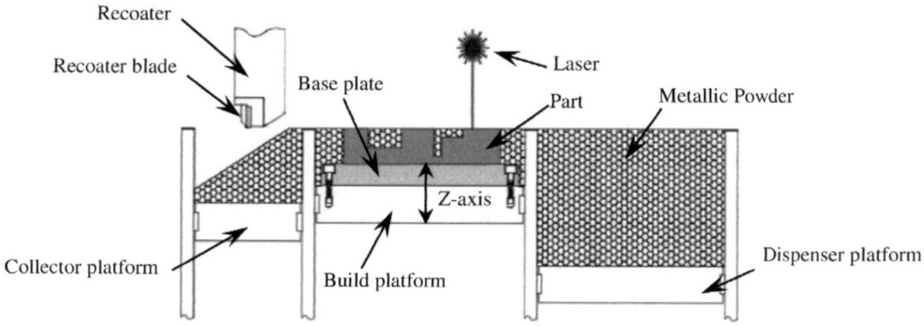

Figure 1. Schematic of SLM Process

2. Powder Characterization

Figure 2(a) shows the scanning electron microscopy (SEM) of AP&C Ti-6Al-4V powder morphology. It is noted that the AP&C powder particles have the similar spherical shape, compared to the EOS provided Ti-6Al-4V powder (figure 2(b)). A certain amount of fine particles are observable. This is confirmed by the particle size analysis (PSA) using a Microtrac S3000 particle analyzer, as shown in figure 3. The PSA of AP&C powder indicates a size distribution between 17.36 μm (D_{10}) and 44.31 μm (D_{90}) with Mean Volume (MV) diameter around 30 μm.

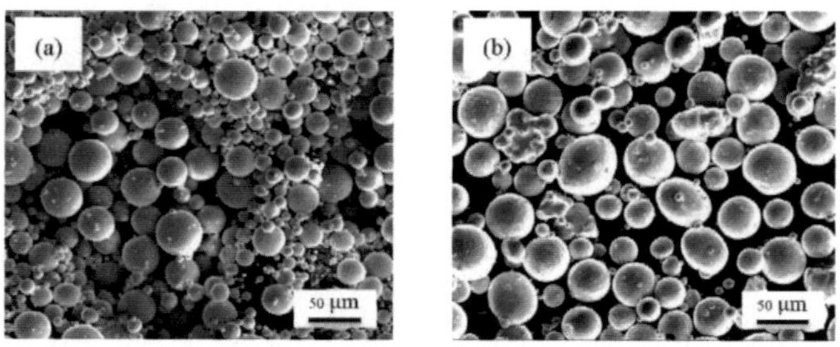

Figure 2. SEM of (a) AP&C Ti-6Al-4V Powder and (b) EOS Ti-6Al-4V Powder.

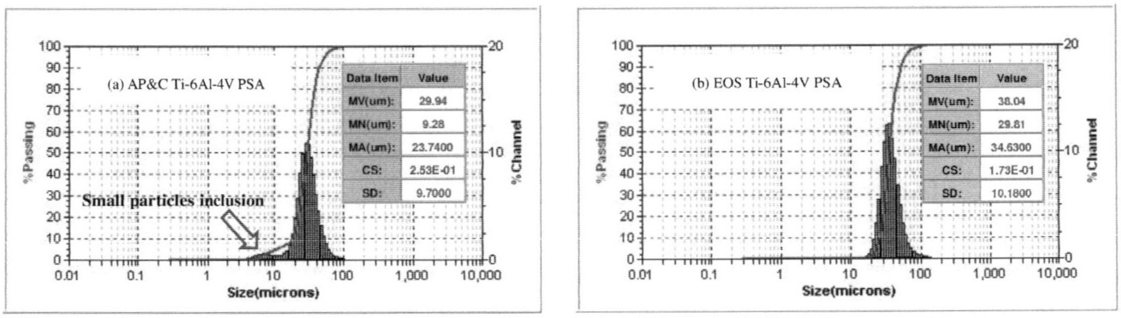

Figure 3. Particle Size Distribution of (a) AP&C Ti-6Al-4V and (b) EOS Provided Ti-6Al-4V Powder

3. Preliminary Test

AP&C Ti-6Al-4V powder was directly loaded into an EOS M270 for a preliminary test, in order to verify the powder adaptability. The machine has a fiber laser (max 200 W). Galvo mirrors and flat field lens are used to conduct laser beam and maintain a constant focal spot size (~100 μm), with scan speed up to 7000 mm/s. Following a standard operation procedure, AP&C powder was melted using the factory default process parameters to fabricate trial parts. After completion, a measurement showed that trail parts have a comparable density to the part fabricated by EOS powder. For further evaluation, microscopy was used to observe the top surface of trial part. As shown in figure 4(a), scan tracks are incoherent on the top surface, deducing voids with small entrapped powder particles inside. So the rough surface for every layer may be expected. The voids could be filled by recoating powder of following layers and melted to minimize porosity inclusion. But there is no apparent evidence that voids could be completely eliminated. A consistent surface topography is desired for using the AP&C powder in SLM process. Thus, development of optimum process parameters for AP&C Ti64 powder is required.

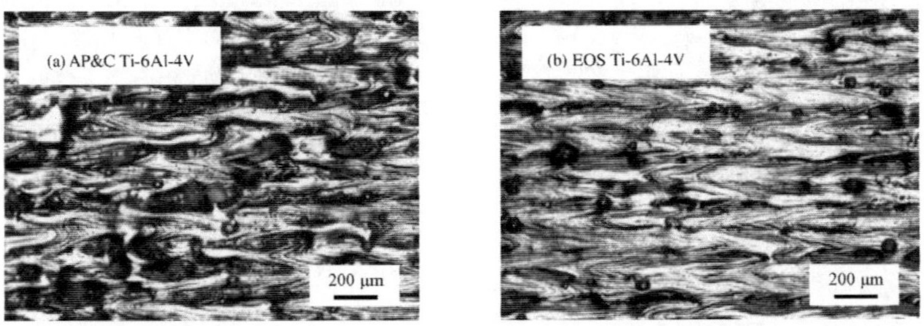

Figure 4. Top surface topography of (a) AP&C Ti-6Al-4V sample and (b) EOS Ti-6Al-4V sample.

4. Experiment Methods

A factorial design of experiment (DOE) is conducted for investigating the optimum parameters of AP&C powder in the EOS M270 machine to make fully dense parts. The laser power and scan speed have a critical impact on the energy density [8]. Thus, these two parameters were varied, as shown in Table 1, with constant hatch spacing (100 μm) and layer thickness (30 μm).

Table 1. Factors and levels of factorial DOE of AP&C Ti-6Al-4V Powder.

Factor	Level
Laser Power (W)	40, 80, 120, 160
Scan Speed (mm/s)	120, 240, 360, …, 1560

IOP Conf. Series: Materials Science and Engineering **272** (2017) 012024 doi:10.1088/1757-899X/272/1/012024

After samples were fabricated, each specimen's density was measured using the Archimedes method (ASTM B962-08) [9] to estimate porosity and determine optimum parameters. As-built parts were sectioned, abrasively grinded and polished, and etched for metallography. It is found that the laser power 120 W and scan speed 960 mm/s are eligible to produce fully dense parts, in conjunction with other default process parameters. Cylindrical bars were built in the *Z* orientation (ISO/ASTM 52921, 2013) using this optimum parameter set, and then machined to specimens conforming to ASTM E8 for tensile testing and ASTM E466 for fatigue testing. Tensile tests were carried out using an Instron 5569A tensile testing machine. High cycle fatigue tests were performed on a 10kN Instron Electropulse 10000 fatigue testing machine. Sinusoidal load was applied to fatigue specimens axially (50 Hz, stress ratio R=0.1). Fatigue testing was stopped when the specimens broke or the fatigue cycles reached 10^7 cycles. Fracture surface was examined by a FEI Nova NanoSEM 600 SEM.

5. Results and Discussion

5.1 Microstructural Analysis

Microstructural evolution is primarily a function of cooling rate. The SLM process undergoes a very high cooling rate which results in a lenticular martensitic (α') microstructure of Ti-6Al-4V, as shown in figure 5. The martensitic phase of AP&C Ti-6Al-4V sample shows similar microscopic morphology in comparison with EOS Ti-6Al-4V sample. Martensitic laths originated from the prior β grain boundaries. All β phases (BCC) are transformed to hexagonal martensite. The microstructure indicates that the small particles inclusion does not influence the formation of crystalline structures in the SLM process.

Figure 5. Optical microscopy of microstructure of SLM samples

5.2 Composition Analysis

The chemical composition of SLM samples was analyzed using energy dispersive X-ray analysis (EDAX). The EDAX is a commonly used analytical technique for the elemental analysis, which allows a unique set of peaks on its electromagnetic emission spectrum for each element. The EDAX spectrum of AP&C sample and EOS sample are shown in figure 6.

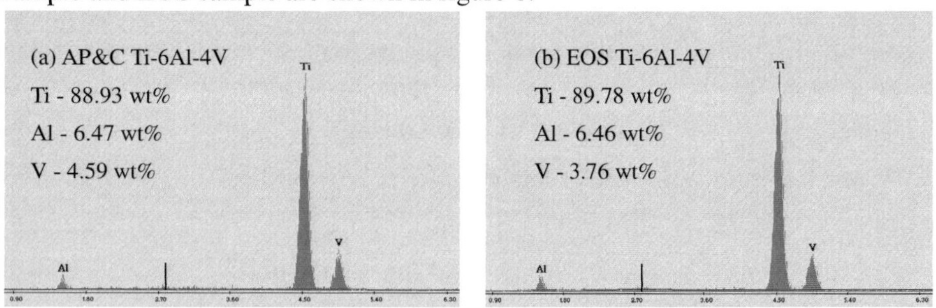

Figure 6. EDAX spectrum of (a) AP&C Ti-6Al-4V sample and (b) EOS Ti-6Al-4V sample.

There is no apparent difference in the composition between Ti-6Al-4V samples. The chemical composition is primarily determined by the composition of pre-alloyed Ti-6Al-4V powder. The melting and solidification process of SLM have few influence on the elemental composition. The variation of particle size distribution could have an impact on the mass transportation during the laser melting process. The small particles may be more easily ejected owing to the recoil force of metallic vapor. But the chemical composition of Ti-6Al-4V sample is not influenced.

5.3 Tensile Property

Six duplicate specimens were performed for the tensile test. Corresponding tensile stress-strain curves were obtained directly from the Instron system, as shown in figure 7. Tensile properties are summarized in Table 2, in comparison with EOS Ti-6Al-4V. Generally, the material properties of AP&C specimens and EOS specimens are comparable. Both SLM Ti-6Al-4V materials show a higher yield strength and tensile strength compared to the wrought Ti-6Al-4V samples, due to the observed martensitic microstructure. The martensite grains nucleate and grow at a rapid rate, resulting in a very small grain size (a greater total grain boundary area). The grain boundaries impede the slip process and increase the strength of the material. No significant influence of small particles inclusion is observed on the tensile property.

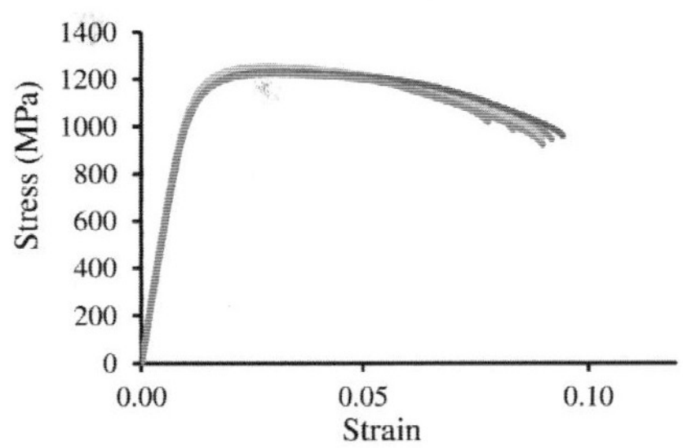

Figure 7. Stress-strain curve of SLM AP&C Ti-6Al-4V samples.

Table 2. Tensile Properties of SLM Ti-6Al-4V samples.

	SLM AP&C Ti-6Al-4V	SLM EOS Ti-6Al-4V[a]
Yield Strength (MPa)	1098 ± 45	1150 ± 80
UTS (MPa)	1237 ± 39	1290 ± 80
Strain at Break (%)	8.8 ± 1.8	8 ± 4
Young's Modulus (GPa)	109 ± 6.3	110 ± 5

[a] www.eos.info

5.4 Fatigue Performance

As for the high-cycle fatigue test, materials performance is shown in figure 8(a), comparing AP&C and EOS specimens. It is noted that the AP&C specimens show considerable scatter in fatigue life. The fatigue limit is also lower than the EOS specimens, which was reported by Rafi et al [5]. The AP&C specimens appear to have an inferior fatigue behavior, as opposed to the tensile properties. Although it is hard to assert that the inclusion of small particles deteriorate the fatigue performance, the variation of powder and process parameters undoubtedly have an influence on the Ti-6Al-4V parts. The small particles inclusion may play a critical role during the SLM process. Due to the small particles inclusion, the absorption is greatly increased on powder content resulted by the multiple scattering of laser [10]. A higher temperature and larger melt pool size can be expected. The drastic phase transformation may cause an unsteady melt pool and scan tracks. Keyhole defects, volumetric voids, and impurities have a great potential to be included in the SLM specimens. The fatigue behavior is sensitive to these imperfections, especially when a high frequency cyclic load is applied. The fractography in figure 8(b) shows the crack initiation site on the fracture surface of an AP&C fatigue specimen. It can be clearly seen that small particles are entrapped in a defective site, which cause stress concentration.

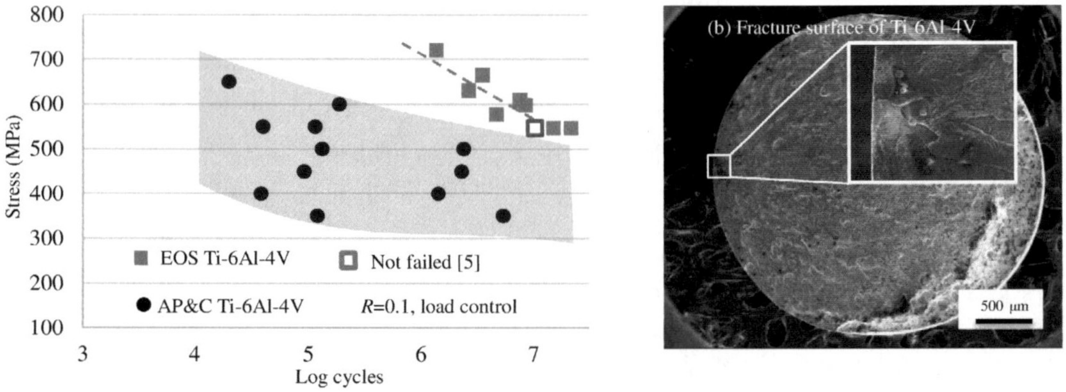

Figure 8. SLM Ti-6Al-4V fatigue specimens (a) *S-N* curve and (b) fractography.

6. Conclusion

This study introduces a new Ti-6Al-4V powder for SLM process. The influence of small particles inclusion is investigated. It is found that process parameters need to be tuned for fabricating fully dense parts if using new powder. The small particle inclusion does not influence the microstructure and elemental composition of SLM Ti-6Al-4V parts. The tensile property of AP&C Ti-6Al-4V specimens is also comparable to the specimen fabricated by EOS powder. However, the fatigue performance of AP&C specimens is poor. The small particles inclusion may influence the laser absorption on the powder bed, resulting in imperfections in the SLM material. Thus, future research is highly desired to conduct quantative analysis of small particles in the powder system.

7. Reference

[1] Soboyejo W O and Srivatsan T S 2006 *Advanced Structural Materials: Properties, Design Optimization, and Applications* (Boca Raton: CRC Press) p 359-400

[2] Facchini L, Magalini E, Robotti P, Molinari A, Höges S and Wissenbach K 2010 *Rapid Prototyping J.* **16** 450-9

[3] Ramosoeu M K E, Booysen G and Ngonda T N 2011 *Proc. Materials Science and Technology Conference (MS&T)* (Materials Park: ASM International) p 1460-8

[4] Wycisk E, Emmelmann C, Siddique S and Walther F 2013 *Adv. Mater. Res.* **816-817** 134-9

[5] Rafi K, Karthik N V, Gong H, Starr T and Stucker B 2013 *J. Mater. Eng. Perform.* **22** 3872-83

[6] Caiazzo F, Cardaropoli F, Alfieri V, Sergi V and Cuccaro L 2013 *Proc. of SPIE* 8677

[7] Cardaropoli F, Alfieri V, Caiazzo F and Sergi V 2012 *Adv. Mater. Res.* **535-537**, 1222-9

[8] Thijs L, Verhaeghe F, Craeghs T, Van Humbeeck J and Kruth J P 2010 *Acta Materialia* **58**, 3303-12

[9] Spierings A B and Schneider M 2011 *Rapid Prototyping J.* **17**, 380-6

[10] Boley C D, Khairallah S A and Rubenchik A M 2015 *Applied Optics* **54**, 2477-82

Cream concentrated latex for foam rubber products

R Suksup[1], C Imkaew[1] and W Smitthipong[1,2 a)]

[1] Specialized center of rubber and polymer materials in agriculture and industry (RPM), Department of Materials Science, Faculty of Science, Kasetsart University, Bangkok, Thailand.
[2] Office of natural rubber research program, The Thailand Research Fund, Bangkok, Thailand.

Corresponding author: fsciwssm@ku.ac.th

Abstract. Fresh natural latex (around 40% rubber and 60% water) can be transformed to concentrated natural latex (around 60% rubber and 40% water) in order to realise economical transportation and easier latex product's preparation. The concentrated natural latex is an extremely valuable material. It can be applied for many types of products, for example, foam rubber as pillow and mattress, elastic band, etc. Industrially, the concentrated natural latex can be prepared by centrifugation which requires an enormous expensive machine. From the eco-friendly products point of view, most of rubber entrepreneurs in the world try to develop a green rubber product. So, the main objective of this study is to prepare the cream concentrated latex without any sophisticated machine. Thus, we work on a simple, cheap and green method that does not use any expensive machine but uses water-based chemical as sodium alginate to prepare the cream concentrated latex. The optimal amount of sodium alginate in the latex was studied. The main characteristics of the cream concentrated latex were tested by various technics, such as alkalinity, total solid content (TSC), dry rubber content (DRC), etc. We found that there are no significant differences of results between fresh natural latex and cream concentrated latex, except for the TSC and DRC. The TSC and DRC of cream latex are higher than those of fresh natural latex. Finally, we propose a model of natural rubber particle and sodium alginate to form the cream concentrated latex.

1. Introduction

Natural rubber is made from runny milky white liquid called latex. Although there are something like 200 plants in the world that produce latex, however, the world's natural rubber is industrially made from the latex that comes from a specific tree species called *Hevea brasiliensis*. Fresh natural latex can be collected from a tapped rubber tree. Fresh natural latex is a colloidal system of natural rubber particle in micron size dispersed in an aqueous medium and contain 30-40% of dry rubber content (Table 1) [1-2].

Composition of fresh natural latex depends on factors such as soil condition, fertilizer quality, tapping system, season, etc. Fresh natural rubber latex can be contaminated by micro-organism because natural latex contains various nutritious substances, otherwise known as non-rubber components [3-4]. The micro-organism consumes these non-rubber constituents and then generates the volatile fatty acid

(VFA) which enables the destabilisation of latex particles. Normally, we use ammonia solution to neutralise the VFA in the latex that keep the stabilization of latex. So, ammonia solution is used to preserve the natural latex. Fresh natural latex can be transformed to concentrated natural latex in order to maintain the constant quality of concentrated natural latex and to generate the economic value for latex's transportation.

Table 1. Composition of fresh natural latex.

Constituent	% by weight of latex
Rubber particles	30-40
Protein	2-3
Lipids	0.1-0.5
Sugars	1-2
Ash	0.5-1.0
Others	1.5-3.5
Water	55-65

Concentrated natural latex contains 60% rubber, which is essential in making products such as foam rubber as pillow and mattress, elastic band, etc [5]. The concentrated natural latex can be produced by various processes including evaporation, electrocantation, centrifuging and creaming. Evaporation refers to the removal of water by floating a double-walled cylindrical vessel and a hollow iron roller in the latex and spray hot latex into partially evacuated chamber. Electrocantation immerses positive electrode into latex. As a result, the positive was combined and would float up on the surface of latex. Centrifuging, in which the latex is placed into a centrifuge and later it is centrifuged to remove some of the water in order to increase the rubber content of latex.

The centrifugation process for concentrated natural latex is very well known and is used widely for commercial purpose, but it uses a big, expensive machine and adding chemical agent (TMTD) to preserve the natural latex that cause the cancer from nitrosamine. One of the methods for preparing concentrated natural latex is creaming, the creaming method is a chemical process that mixes creaming agent and fresh natural latex and facilitates phase separation (rubber fraction and serum). Creaming agent helps in increasing the creaming efficiency. Cream concentrated latex was successfully prepared using a creaming agent include sodium carboxymethyl cellulose, poly(vinyl alcohol), methyl cellulose, carboxymethyl cellulose, locust bean gum, tamarind seed powder, casein and sodium alginate [6-7]. This method is environmentally friendly for it requires simple equipment and incurs lower power consumption. The main objective of this paper is to prepare the cream concentrated latex using sodium alginate as a creaming agent. Various properties of the cream concentrated latex were characterized and compared with the fresh natural latex. This aspect has not been reported in the previous literature reviewed.

2. Experimental

2.1. Materials

Fresh natural latex was kindly collected from plantation in the southern of Thailand (Tha chang industries group, Suratthani). The characteristics of latex are in the table 2. Sodium alginate was supplied from Thanodom trading, Thailand. Ammonia solution was supplied from RCL Labscan, it was used as a latex preservative against bacteria attack and yo provide long-term stability..

Table 2. Characteristics of latex.

Materials	Characteristics		Sources
Fresh natural latex	TSC	37.17%	Tha chang industries, Thailand
	DRC	27.28%	
Centrifuged concentrated latex	TSC	61.71%	Tha chang industries, Thailand
	DRC	60.13%	

2.2. Preparation of cream concentrated latex using sodium alginate

Dissolved 3 g of sodium alginate (creaming agent) in 100 ml of distilled water. Varies of 1, 2, 3, 4 and 5 ml of the sodium alginate solution was transferred into beaker containing 25 ml of latex and then agitation for 30 min. Incubation at room temperature and leave it undisturbed for 3 day, observing that the latex separated into two layers after 1 day and allowed the rubber particles (cream phase) to move up through a period of several weeks. When the separation is completed, the aqueous serum phase is run off leaving only the upper rubber fraction.

2.3. Characterizations of the fresh natural latex and cream concentrated latex

Analysis and testing of latex properties include total solid content (TSC), dry rubber content (DRC), Alkalinity (%), Volatile fatty acid (VFA) and Mechanical stability (MST) as shown below.

2.3.1. Total solid content (TSC) (ISO, 124). Pour 2 g of latex into the dish and determine the exact mass by weighing, approximate to the nearest 0.1 mg, then place the dish in the oven at 105 ± 5°C for 2 h, or until the test portion has lost its whiteness. Remove the dish from the oven and allow it to cool to ambient temperature.

2.3.2. Dry rubber content (DRC) (ISO 126). 10 ± 1 g of latex concentrate into dish. Add of 20 g/dm^3 acetic acid solution, pouring it down along the inner edge of dish and slowly rotating the dish while the acid is being added. When the serum is clear, collect any small particles of coagulated rubber by rubbing with the main bulk. Soak the coagulated rubber in several change of water until the water is no longer acidic to litmus, after that running of water for at least 5 min. Dry the sheet at 70 ± 5°C until it has no white patches.

2.3.3. Alkalinity (%) (ISO, 125). A test portion of latex is titrated with acid to pH 6 in the presence of a stabilizer to prevent coagulation, either electrometrically or with methyl red as a visual indicator. The alkalinity is calculated from the quantity of acid required.

2.3.4. Volatile fatty acid or VFA (ISO, 506). A test portion is coagulated with ammonium sulphate and a portion of the resultant serum is separated and acidified with sulphuric acid. The acidified serum is steam-distilled and the volatile acids present in the test portion are determined by titration of the distillate with a standard volumetric barium hydroxide solution

2.3.5. Mechanical stability (MST) (ISO, 35). A test portion of latex is diluted to 55% by mass total solids content and stirred at high speed. The time required to initiate visible flocculation is recorded, this regarded as a measure of the mechanical stability.

3. Results and discussion

When fresh natural latex was added the creaming agent, two layers of latex appeared or we can call it "the creaming phenomenon". The upper layer is cream concentrated latex, it was then added ammonia solution with a concentration of 0.6% for its preservation. On the other hand, the lower fraction is serum that is separated from the cream latex by gravity with the activity of the added creaming agent.

3.1. Quantity of creaming agent for preparation of cream concentrated latex

Figure 1 presents the quantity creaming agent at 1, 2, 3, 4, and 5 ml in the fresh natural latex at 25 ml. The creaming phenomenon was observed at different incubation times. This figure shows the relationship between incubation time and quantity of creaming agent. We found that incubation time is decreased with increasing of creaming agent then reached at a constant value of incubation time for 4 ml of creaming agent. So, the optimal quantity of creaming agent is 4 ml in fresh natural latex 25 ml, this optimal ratio will be applied for the other sections.

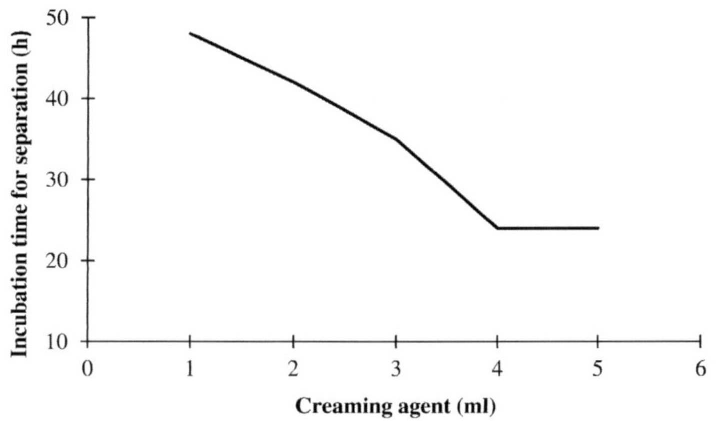

Figure 1. Relationship of quantity of creaming agent and incubation time (h)

3.2. Properties of cream concentrated latex

Characterizations of fresh natural latex and cream concentrated latex were compared in table 3. When fresh natural latex is transformed to concentrated natural latex by creaming process using sodium alginate as a creaming agent, we found that TSC and DRC values of the cream concentrated latex are clearly higher than those of fresh natural latex. This proves that the sodium alginate provides efficiency to separate rubber particles from the serum solution. However, there are no significant differences of alkalinity and VFA between fresh natural latex and cream concentrated latex. The cream concentrated latex was still incubated just 3 days with 4 ml of creaming agent. The alkalinity indicates to the ammonia level in latex and the VFA indicates to the degradation of latex by micro-organism. The results from table 3 mean that both types of latex are in good condition and are not yet degraded.

Table 3. Properties of fresh natural latex and cream concentrated latex.

Properties	Fresh natural latex	Cream concentrated latex
Total solid content (%)	37.17	46.31
Dry rubber content (%)	27.28	44.61

Alkalinity (%)	0.67	0.14
Volatile fatty acid or VFA	0.042	0.031

3.3. Effect of incubation time on TSC and DRC

Figure 2 shows TSC and DRC values of cream concentrated latex with incubation time. The TSC and DRC values of cream concentrated latex increased in the first month and then reached the plateau constant value with longer period of incubation time. The TSC value is always higher than DRC value because TSC includes rubber and non-rubber components, while the DRC presents only rubber component.

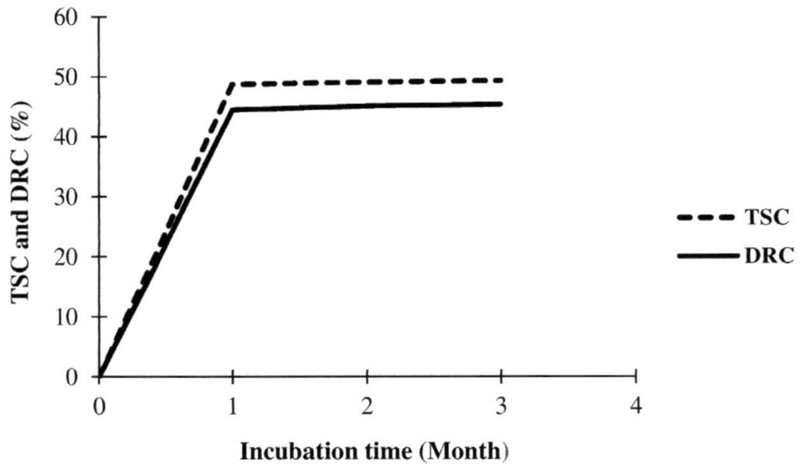

Figure 2. Relationship of TSC and DRC of cream concentrated latex and incubation time (Month).

3.4. Effect of incubation time on MST of cream concentrated latex.

Mechanical stability or MST level relates to incubation time as in the figure 3, indicates that cream concentrated latex has higher MST with increasing incubation time or storage time. This is attributed to the increase fatty acid soap from hydrolysis of the phospholipid and ammonia solution [6]. It has been known that the fatty acid soap is adsorbed at the surface of rubber particle to form hydration layer to cover rubber particle [8]. This affects the enhancement of the colloid stability of the latex and thus increasing MST.

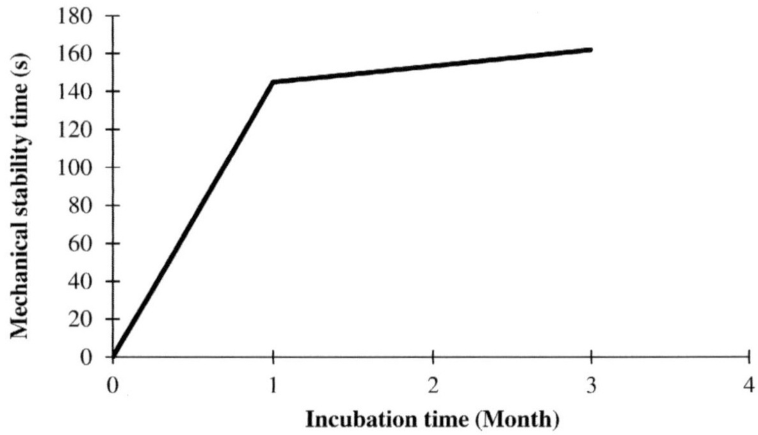

Figure 3. MST of cream concentrated latex and incubation time (Month).

3.5 Model of relation between sodium alginate

Sodium alginate is used as a creaming agent with high degree of polarity (hydroxyl group, -OH) and sequences of the mannuronate and glucuronate residues (figure 4). The sodium alginate can become hydrophilic colloids dispersed in the latex and interacts with the rubber particles through the negative charges at the surface of particles (figure 5).

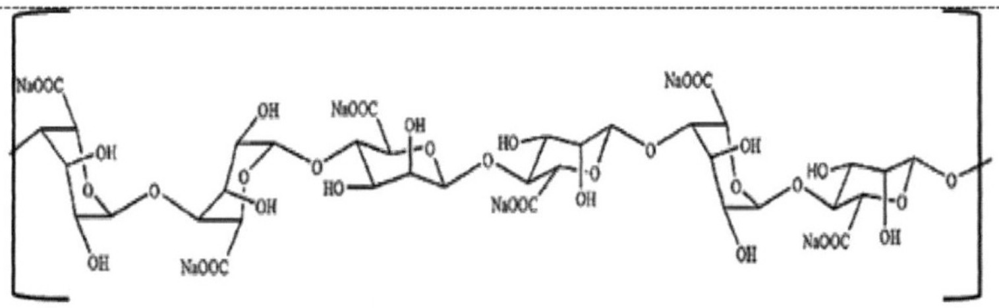

Figure 4. Molecular structure of sodium alginate.

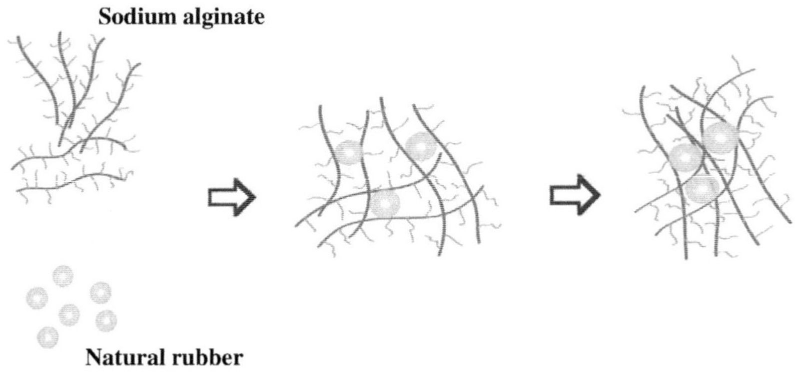

Figure 5. Ideal model structure of cream concentrated formation (a) surface of rubber particles interaction with sodium alginate molecule and (b) entanglement of sodium alginate molecules with rubber particles.

The sodium alginate molecules cover the surface of rubber particles (figure 5), more adsorption of sodium alginate molecules at the surface of rubber particles with longer incubation time. The branched segments of the sodium alginate molecule could be entangled with the sodium alginate in neighbouring rubber particles, this causes larger size of rubber particle. High molecular weight sodium alginate with a number of hydroxyl groups (OH) led to higher viscosity of the latex and cause to slowly movement of rubber particles [6].

4. Conclusions

We have succeeded in preparing the concentrated natural latex using creaming technic with the optimal amount of creaming agent. Cream concentrated natural latex using sodium alginate as a creaming agent represented higher TSC and DRC compared to the fresh natural latex. The other properties of both types of latex show no significant differences, except for the MST of cream concentrated latex increasing with the increase of incubation time. We also propose a model to explain the cream concentrated latex between sodium alginate and rubber particles. This type of concentrated latex could be applied for the foam rubber products, especially green rubber products.

5. Acknowledgments

The authors gratefully acknowledge the financial support of the Graduate School Kasetsart University, Bangkok, Thailand, and National Science and Technology Development Agency (NSTDA) from the STEM Workforce, Thailand.

6. References

[1] Smitthipong W, Nardin M, Schultz J, Nipithakul T and Suchiva K 2004 *J. Adhesion Sci. Technol.* **18** 1449.
[2] Smitthipong W, Nardin M and Schultz J 2008 *Kasetsart J. (Nat. Sci.)* **42** 318.
[3] Vallat M-F, Radabutra S, Thanawat S, Bresson B, Chollakup R, Rungsanthien K, Suwanruji P, Sriroth K, Smitthipong W and Mougin K 2014 *Adv. Mat. Res.* **844** 369.
[4] Smitthipong W, Tantatherdtam R, Rungsanthien K, Suwanruji P, Sriroth K, Radabutra S, Thanawat S, Vallat M-F, Nardin M, Mougin K and Chollakup R 2014 *Adv. Mat. Res.* **844** 345.
[5] Smitthipong W, Suethao S, Shah D and Vollrath F 2016 *Polym. Testing* **55** 17.

[6] Yimae N, Kaesaman A, Rungvichaniwat A, Thepchalerm C and Nakason C 2010 *J. Elast. Plast.* **42** 453.
[7] Ochigbo S S, Lafia-Araga R A and Suleiman M A T 2011 *African J. Agricul. Res.* **6** 2916.
[8] Blackly D C 1997 *Polymer Lattices*, London, Chapman & Hall **2** 106.

IOP Conf. Series: Materials Science and Engineering **272** (2017) 012026 doi:10.1088/1757-899X/272/1/012026

Existence of Nonsteady Planar Ideal Flows in Anisotropic Plasticity

S Alexandrov[1], A Pirumov[2] and P P Date[3]

[1] Laboratory of Mechanics of Strength and Fracture of Materials and Structures, Institute for Problems in Mechanics, 101-1 Prospect Vernadskogo, Moscow 119526, Russia

[2] Department of Elektrotechnics and Mechanics, Moscow Technological University, 78 Prospect Vernadskogo, Moscow 119526, Russia

[3] Department of Mechanical Engineering, IIT Bombay, Powai Mumbai 400076, India

Abstract. Ideal plastic flows are those for which all material elements follow minimum work paths. The general equations for steady and nonsteady planar ideal flows in Tresca solids have been given elsewhere. The present paper focuses on nonsteady planar ideal flows in anisotropic plasticity. In particular, the existence of such flows is proven under a certain assumption concerning the orientation of principal stress trajectories at the initial instant. It is also shown that the system of kinematic equations is hyperbolic. This system can be treated separately from the stress equations. The original ideal flow theory is widely used as the basis for inverse methods for the preliminary design of metal forming processes driven by minimum plastic work. The new theory extends this area of application to anisotropic materials.

1. Introduction

The ideal flow theory has long been associated with solids satisfying Tresca's yield criterion and its associated flow rule [1]. The ideal flow condition is that trajectories of the major principal stress are fixed in the material. This condition is an additional equation to the standard system of equations of plasticity theory. However, in many cases it is possible to show that a large class of solutions exists for this over-determined system of equations. In particular, a proof of the existence of steady ideal flows in Tresca solids has been provided in [2]. This result has been extended to nonsteady flows in [1]. Ideal flows result in maximum uniformity and minimum resistance and, therefore, are useful for the preliminary design of material forming processes [3]. A review of the ideal flow theory and ideal flow solutions has been provided in [4]. Solutions for anisotropic materials are only available in the case of sheet metal forming (for example, [5]). On the other hand, plastic anisotropy is a common property of many materials in bulk material forming. It is therefore of interest and importance to extend the bulk ideal flow theory to such materials. A proof of the existence of steady planar ideal solutions for orthotropic materials has been given in [6]. This ideal flow theory is based on the model of anisotropic

IOP Conf. Series: Materials Science and Engineering **272** (2017) 012026 doi:10.1088/1757-899X/272/1/012026

plasticity proposed in [7] and the yield criterion proposed in [8]. The objective of the present paper is to extend the proof given in [6] to nonsteady flow.

2. Material model

The plane strain yield criterion of any incompressible anisotropic material which complies with the principle of maximum plastic dissipation is expressed solely in terms of the stress variables $s=(\sigma_{xx}-\sigma_{yy})/2$ and $\tau=\sigma_{xy}$ [8]. Here σ_{xx}, σ_{yy} and σ_{xy} are the components of the stress tensor in a Cartesian coordinate system (x, y). This yield criterion is represented as a unique contour in the Mohr stress space (Fig. 1[a]). Let us introduce another Cartesian coordinate system (x', y') in the plane of deformation. The corresponding stress components are denoted as σ'_{xx}, σ'_{yy} and τ'_{xy}. Consequently, $\tau'=\tau'_{xy}$ and $s'=(\sigma'_{xx}-\sigma'_{yy})/2$. Let γ be the angle between the $s'-$ axis and the outward normal to the yield contour (Fig. 1[a]). It is evident that there are primed coordinate systems in which $\gamma=0$. One of such systems is shown in Fig.1[b]. Assume that the axes of the primed system are tangent to the principal stress trajectories at a given point *M*. Then, point *M* corresponds to point *P* in the Mohr stress space and, according to the associated flow rule, the principal stress and principal strain rate directions coincide at *M*. In this case the yield criterion can be written as

$$\sigma_1-\sigma_2=K \tag{1}$$

where σ_1 and σ_2 are the principal stresses and *K* is a material constant. The subsequent analysis is valid for deformation processes in which the axes of the primed system are tangent to the principal stress trajectories at each point of the plastic region at the initial instant. Therefore, the yield criterion (1) is also valid in the entire plastic region at the initial instant. The evolution of plastic anisotropy is governed by the law proposed in [7].

3. Existence of ideal flow

Let us introduce two coordinate systems; namely, a Cartesian coordinate system (x,y) and a principal lines system (ξ,η) (i.e. the coordinate curves of this coordinate system coincide with trajectories of the principal stress directions). In the principal lines coordinate system $\sigma_\xi\equiv\sigma_1$ and $\sigma_\eta\equiv\sigma_2$. Therefore, equation (1) becomes

$$\sigma_\xi-\sigma_\eta=K \tag{2}$$

at the initial instant. According to the ideal flow condition the principal lines coordinate system is Lagrangian. In this case, the evolution law for plastic anisotropy proposed in [7] shows that equation (2) is valid throughout the process of deformation. Thus, in order to demonstrate the existence of ideal flow, it is necessary to prove that the yield criterion (2) and its associated flow rule are compatible with the ideal flow condition. The associated flow rule is equivalent to two equations: (i) the incompressibility equation and (ii) zero shear strain rate in the (ξ,η) coordinate system. Let h_ξ and

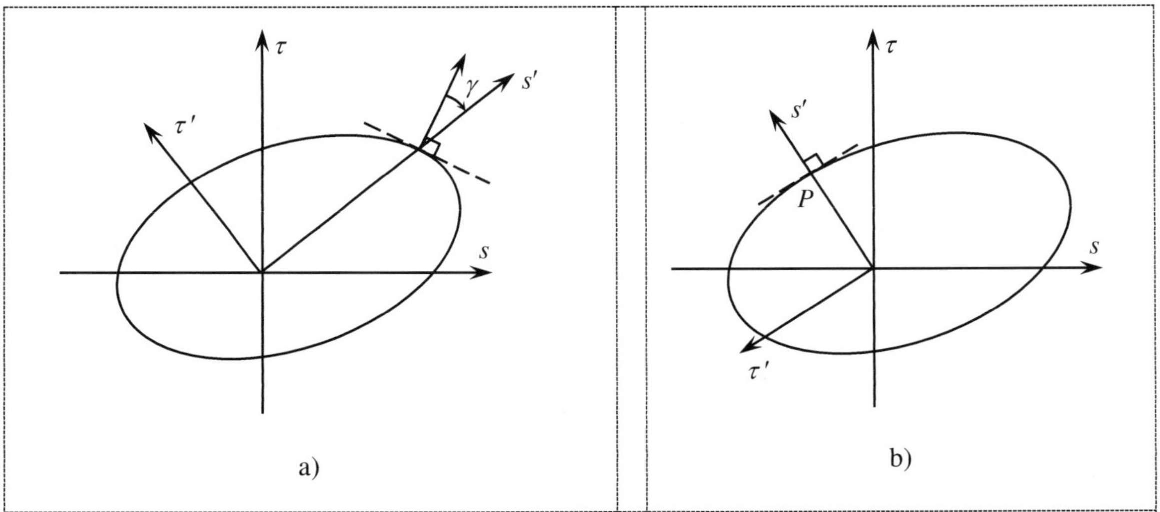

Figure 1. Yield contour in Mohr plane.

h_η be the scale factors for the coordinate curves of the (ξ,η) coordinate system. It has been shown in [6] that in the case of the yield criterion (2) it is always possible to choose the principal lines coordinate system such that

$$h_\xi h_\eta = 1. \tag{3}$$

This equation generalizes the corresponding equation in isotropic plasticity derived in [9]. Since the (ξ,η) coordinate system is Lagrangian, the incompressibility equation requires that equation (3) is satisfied throughout the process of deformation. Since the shear strain rate vanishes in the (ξ,η) coordinate system, the $\xi-$ and $\eta-$ material lines are orthogonal throughout the process of deformation. Then, it follows from the geometry of Fig.2 that

$$\partial x/\partial\xi = h_\xi \cos\psi, \quad \partial x/\partial\eta = -h_\eta \sin\psi, \quad \partial y/\partial\xi = h_\xi \sin\psi, \quad \partial y/\partial\eta = h_\eta \cos\psi. \tag{4}$$

The compatibility equations are

$$\frac{\partial^2 x}{\partial\xi\partial\eta} = \frac{\partial^2 x}{\partial\eta\partial\xi}, \quad \frac{\partial^2 y}{\partial\xi\partial\eta} = \frac{\partial^2 y}{\partial\eta\partial\xi}. \tag{5}$$

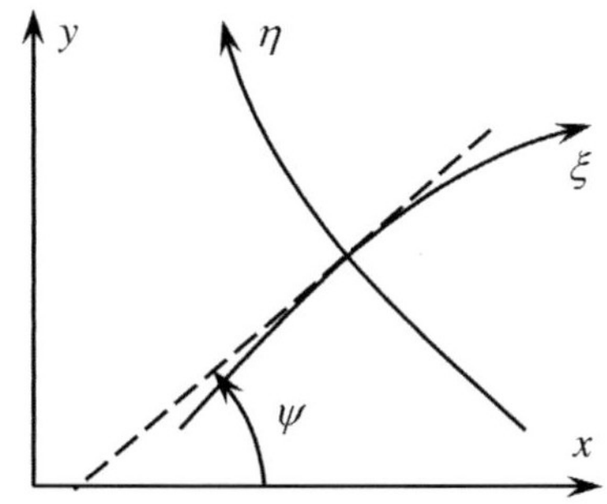

Figure 2. Cartezian (x,y) and principal lines-based (ξ,η) coordinate systems.

Substituting (4) into (5) yields

$$\begin{aligned}\cos\psi\,\partial h_\xi/\partial\eta - h_\xi\sin\psi\,\partial\psi/\partial\eta &= -\sin\psi\,\partial h_\eta/\partial\xi - h_\eta\cos\psi\,\partial\psi/\partial\xi,\\ \sin\psi\,\partial h_\xi/\partial\eta + h_\xi\cos\psi\,\partial\psi/\partial\eta &= \cos\psi\,\partial h_\eta/\partial\xi - h_\eta\sin\psi\,\partial\psi/\partial\xi.\end{aligned} \tag{6}$$

It is always possible to choose the Cartesian coordinate system such that $\psi = 0$ at a given point. Moreover, h_η can be eliminated by means of (3). Then, equation (6) becomes

$$h\partial h/\partial\eta + \partial\psi/\partial\xi = 0, \quad h^3\partial\psi/\partial\eta + \partial h/\partial\xi = 0 \tag{7}$$

where $h \equiv h_\xi$. Using a standard technique it is possible to show that this is a hyperbolic system of equations. Its solution can be found independently of the stress equations if appropriate boundary conditions are prescribed. Therefore, it remains to show that the stress equations are compatible with (7). The system of stress equations comprises the equilibrium equations and yield criterion (2). Since the shear stress vanishes in the (ξ,η) coordinate system, the equilibrium equations are [10]

$$h_\eta\,\partial\sigma_\xi/\partial\xi + (\sigma_\xi - \sigma_\eta)\partial h_\eta/\partial\xi = 0, \quad h_\xi\,\partial\sigma_\eta/\partial\eta + (\sigma_\eta - \sigma_\xi)\partial h_\xi/\partial\eta = 0. \tag{8}$$

Equations (2), (3) and (8) combine to give $h\partial\sigma_\xi/\partial\xi - K\,\partial h/\partial\xi = 0$ and $h\partial\sigma_\eta/\partial\eta - K\,\partial h/\partial\eta = 0$. These equations can be immediately integrated to arrive at

$$\sigma_\xi/K - \ln h = f_1(\eta), \quad \sigma_\eta/K - \ln h = f_2(\xi). \tag{9}$$

Here $f_1(\eta)$ is independent of ξ and $f_2(\xi)$ is independent of η. Eliminating $\ln h$ in (9) leads to

$$(\sigma_\xi - \sigma_\eta)/K = f_1(\eta) - f_2(\xi). \tag{10}$$

It is evident that (10) is compatible with (2) if $f_1(\eta) - f_2(\xi) = 1$. Therefore, $f_1(\eta) = C$ and $f_2(\xi) = 1 + C$ where C is independent of both ξ and η. Then, equation (10) becomes

$$\sigma_\xi/K - \ln h = C, \quad \sigma_\eta/K - \ln h = C + 1. \tag{11}$$

This equation connects the kinematic quantity h and the stress field in ideal flows.

4. Conclusion

It has been shown that non-trivial nonsteady planar ideal flow solutions exist in anisotropic plasticity assuming that the constitutive equation proposed in [7] is valid. An additional requirement, as compared to ideal flow in isotropic plasticity, is that the principal axes of anisotropy coincide with principal stress trajectories at the initial instant. If this condition is satisfied then equation (11) connects the kinematic quantity h and the stress field. The theory developed is important for training of high skill professionals in Russia [11].

5. Acknowledgment

This research was made possible by the grant RSCF-16-49-02026 (Russia) and INT/RUS/RFBR/P-214 (India).

6. References

[1] Richmond O and Alexandrov S 2002 *Acta Mech.* **158** 33
[2] Hill R J 1967 *J. Mech. Phys. Solids* **15** 223
[3] Richmond O and Morrison H L 1967 *J. Mech. Phys. Solids* **15** 195
[4] Chung K and Alexandrov S 2007 *Appl. Mech. Rev.* **60** 316
[5] Park S H, Yoon J W, Yang D Y and Kim Y H 1999 *Int. J. Mech. Sci.* **41** 1217
[6] Alexandrov S, Mustafa Y and Lyamina E 2016 Meccanica **51** 2235
[7] Collins I F and Meguid S A 1977 *ASME J. Appl. Mech.* **44** 271
[8] Rice J 1973 *Mech. Phys. Solids.* **21** 63
[9] Sadowsky M A 1941 *ASME J. Appl. Mech.* **63** A74
[10] Malvern L E 1969 *Introduction to the Mechanics of a Continuous Medium* (Englewood Cliffs:Prentice-Hall)
[11] Bodrova E V and Pirumov A R 2016 *Russ. Technol. J.* **4** 62

IOP Conf. Series: Materials Science and Engineering **272** (2017) 012027 doi:10.1088/1757-899X/272/1/012027

A new approach to generate local resonator for the application of acoustic metamaterials

Sheng Sang[1], Eric Sandgren[1] and Ziping Wang[2]

[1]Department of System Engineering, University of Arkansas at Little Rock, Little Rock, AR, USA 72204

[2]Faculty of Civil Engineering and Mechanics, Jiang Su University, Zhenjiang, China 212013

Corresponding author: sxsang@ualr.edu

Abstract. In this paper, we proposed a type of local resonator which works by introducing more than one vibration modes, and thus may provide a new approach to formulate acoustic/elastic metamaterial. Such new concept of resonator is fundamentally different from traditional translational and rotational resonators. The potential application and design of acoustic/elastic metamaterial based on this kind of structures are also illustrated. The acoustic metamaterial filter we developed has unique property that only the waves with specific frequency can pass the material while other waves will be attenuated. We also proposed a kind of plane wave lens which can transfer plane wave with uniformly distributed amplitude profiles into pure plane wave. With these novel resonators, the research and application regarding it are highly anticipated in the near future.

Key words: Acoustic Metamaterials, Resonator, vibration Mode, Wave filter, Plane wave lens

1. Introduction

Acoustic metamaterial is a kind artificial material with periodic local microstructures [1]. Many acoustic metamaterials with local resonators of various topologies have been proposed and studied. Normally, acoustic/elastic metamaterials can be classified as 1-D [2], 2-D [3][4] and 3-D [5] metamaterials according to their dimensions. They can also be divided into groups of single phase and multiphase metamaterials by considering their building components.

To our best knowledge, basically, there are only two types of local resonators, that is translational resonator and rotational resonator, which can introduce negative effective mass density and negative effective elastic modulus, respectively. By the combination of using those two resonators, one can obtain double negative systems [6] [7], which in return attribute to negative refractive index. A variety of designs [8]-[10] based on those resonators have been proposed and implemented to explore and expand its applications.

The objective of this work is to propose a new type of resonator which is based on the second or high vibration modes of substructures. This work is organized into four sections including this instruction: in section 2, two typical concepts of resonators are reviewed and the concept of multi-mode resonator is for the first time proposed and studied analytically; in section 3 a simplified acoustic/elastic wave filter containing beams-masses resonator is formulated. Numerical modelling is carried out to obtain a better understanding of its working mechanism in section 4.

2. Concept of multi-mode resonator

In order to understand the difference between resonator that we suggested and other resonators, lattice models of two normal resonators is illustrated first. As figure 1(a) shows, negative effective mass density can be perfectly explained in a 1D mass-in-mass model, which here we defined as translational resonator, those metamaterials which can be explained based by translational resonator, such as mass-spring system and matrix-core system are actually have similar working mechanics, and thus we classify them as one type of metamaterials. Likewise, as it is shown in figure 1(b) negative effective modulus is mainly achieved by rotational resonator, chiral structure has been proved to be a realistic way to obtain rotational oscillations. Also some researchers have proposed plate-type acoustic/elastic metamaterials which combine those two resonators and successfully archive negative refractive index. For the sake of brevity, detailed explanation of the governing equation and working mechanism of those two types of resonators are not presented here.

Basically those two types of resonators have only one vibration mode, high frequency vibration modes are hard to be activated. As it is well known that structures such as beams have more than just one vibration mode, and each vibration mode is corresponding to a constant eigenfrequency. However those eigenfrequencies are usually very high and not practically for metamaterial design which mainly focuses on low frequency wave manipulation. In order to make use of more than one vibration modes at low vibration frequency, as it is shown in figure 1(c), a beam-masses structures which compromising a continues thin slender massless beam and two concentrated masses are proposed. To obtain a simplified dynamic model, the thickness of the beam and the dimension of the masses are ignored. The Young modulus of the beam is denoted by E, while the length of the whole beam is l, the distance between the concentrated masses and ends of beam is $l/3$, the mass is represented by m. The reaction force at the left and right end of the beam are denoted by $F_l(t)$ and $F_r(t)$ respectively. The first and second vibration modes are demonstrated in the figure by dashed lines, so that the vibration frequency corresponding to the first and second vibration modes are respectively. If the first vibration mode is activated, then the two concentrated masses moves in phase, then the it behave like a simple spring-mass model, thus it functions as an translational resonator. However when the second vibration mode is activated, the motions of the two masses are 180° out of phase, and hence, $F_l(t)=-F_r(t)$. Then a harmonic moment of force will be generated, this character, we believe, can be used for the purpose of shear wave manipulation.

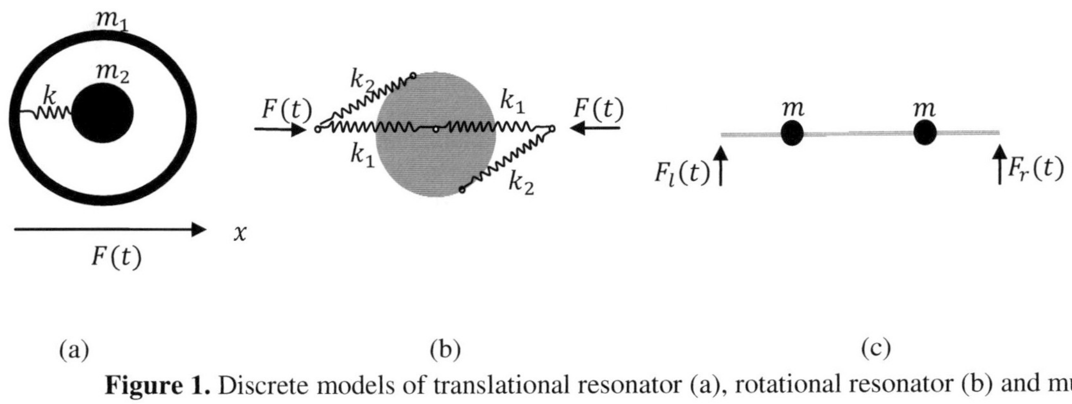

Figure 1. Discrete models of translational resonator (a), rotational resonator (b) and multi-mode resonator (c).

3. A design of elastic wave filter based on multi-mode resonator

To fully understand the dynamic characteristic and demonstrated its potential application, a passive elastic wave band-pass filter based on multi-mode resonator is proposed in figure 2. The basic function of it is to pass signals with frequencies within a specified band, the pass-band, and to reject signals outside this band. This type of filter is fundamentally different from the other acoustic filters which only prohibit signals with frequencies within certain band. The beam-masses resonator presented in previous section is adapted here, and two ends of the slender beam are clamped to the upper and lower bars respectively which serve as media for longitudinal wave propagation. For convenience, we assume that the disks are taken as rigid and the radiuses of them are neglected. Only the horizontal displacements of those components are considered here.

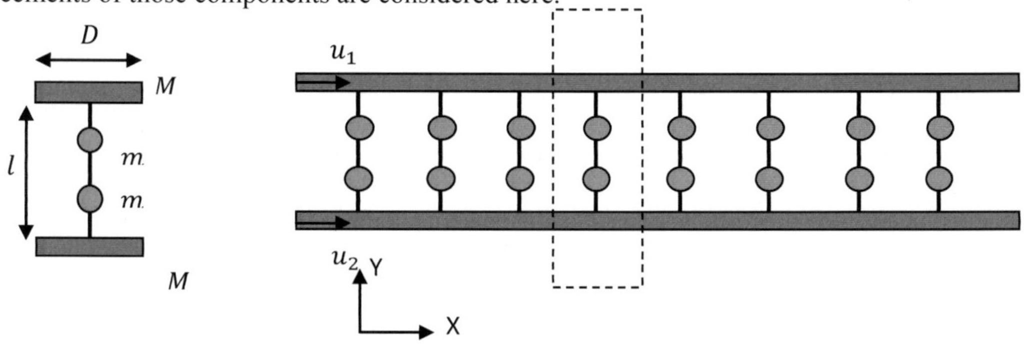

Figure 2. An acoustic metamaterial bar based on beam-masses resonator.

If a single frequency elastic longitudinal wave propagating through the upper and down bar, then it can be assumed to have:

$$u_1(x,t) = p_1 e^{i(\alpha x - wt)}, \tag{1}$$

$$u_2(x,t) = p_2 e^{i(\alpha x - wt)} \tag{2}$$

$$v_1(t) = q_1 e^{-iwt}, \tag{3}$$

$$v_2(t) = q_2 e^{-iwt}, \tag{4}$$

where p_1, p_2, q_1 and q_2 denote the vibration amplitudes, α and w are the wave number and radian frequency respectively. Thus the longitudinal stiffness of the beam segments are approximately estimated as $k = 81EI/l^3$, E, I and l are elastic modulus, moment of inertia and length of the beam. Since the wavelength λ is assumed to be much larger than the width of unit cell d, and the structure consists of repeated cells, thus only one cell needs to be considered in the analysis. The governing equation can be obtained based on Newton's law:

$$EADu_1'' + k(v_1 - u_1) = M\ddot{u}_1 \tag{5}$$

$$k(u_1 - v_1) + \frac{k}{2} * (v_2 - v_1) = m\ddot{v}_1 \tag{6}$$

$$k(u_2 - v_2) + \frac{k}{2} * (v_1 - v_2) = m\ddot{v}_2 \tag{7}$$

$$EADu_2'' + k(v_2 - u_2) = M\ddot{u}_2 \tag{8}$$

Assume that $p_1 = p_2 = 1$, then q_1 and q_2 can be represented by $(Mw^2 - EA\alpha^2 - k)/k$,

$$\begin{bmatrix} Mw^2 - EAD\alpha^2 - k & k & 0 & 0 \\ k & -1.5 * k + mw^2 & k/2 & 0 \\ 0 & k/2 & -1.5 * k + mw^2 & k \\ 0 & 0 & k & Mw^2 - EAD\alpha^2 - k \end{bmatrix} \begin{bmatrix} p_1 \\ q_1 \\ q_2 \\ p_2 \end{bmatrix} = 0 \tag{9}$$

In order to have nontrivial solution to this eigenvalue problem, the determination of the matrix should be zero, and thus the dispersion equation is achieved as:

$$(Mw^2 - EAD\alpha^2 - k)^2 * \left(\frac{3}{2} * k - mw^2\right)^2 - k^4 = 0 \tag{10}$$

Thus the stiffness of the beam segment is approximately estimated as $k = 81EI/l^3$, E, I and l are elastic modulus, moment of inertia and length of the beam. To demonstrate an example, a unit cell having the following geometric and material parameters is listed below:

Width of unit cell: $d = 10$ mm

Length of the beam: $l = 12$ mm

Radius of the mass disk: $r = 2\ mm$

Beam thickness: $h = 0.2\ mm$

Bar thickness: $H = 4\ mm$

Young's Modulus: $E = 70\ Gpa$

Poisson's ratio: $v = 0.33$

Mass density: $\rho = 2800kg/m^3$

Width of unit cell: $D = 10$ mm

Cross-sectional area: $A = H * 1$

Mass of bar segment: $M = \rho HD * 1$

Mass of Disk: $\mathrm{m} = \rho\pi r^2 * 1 = 2800 * 3.14 * 8 * 10^{-9} = 7.03 * 10^{-5}$

Moment of inertia of beam section: $\mathrm{I} = \frac{h^3}{12} = \frac{2}{3} * 10^{-12}$

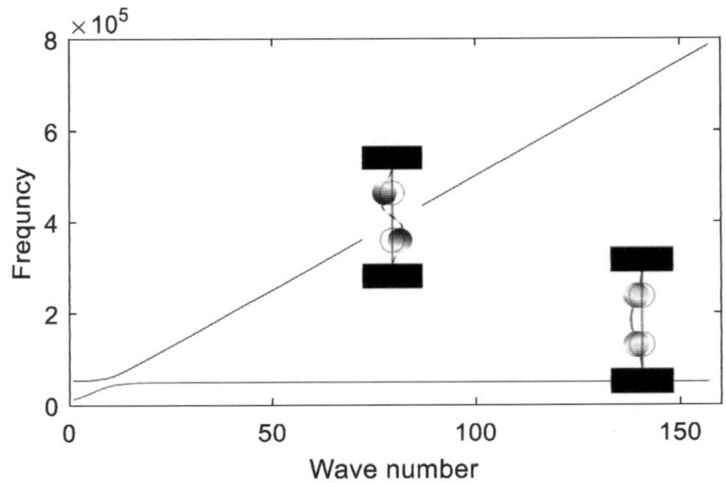

Figure 3. Band diagram.

The band diagram of a single cell is shown in Fig. 3. For an elastic wave having a specific wave number, it can propagate at two different frequencies. The lower branch can be named as in-phase mode because the displacements $v_1(t)$ is same to $v_2(t)$, and the upper branch is called the out-phase mode is due to the fact that $v_1(t) = -v_2(t)$. It should be pointed out that p_1 can be different than p_2 here. If we assume that the left end of the upper and bottom bars are subject to the same excitations with same phase and amplitude, then two bars can be seen as an single bar, thus similar to single bar model, when the excitation frequency is above the vibration frequency of in-phase mode the wave will be attenuated, the effect of the resonator is same to translational resonator. This is not the concentration of this work, so the mechanism is not demonstrated here in detail. We mainly focus on the out of phase mode, which brings in the concept of so called new approach to generate local resonator.

In order to active the second vibration mode of the resonator, we assume that $p_1$is larger than p_2, and $p' = p_1 - p_2$ is defined. For the analysis of a single unit, the lower bar segment is assumed to be fixed on the ground, thus the matrix of the governing equation can be reduced as:

$$\begin{bmatrix} Mw^2 - EAD\alpha^2 - k & k & 0 \\ k & -1.5 * k + mw^2 & k/2 \\ 0 & k/2 & -1.5 * k + mw^2 \end{bmatrix} \begin{bmatrix} p' \\ q_1 \\ q_2 \end{bmatrix} = 0 \tag{11}$$

Dispersion equation is obtained as:

$$(Mw^2 - EAD\alpha^2 - k)[(-1.5 * k + mw^2)^2 - k^2/4] - k^2(-1.5 * k + mw^2) = 0 \tag{12}$$

4. Numerical simulation

In this section, wave propagation and working mechanism of the proposed acoustic wave filter and plane wave lens are numerically studied. The free wave motion properties are reported in the form of band diagrams. In the following, results are presented in normalized frequency.

1. Filtering properties of the acoustic wave filter with finite units.

Previous subsection analyzed the free wave motion characteristics of the acoustic wave filter with infinite subunits, however it is not available in real industry field. In this subsection, a finite length acoustic wave filter is proposed as shown in Fig.4. The parameters of the bar used here are the same as those of the acoustic bar in the previous section, and it contains 8 unit cells in the length direction. In order to get rid of the influence of the reflection of waves at the boundary, Perfect Matched Layer (PML) is applied. A horizontal displacement field which serves as input source is applied at the left edge of the upper bar, and the output is the right edge of the lower bar. The excitation frequency is swept from 0 to $10000Hz$, and then it will cover the first two eigenfrequency of the local resonator. The wave filter can be characterized by the frequency response functions of the displacement at the output. Fig.5 shows calculated response spectra at the output. Significant high transmittance appears when the excitation frequency is close to the resonance frequency $(1600Hz, 6250Hz)$ of the local resonator especially the second eigenfrequency. In order to understand the working mechanism, a snap shot of the deformation of the wave filter when the excitation frequency is and is shown in Fig.5, we can observe that the formation of the high transmittance ratio is due to the local resonance of the resonator.

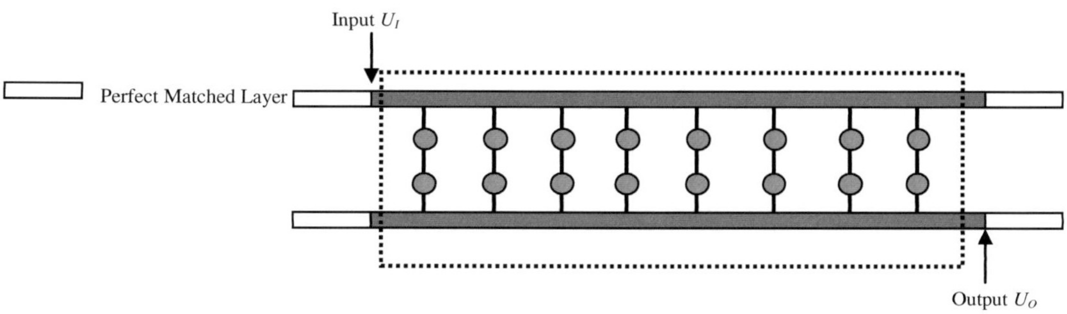

Figure 4. Simulation model of a simple Acoustic wave filter consists of beam-masses resonator.

5. Plane wave lens design and its characterization.

Fig. 6 shows a design of acoustic plane wave lens composed of horizontal bars and proposed local resonators, 5 bars are placed parallel to each other, and 8 beam-masses resonators are placed periodically between and welded to each long bar. The 5 edges of the left ends of the bars serve as inputs, and the 5 edges of the right ends serve as outputs. The microstructure parameter used here is same to that in the design of the wave filter. Obviously, the uneven input wave source would not change at the output without the embedded local resonator. In the numerical simulation, PFL is also adapted here to avoid wave reflection from the boundary. The uneven input and the outputs at two eigenfrequencies are plotted in Fig.7, an almost pure plane wave are generated at the second resonance frequency. This design can certainly be expanded to large scale, which means a perfect elastic pane wave generator can be obtained with similar designs.

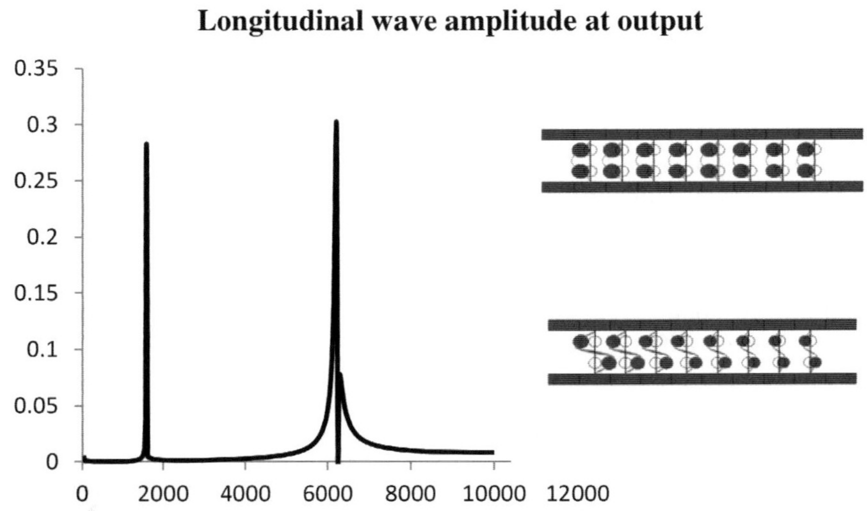

Figure 5. Frequency responses at the output.

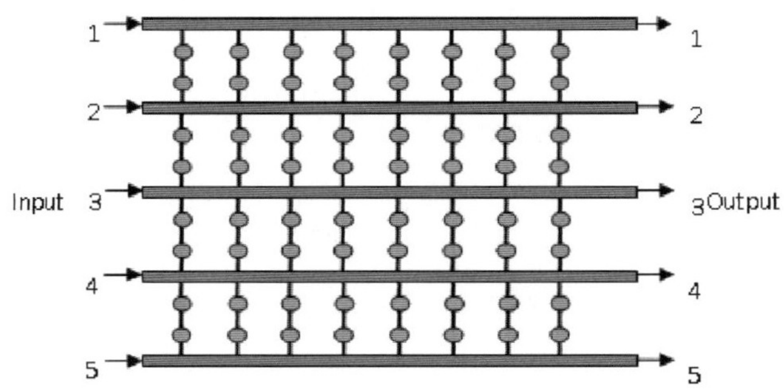

Figure 6. A simple Acoustic plane wave lens consists of beam-masses resonator.

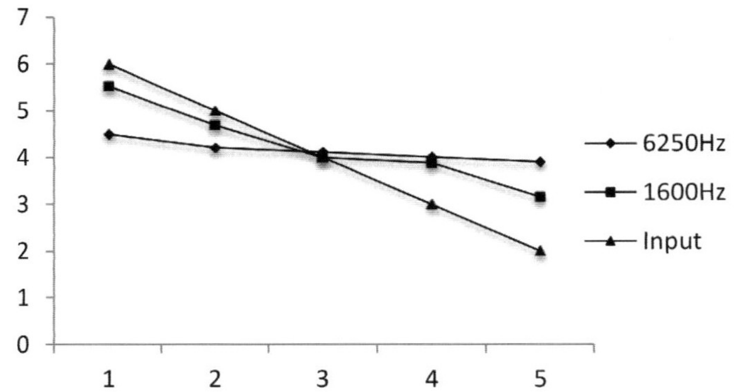

Figure 7. Amplitude of input and output of the plane wave generator.

6. Concluding remarks

In this work, a new approach to generate local resonator is suggested by introducing a beam-masses resonator. The second vibration mode of the local resonator is for the first time used in the design of local resonator. The special Based on this resonator, simple models of wave filter and plane wave generator are proposed and studied analytically and numerically. The reason of high transmittance of the wave filter and the working mechanism of the plane wave generator are discussed. There are more applications such as wave guide and wave hyper-lens for shear wave, we believe, can be obtained based on this type of resonator. This a new approach expanded the way we have to design the local resonator and family of acoustic metamaterials.

7. Acknowledgments

We appreciate the financial support from the National Natural Science Foundation of China (Grants No.11402101).

8. References

[1] Li J and Chan C 2004 Phys. Rev.E 70
[2] Pai PF 2010 J. Intell. Mater. Syst. Struct. 21 517-528
[3] Pai PF, Peng H and Jiang S 2014 Int. J. Mech. Sci. 79 195-205
[4] Peng H amd Pai PF Int. J. Mech. Sci. 89 350-361
[5] Lu M, Feng L and Chen Y 2009 Mater Today 12 34-42
[6] Lai Y, Wu Y, Sheng P and Zhang Z 2011 Nat. Mate. 10 620-624
[7] Zhu, R. et al. 2014 Nat. Commun. 5 5510
[8] Cheng Y, Xu J and Liu X 2008 Phys. Rev. B 77
[9] Chan C, Li J and Fung K 2006 Journal of Zhejiang University-SCIENCE A 7 24-28
[10] Yao S, Zhou X, and Hu G. 2008 New J. Phys.10 043020.

IOP Conf. Series: Materials Science and Engineering **272** (2017) 012028 doi:10.1088/1757-899X/272/1/012028

Corrosion behaviour of Al-Fe-Ti-V medium entropy alloy

M O Bodunrin[1,2], B A Obadele[3], L H Chown[2] and P A Olubambi[3]

[1] Department of Metallurgical and Materials Engineering, Federal University of Technology Akure, Ondo State, PMB 704, Nigeria
[2] School of Chemical and Metallurgical Engineering, University of Witwatersrand, Johannesburg, 2050, South Africa
[3] Centre for Nanoengineering and Tribocorrosion, School of Mining, Metallurgy and Chemical Engineering, University of Johannesburg, Doornfontein Campus, Johannesburg, South Africa

Abstract. Alloys containing up to four multi-principal elements in equiatomic ratios are referred to as medium entropy alloys (MEA). These alloys have attracted the interest of many researchers due to the superior mechanical properties it offers over the traditional alloys. The design approach of MEA often results to simple solid solution with either body centered cubic; face centered cubic structures or both. As the consideration for introducing the alloys into several engineering application increases, there have been efforts to study the corrosion behaviour of these alloys. Previous reports have shown that some of these alloys are more susceptible to corrosion when compared with traditional alloys due to lack of protective passive film. In this research, we have developed AlFeTiV medium entropy alloys containing two elements (Ti and Al) that readily passivate when exposed to corrosive solutions. The alloys were produced in vacuum arc furnace purged with high purity argon. Open circuit potential and potentiodynamic polarisation tests were used to evaluate the corrosion behaviour of the as-cast AlFeTiV alloy in 3.5 wt% NaCl and 1 M H_2SO_4. The corrosion performance of the alloy was compared with Ti-6Al-4V alloy tested under similar conditions. The results show that unlike in Ti-6Al-4V alloy, the open circuit potential of the AlFeTiV alloy move towards the negative values in both 3.5 wt% NaCl and 1 M H_2SO_4 solutions indicating that self-activation occurred rapidly on immersion. Anodic polarisation of the alloys showed that AlFeTiV alloy exhibited a narrow range of passivity in both solutions. In addition, the alloys exhibited lower E*corr* and higher I*corr* when compared with traditional Ti-6Al-4V alloy. The traditional Ti-6Al-4V alloy showed superior corrosion resistant to the AlFeTiV alloy in both 3.5 wt.% NaCl and 1 M H_2SO_4 solutions.

1. Introduction

About twelve years ago, a paradigm shift in the design approach of metallic alloys led to the emergence of high entropy alloys (HEA) [1]. Unlike conventional alloy systems with a parent element, HEA are alloys with equimolar or equiatomic concentration of constituent elements [2]. The alloys have been reported to have superior mechanical properties and sometimes excellent corrosion resistance over the conventional alloys [3-4]. These properties have been attributed to the strong

binding energy among the constituent elements and the simple solid solution phase that is formed in the alloys due to high configurational entropy of mixture that was hypothesized by Boltzmann [5].
The unique properties showed by these alloys have continued to attract the interest of researchers since the mechanisms driving the behaviour of these alloys are not well understood. Consequently, there have been a lot of contradictions and controversies in the reports of scientists in open literature. For example, the general claim that HEA forms simple solid solution have been found to be inconsistent in some other alloys where intermetallic compounds and complex phases were found. Also, the initial definition of HEA as alloys containing at least five elements with equiatomic concentrations has been questioned by some researchers, they reported that some HEAs can be developed with just four elements with equimolar concentrations. These sets of alloys have been referred to as medium entropy alloys (MEA).
While the controversies on the definition and properties of medium and high entropy alloy continue, the alloys exhibit comparable hardness and tensile properties. With a lot of potential medium and high entropy alloys yet to be explored, we have developed an experimental Al25Fe25Ti25V25 medium entropy alloy with the aim of examining its corrosion behaviour.
In this study, the AlFeTiV medium entropy alloy was tailored to conventional Ti-6Al-4Valloy which exhibit excellent corrosion resistance in several corrosive environments. The alloy contains Al and Ti that readily passivate when exposed to some corrosive solution, the corrosion behaviour of these alloys were then investigated in two common industrial solutions in comparison with commercial Ti-6Al-4V alloy.

2. Experimental procedure

The experimental powders were produced from elemental powders of aluminium, iron, titanium and vanadium. A total of 20 g of mixed Al25Fe25Ti25V25 powders were pressed and then melted three times in an electric arc furnace purged with high purity argon to produce the Al25Fe25Ti25V25 ingot. The ingot was turned in between melts to ensure homogenization of the alloy. The as-cast alloy was then sectioned and mounted for microstructural examination, hardness and corrosion testing.
The Al25Fe25Ti25V25 and commercial Ti-6Al-4V samples used for microstructural examination was mounted, grounded and polished following standard metallographic procedures. The samples were etched for 15second in standard Kroll's reagent, rinsed with acetone in an ultrasonic bath and dried in compressed air. The elemental composition of the Al25Fe25Ti25V25 was taken prior to etching of the alloy to avoid the dissolution of some of the elements in the alloy.
Back-scattered electron images of the etched sample were taken using field emission scanning electron microscope (Jeol FE-SEM JSM-7600F) while phase analysis of the as-cast AlFETiV was carried out using the PW1710 Philips X-ray diffractometer with Cu Kα radiation at 40 kV and 40 mA to identify the constituent phases. Diffractograms were collected over a range of 2θ between 10 and 90º, with a step size of 0.02. A search-match routine was performed using X'Pert High Score Plus software for phase identification.
The hardness of the sample was measured on FM700 microhardness testing machine using a load of 300 kgf and a dwell time of 15 s.
Open circuit potential (OCP) and potentiodynamic polarisation techniques were used in investigating the corrosion performance of the MEA in 3.5 wt% NaCl and 1 M sulphuric acid solutions. The solutions were prepared using analytical grade chemicals and distilled water. A three-cell set-up which consisted of silver-silver chloride as reference electrode, graphite rod as counter electrode and both Al25Fe25Ti25V25 and Ti-6Al-4V alloys as working electrodes was connected to an AUTOLAB potentiostat (302N model) with general purpose electrochemical system (GPES 4.9) software. Before potentiodynamic measurement, the OCP of the alloys was traced for 3600 s in 1 M sulphuric acid and 3.5 wt% sodium chloride solutions to allow stability. The potentiodynamic scan was traced from -1.5 to 2.0 V at a scan rate of 0.2 mV/s in both solutions. Electrochemical parameters such as Ecorr, Icorr and corrosion rate were calculated using Tafel extrapolation technique.

3. Results and discusssion

The microstructure of the as-cast Al25Fe25Ti25V25 alloy (nominal composition) is shown in figure. 1. The alloy contains typical as-cast dendritic and interdendritic regions marked as A and B respectively. Interdendritic and dendritic regions constituting FCC and BCC phases have been reported in some HEA alloys.

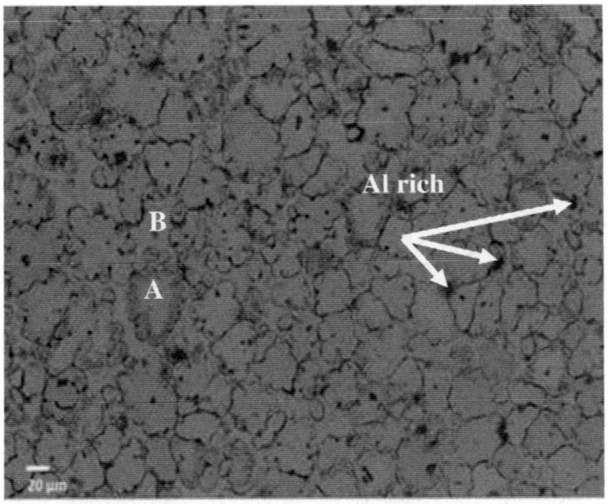

Figure 1. SEM-BSE of as-cast Al25Fe25Ti25V25 medium entropy alloy.

XRD peaks in figure 2 indicate that a mixture of FCC and BCC solid solutions are present in the alloy. The segregation of dark Al rich spots is also observed in the SEM micrograph. The EDX area analysis (figure 3) shows that there is a loss in aluminium during melting of the alloy. This led to the reduction in the aluminium concentration in the alloy. The hardness of the alloy was 692.5 ±21 HV (6.79 ±0.2 GPA) which is comparable to hardness values of some HEAs [6].

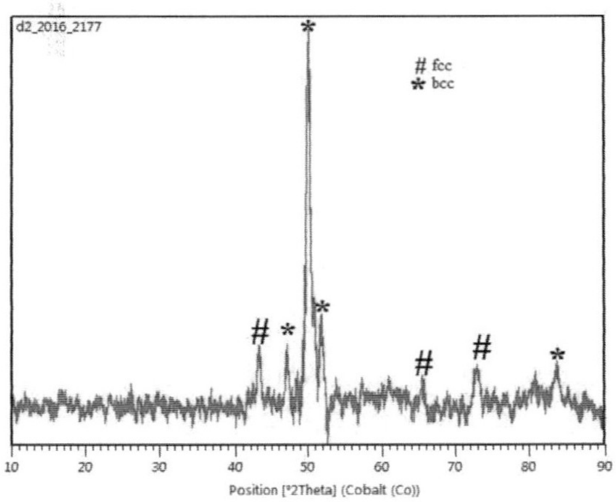

Figure 2. XRD pattern of as-cast Al25Fe25Ti25V25 medium entropy alloy.

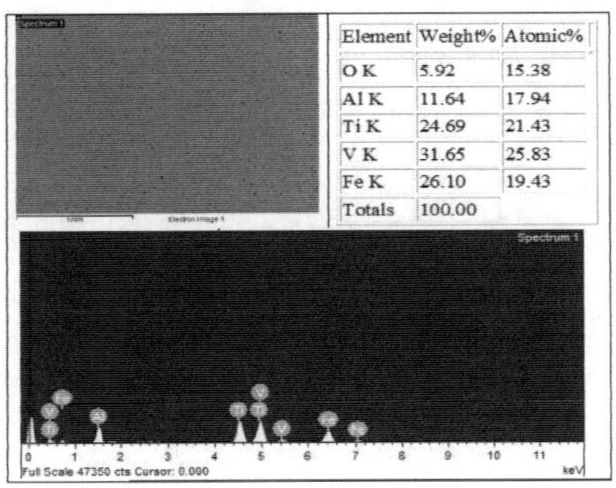

Figure 3. EDX spectra of $Al_{25}Fe_{25}Ti_{25}V_{25}$ medium entropy alloys.

The high hardness of the alloy can be attributed to the presence of BCC solid solutions in the alloys. Increasing aluminium content has been reported to promote the transition of FCC to BCC phase in some HEAs due to lattice distortions [7]. In addition, it was reported that refractive elements such as Ti and V impacts high hardness in high entropy alloys [8].

The linear polarization curves for alloys immersed in 3.5 wt% sodium chloride and 1 M sulphuric acid solutions are presented in figure 4. In both solutions, the traditional Ti-6Al-4V alloys showed superior corrosion resistance in comparison with the Al25Fe25Ti25V25 medium entropy alloys. Ti-6Al-4V alloy has higher Ecorr values and a wider range of passivity in the anodic region in both solutions. An active to passive behaviour could be observed in 1 M H_2SO_4. The range of the passivity exhibited by the Al25Fe25Ti25V25 alloy was narrower than that of the Ti-6Al-4V alloy. Narrow range of passivity was also reported in CrMnFeCoNi high entropy alloys [9]. The Al25Fe25Ti25V25 alloy showed a lower breakdown potential in sulphuric acid than in sodium chloride solution whereas the traditional Ti-6Al-4V did not show any breakdown potential up to 2 V.

In sodium chloride solution (figure 4a), both alloys displayed spontaneous passivation since no evidence of active to passive transition nose is observed. However, the Al25Fe25Ti25V25 alloy showed transient currents in the passive region. These transient currents are signatures of unstable passive film which indicates the dissolution and building of the films in the anodic region. The passivation current of the Al25Fe25Ti25V25 alloy is slightly lower than the conventional alloy in sodium chloride solution. In sulphuric acid medium (figure 4b), the traditional alloy displayed an active to passive transition nose while the Al25Fe25Ti25V25 alloy showed transient current close to the active nose. The passivation current and critical current density of the conventional Ti-6Al-4V alloys is slightly lower than that of the Al25Fe25Ti25V25 alloy.

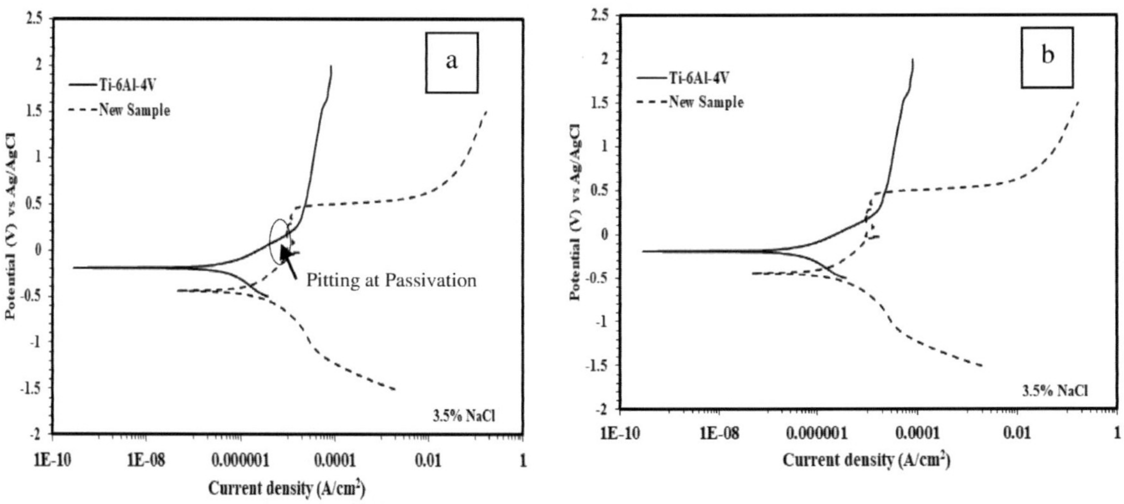

Figure 4. Polarisation curves for Al25Fe25Ti25V25 and Ti-6Al-4V in (a) 3.5 wt% sodium chloride solution and (b) 1 M sulphuric acid.

Table 1. Corrosion data obtained from potentiodynamic polarization curves.

Sample	E_{corr} (V)	I_{corr} (μA/cm^2)	Corrosion rate
		3.5% NaCl	
Ti-6Al-4V	-0.283	0.33	2.828E-3
Al25Fe25Ti25V25	-0.442	0.49	4.010E-3
		1 M H_2SO_4	
Ti-6Al-4V	-0.038	0.177	1.517E-2
Al25Fe25Ti25V25	-0.453	49.19	3.997E-1

Table 1 presents the E_{corr}, I_{corr} and Corrosion rate value derived from the Tafel extrapolation. The E_{corr} of Ti-6Al-4V alloy in both media are nobler when compared to Al25Fe25Ti25V25 alloy. Furthermore, the corrosion current density, I_{corr}, measured for Ti-6Al-4V are lower as shown in Figure 4.

Generally, the presence of Cl^- in the chloride containing solution significantly affect the pitting behaviour of Al25Fe25Ti25V25 alloy and thus, pitting within passivation range. In H_2SO_4 solution, the presence of H^+ influences the oxide film formation, lowering the pitting potential of Al25Fe25Ti25V25 alloy, thus affecting the corrosion behaviour (I_{corr} = 49.19 μA/cm^2). This explains the high corrosion rate reported for Al25Fe25Ti25V25 alloy (3.997E-1).

4. Conclusion

The corrosion behaviour of Al25Fe25Ti25V25 medium entropy in comparison with commercial Ti-6Al-4V alloy was evaluated in sodium chloride and sulphuric acid media. The Al25Fe25Ti25V25 medium entropy alloy exhibited inferior corrosion resistance in comparison with the conventional Ti-6Al-4v alloy due to unstable protective film. The substantial amount of the principal elements such as Al, V and Fe must have promoted defects in the passive film thereby causing severe attack of the corrosive solution on the alloy. Additionally, the segregated Al rich spots seen in the microstructure of the Al25Fe25Ti25V25 alloy could serve as a preferred site for corrosion attack to occur.

5. Acknowledgement

Dr B.A. Obadele would like to thank the NRF (KIC-NRF unique grant number 110134) for providing funding to present this paper.

6. References

[1] J.W. Yeh, et al, "Nano-structured high-entropy alloys with multiple principal elements: novel alloy design concepts and outcomes," Adv. Eng. Mater., vol. 6, no. 6, pp. 299–303, 2004.
[2] B. Cantor, I. Chang, P. Knight, A. Vincent, "Microstructural development in equiatomic multicomponent alloys," Mater. Sci. Eng. A, vol. 375, pp. 213–218, 2004.
[3] X.W. Qiu, Y.P. Zhang, L. He, C.G. Liu, "Microstructure and corrosion resistance of AlCrFeCuCo high entropy alloy," J. Alloys Compd, vol. 549, pp. 195–199, 2013.
[4] Z.X. Liu, D.M. Li, L. Shi, B. Cai, M.X. Wang, "Corrosion behavior of CuCrFeNiMn high entropy alloy system in 1 M sulfuric acid solution," Mater. Corros., vol. 63, pp. 828–834, 2012.
[5] D.H. Xiao, P.F. Zhou, W.Q. Wu, H.Y. Diao, M.C. Gao, M. Song, P.K. Liaw, "Microstructure, mechanical and corrosion behaviors of AlCoCuFeNi-(Cr,Ti) high entropy alloys," Mater. & Des., vol. 116, pp. 438-447, 2017.
[6] M.R. Chen, S.J. Lin, J.W. Yeh, M.H. Chuang, S.K. Chen, and Y.S. Huang, "Effect of vanadium addition on the microstructure, hardness, and wear resistance of Al0.5CoCrCuFeNi high-entropy alloy," Met. Mater. Trans. A, vol. 37, no. 5, pp. 1363-1369, 2006.
[7] W. Chen, Z. Fu, S. Fang, H. Xiao, and D. Zhu, "Alloying behavior, microstructure and mechanical properties in a FeNiCrCo0.3Al0.7 high entropy alloy," Mater. Des., vol. 51, pp. 854-860, 2013.
[8] K.E. Easterling, Phase Transformations in Metals and Alloys. Chapman & Hall, New York, 1993, pp. 314.
[9] Q. Ye, K. Feng, Z. Li, F. Lu, R. Li, J. Huang, and Y. Wu, "Microstructure and corrosion properties of CrMnFeCoNi high entropy alloy coating," Appl Surf. Sci., vol. 396, pp.1420-1426, 2017.

Synthesis of Ti-6Al-4V alloy with nano-TiN microstructure via spark plasma sintering technique

O E Falodun[1], B A Obadele[1], S R Oke[1], M E Maja[1] and P A Olubambi[1]

[1] Centre for Nanoengineering and Tribocorrosion, School of Mining, Metallurgy and Chemical Engineering, University of Johannesburg, Doornfontein Campus, Johannesburg, South Africa

Abstract. The effect of nano-TiN dispersion strengthened Ti-6Al-4V via spark plasma sintering method has been investigated. Ti-6Al-4V with 4 vol. percent of nano-TiN were mixed in a Turbula shaker mixer for 8 h at a speed of 49 rpm and the admixed powders were sintered at sintering temperature range of 1000 - 1100 °C, holding time of 10-30 mins, heating rate of 100 °C/min under an applied pressure of 50 MPa. The morphology of the as-received and sintered compacts was examined by scanning electron microscopy (SEM) equipped with energy dispersive X-ray spectroscopy (EDS) and phase analysis was done by X-ray diffractometry (XRD). The sintered compacts without nano-TiN reveal lamellar structure while reinforced Ti-6Al-4V with nano-TiN shows a bimodal structure and titanium nitride has a great influence on α grain growth at high temperature. Furthermore, the microstructural formation mechanism was investigated. With the addition of the content of Ti-6Al-4V with 4 vol.% of nano-TiN, the micro-hardness also improved and this was due to homogenous distribution of TiN in Ti-6Al-4V matrix.

1. Introduction

Powder metallurgy (PM) has gained attention over the conventional methods for the fabrication of composites reinforced with particles by applying new powder metallurgy processes such as using spark plasma sintering (SPS) [1-3]. The advantageous of powder metallurgy which includes better control on the microstructure, uniform dispersion of reinforced particle (second) in the sintered composite [3, 4]. The powder metallurgy technique has been proven for the production titanium alloys and their composites [5]. Furthermore, spark plasma sintering (SPS) method has been developed as a newly versatile technique to rapidly fabricate a number of materials including metals, composites and ceramics. [6-8]. Compared with other conventional methods of processing titanium alloys and composites, which takes longer time, required higher temperature and higher density is produced [9, 10]. However, longer time of sintering could cause grain growth, which is detrimental to the properties and microstructure of the composite material [11, 12]. Recent development in technology allows materials to be sintered with a reduction in time, temperature with the positive impact on the economic aspect of the fabrication of composite.

Titanium and its alloys are widely used due to their specific properties they possess like hardness, high specific strength and high corrosion resistance for various applications [13, 14]. Furthermore, Ti-6Al-4V remains the most widely used in various applications due to its extreme combination of mechanical

IOP Conf. Series: Materials Science and Engineering **272** (2017) 012029 doi:10.1088/1757-899X/272/1/012029

properties as well corrosion resistance Among the various titanium alloys, Ti-6Al-4V alloy is still the most commonly used one due to its favourable combination of mechanical properties and corrosion resistance [15]. Reinforcing titanium alloys with hard precipitates in sintered matrix composites such as titanium nitride helps with the improvement of corrosion and wear resistance of the materials. The aim of this research is to investigate the effect of reinforcement in the sintered materials in terms of increase in strength and hardness of titanium materials by developing a novel titanium matrix composite and dispersion-strengthened titanium material.

2. Experimental procedure

The materials used in the present composite production are a commercially pure Ti-6Al-4V powder (99.9%, 25 μm by TLS-Technik, Germany) and TiN (97%, 20 nm supplied by Nanostructured & Amorphous Material Inc., Texas, USA) were used as reinforcement phase. A composite mixture of Ti-6Al-4V was added to TiN while, turbula mixer was used to blend the powders according to varying ratios of 4 vol.% TiN for 8 h at a speed of 49 rpm so as to achieve a homogeneous mixture of titanium matrix composite. The mixed powders were sintered using a spark plasma sintering equipment (HHPD-25 FCT type SPS, Germany), the mixed powders were fed into a graphite die mould and then pressed with a pressure of 10 MPa. The die with an internal diameter of 30 mm and thickness of 5 mm was put into the SPS chamber machine. The sintering temperature range of 1000 - 1100 ^{o}C for 10 and 30 min, while the heating rate was 100 ^{o}C/min, and the applied pressure of 50 MPa used from the start to the end of the sintering operation. Sintered titanium composites produced were then characterized by XRD and SEM analyses.

The sintered density of the samples was carried out by Archimedes principle using an electronic balance with an accuracy of ±0.001 g. The micrograph of the synthesized composites was evaluated using field emission scanning electronic microscope (JEOL JSM-7600F) equipped with energy dispersive X-ray spectroscopy (EDS) at an accelerating voltage of 15 kV. An X-ray diffraction (PW1710 Philips diffractometer, with monochromatic Cu K radiation at 40 kV and 20 mA) was performed to identify the presence of the constituent phases of the sintered composites using an highscore X'Pert Software. Prior to observation, the surface of the sample was prepared by silicon carbide with the grit size of 320 and then polished with diamond paste to 0.2 μm. Kroll's reagent was used as an etching solution for 10–30 s in accordance with ASTM E3-11 ASTM standard [16] in order to reveal the grain boundaries of the sintered composites.

The hardness and fracture surface of the sintered composites were carried out in order to determine the level of mechanical properties. The micro indentation test was done by using a Vickers micro hardness tester (FALCON 500 series) at room temperature upon a polished surface at a load of 100 gf (0.98 N) and dwell time of 15 secs, while the fracture surfaces were observed by using a SEM to check the morphology.

3. Results and discussion

Figure 1a. present SEM micrograph of Ti-6Al-4V while Figure 1b demonstrates SEM morphology of admixed Ti–6Al–4V and TiN powders subjected to intensive, periodically pulsating motions (rotation, translation and inversion) [17]. It could be seen that TiN diffuses into Ti–6Al–4V. Turbula mixing of Ti–6Al–4V with nano-sized titanium nitride powder resulted in even dispersion within the titanium alloy powder. The effectiveness of mixing process is usually described by the homogeneity of the mixed powder [9].

IOP Conf. Series: Materials Science and Engineering **272** (2017) 012029 doi:10.1088/1757-899X/272/1/012029

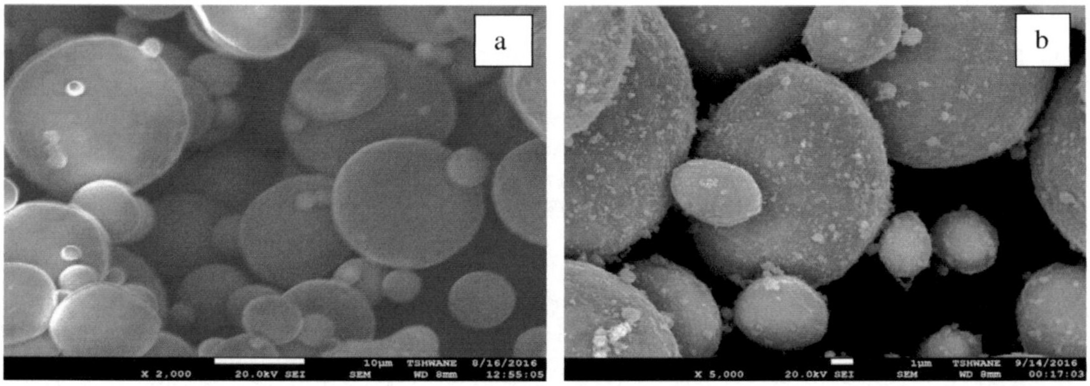

Figure 1. (a) SEM morphology of Ti–6Al–4V powder and (b) SEM morphology of Turbula mixed powders.

Figure 2 present the effect of sintering temperature and holding time on the displacement of punches during sintering and this demonstrates the deformation of the samples at an applied pressure of 50 MPa. Firstly, a sintering temperature of 250 – 300 °C was maintained constantly for 240 to 420 s. Thereafter, it was increased to 1000 °C and 1100 °C respectively and maintained for 600 to 900 s. At the initial stage, the displacement increases from 0.3 to 0.5 mm under the effect of applied pressure exacted on it, although the initial temperature was too low to have an effect on the material. The displacement observed was as a result of cold compaction of the powder. Displacement begins to increase as a result of rising in temperature thereby causing rearrangement of the particles. The effect of pressure becomes stronger which causes the escape of trapped air from the green compact. The densities of samples sintered at 1000 – 1100 °C for 10 and 30 mins respectively resulted in the theoretical density of 97–99% upon the addition of titanium nitride. The increase in densification with increase in temperature can further lead to the grain growth which therefore leads to gradual closure of pores in the material [18].

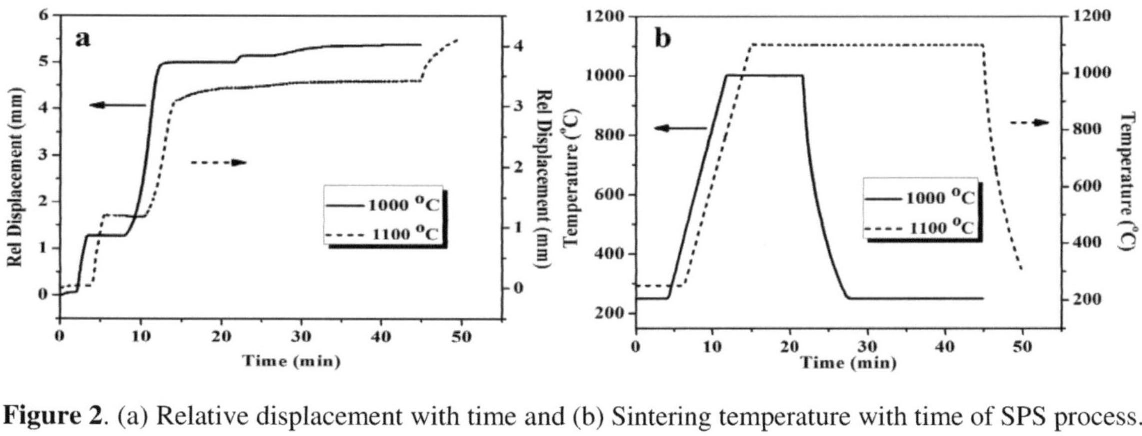

Figure 2. (a) Relative displacement with time and (b) Sintering temperature with time of SPS process.

The hardness value of all the composites was higher irrespective of the sintering temperature and time than Ti–6Al–4V alloy, which indicates that the addition of titanium nitride into titanium alloy really enhanced the hardness of the sintered composites from 389 to 531 $HV_{0.1}$. The microstructure contribution to the observed increase in hardness was as a result of the uniform distribution of TiN particles and Ti_2N hard phases in Ti–6Al–4V matrix. The increase in microhardness with increasing temperature and time may be attributed to decrease in porosity and increases in sintered density. Also, the increase can be explained by the increase in α volume fraction with hardness phase and presence of nitrogen in solid solution.

Figure 3 shows SEM microstructure of Ti–6Al–4V with an equiaxed structure with intergranular β and evidence of prior β grain boundaries, while figure 4 shows Ti–6Al–4V reinforced with 2 and 4 vol.% content of titanium nitride. The presence of TiN shows a relatively refined homogeneous distribution in the composite matrix which was located at the grain boundaries. Moreover, the α-nucleation sites show an increase when TiN particles were incorporated which further acted as a point of nucleation for the primary α-phase. Also obstructions which forestall those lamellae growth development in the microstructural transform. The introduction of nano-sized TiN in the titanium matrix is suggested to be responsible for the enhancement of Ti–6Al–4V alloy transforming to bimodal-structure. The composite were nearly free from porosity.

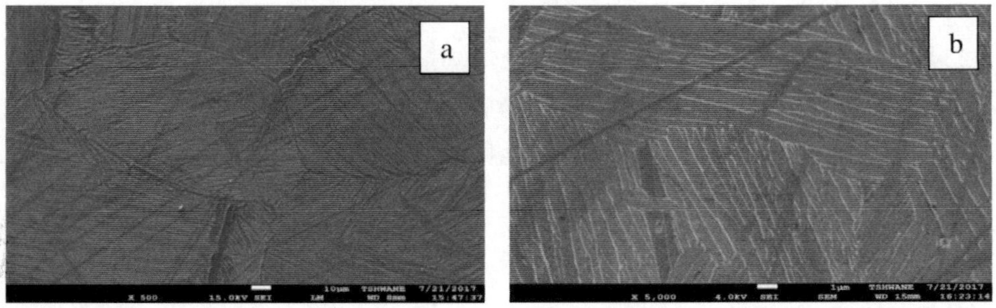

Figure 3. SEM micrograph of sintered Ti-6Al-4V at different magnifications.

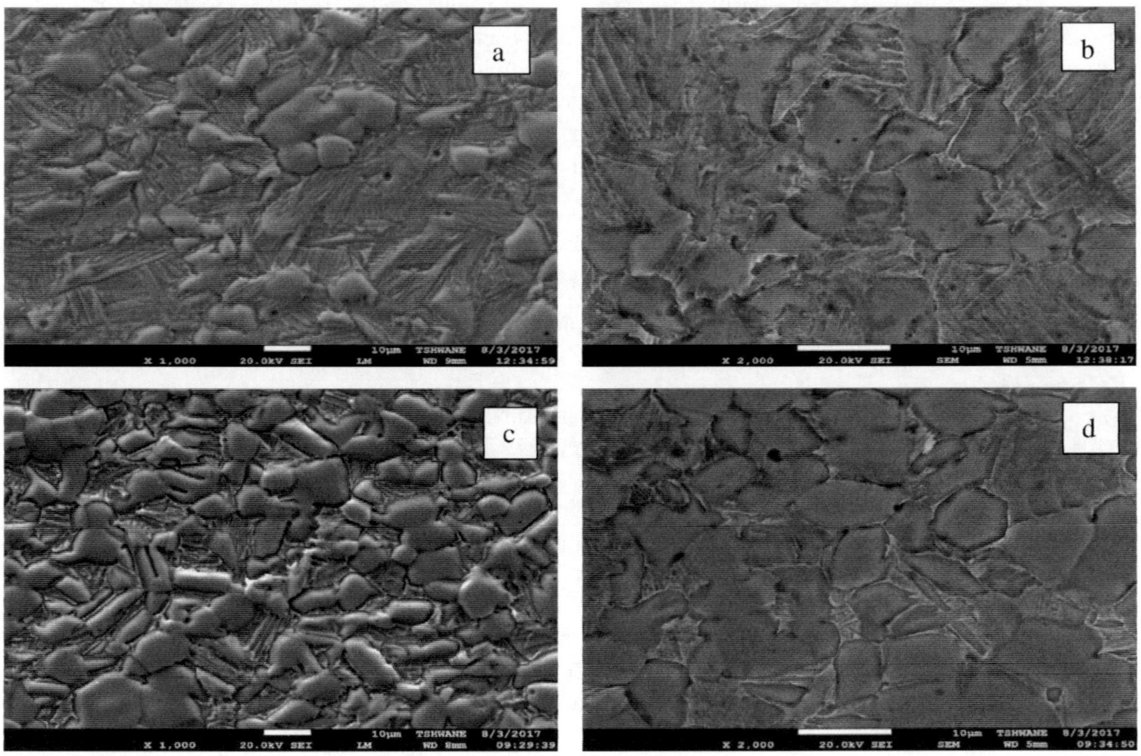

Figure 4. SEM images of Ti-6Al-4V with TiN: (a-b) 2 vol% TiN and (c-d) 4 vol% TiN.

Figure 5 reveals fracture morphology of the sintered Ti-6Al-4V with cleavage feature and dimples. Ti-6Al-4V reinforced with 4 vol.% TiN shows a transgranular pattern with fine dimples features and TiN grains thereby causing cleavage facets with a river pattern. This fracture mode remains a complex one. These features may be as a result of TiN addition and which may improve the fracture hardness of the sintered composites.

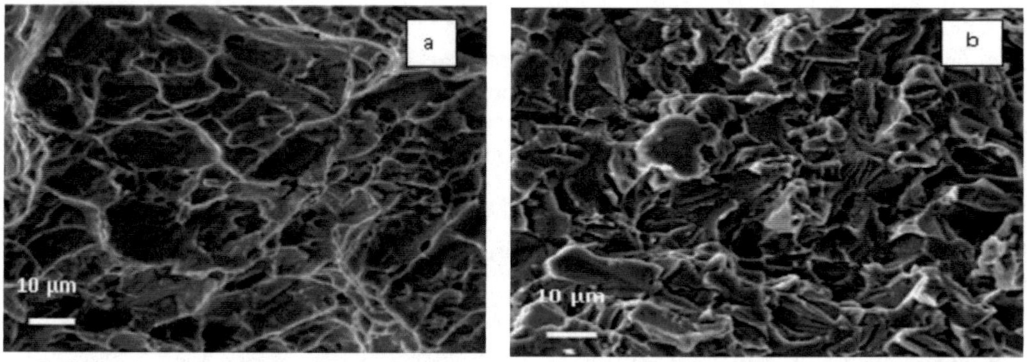

Figure 5. SEM fractographs of sintered (a) Ti-6Al-4V and (b) Titanium matrix composite with 4 vol.% TiN addition.

4. Conclusion

The sintering temperature and time have a critical impact on densification, microstructure, hardness and fracture surface of the sintered titanium matrix composites. Conclusions are as follows:

1. For these materials, the relative density was between 97.53 – 99.89%. Sintering at higher temperatures and time tends to promote the formation and growth of new phases in the sintered matrix composites and microstructural transformation from lamellae to bi-modal structures.
2. The microhardness of the sintered composites significantly increases from 389 HV and upon addition of 4 % volume fraction of TiN in the matrix increases to 531 $HV_{0.1}$.
3. An intergranular was revealed with Ti–6Al–4V which further transform within the titanium matrix composite to transgranular with fine dimples features.

5. Acknowledgement

The authors would like to acknowledge the Global and Excellence stature (GES), University of Johannesburg, South Africa and KIC-National Research Foundation for funding Oluwasegun Falodun.

6. References

[1] Ye LL, Liu ZG, Raviprasad K, Quan MX, Umemoto M, Hu ZQ. Consolidation of MA amorphous NiTi powders by spark plasma sintering. Materials Science and Engineering A. 1998;241:290-3.
[2] Tongsri R, Asavavisithchai S, Mateepithukdharm C, Piyarattanatrai T, Wangyao P. Effect of powder mixture conditions on mechanical properties of sintered Al2O3-SS 316L composites under vacuum atmosphere. Journal of Metals, Materials and Minerals. 2007;17:81-5.
[3] Rahimian M, Ehsani N, Parvin N, reza Baharvandi H. The effect of particle size, sintering temperature and sintering time on the properties of Al–Al 2 O 3 composites, made by powder metallurgy. Journal of Materials Processing Technology. 2009;209:5387-93.
[4] Li B, Liu Y, Li J, Cao H, He L. Effect of sintering process on the microstructures and properties of in situ TiB 2–TiC reinforced steel matrix composites produced by spark plasma sintering. Journal of Materials Processing Technology. 2010;210:91-5.
[5] Wang L, Lang Z-b, Shi H-p. Properties and forming process of prealloyed powder metallurgy Ti-6Al-4V alloy. Transactions of Nonferrous Metals Society of China. 2007;17:s639-s43.

IOP Conf. Series: Materials Science and Engineering **272** (2017) 012029 doi:10.1088/1757-899X/272/1/012029

[6] Zhang Y, Zhang J. Rapid reactive synthesis and sintering of textured Ca 3 Co 4 O 9 ceramics by spark plasma sintering. journal of materials processing technology. 2008;208:70-4.

[7] Feng K, Yang Y, Hong M, Wu J, Lan S. Intensified sintering of iron powders under the action of an electric field: Effect of technologic parameter on sintering densification. Journal of Materials Processing Technology. 2008;208:264-9.

[8] Feng H, Zhou Y, Jia D, Meng Q. Rapid synthesis of Ti alloy with B addition by spark plasma sintering. Materials Science and Engineering: A. 2005;390:344-9.

[9] Tang C, Wong C, Zhang L, Choy M, Chow T, Chan K, et al. In situ formation of Ti alloy/TiC porous composites by rapid microwave sintering of Ti6Al4V/MWCNTs powder. Journal of Alloys and Compounds. 2013;557:67-72.

[10] Chaudhari R, Bauri R. Microstructure and Mechanical Properties of Titanium Processed by Spark Plasma Sintering (SPS). Metallography, Microstructure, and Analysis. 2014;3:30-5.

[11] Yang F, Kong F, Chen Y, Xiao S. Effect of spark plasma sintering temperature on the microstructure and mechanical properties of a Ti 2 AlC/TiAl composite. Journal of Alloys and Compounds. 2010;496:462-6.

[12] Chen Y, Yu H, Zhang D, Chai L. Effect of spark plasma sintering temperature on microstructure and mechanical properties of an ultrafine grained TiAl intermetallic alloy. Materials Science and Engineering: A. 2009;525:166-73.

[13] Garbiec D, Siwak P, Mróz A. Effect of compaction pressure and heating rate on microstructure and mechanical properties of spark plasma sintered Ti6Al4V alloy. Archives of Civil and Mechanical Engineering. 2016;16:702-7.

[14] Koo MY, Park JS, Park MK, Kim KT, Hong SH. Effect of aspect ratios of in situ formed TiB whiskers on the mechanical properties of TiB w/Ti–6Al–4V composites. Scripta materialia. 2012;66:487-90.

[15] Wang X, Ma X, Nie Q, Wang M. Effects of Y addition on microstructure and mechanical properties of TiC/Ti6Al4V composites. Intermetallics. 2012;31:242-8.

[16] Standard A. E384-11E1,—Standard test method for Knoop and Vickers hardness of materials,‖ in ASTM International. West Conshohocken, PA. 2007.

[17] Obadele BA, Masuku ZH, Olubambi PA. Turbula mixing characteristics of carbide powders and its influence on laser processing of stainless steel composite coatings. Powder Technology. 2012;230:169-82.

[18] Gülsoy H, Özbey S, Pazarlioglu S, Çiftci M, Akyurt H. Sintering and Mechanical Properties of Titanium Composites Reinforced Nano Sized Al2O3 Particles 2015.

IOP Conf. Series: Materials Science and Engineering **272** (2017) 012030 doi:10.1088/1757-899X/272/1/012030

Influence of Sintering Temperature on Hardness and Wear Properties of TiN Nano Reinforced SAF 2205

S R Oke[1,2], O O Ige[1,3], O E Falodun[1], B A Obadele[1], M R. Mphalele[1] and P A Olubambi[1]

[1] Centre for Nanoengineering and Tribocorrosion, School of Mining, Metallurgy and Chemical Engineering, University of Johannesburg, Doornfontein Campus, Johannesburg, South Africa

[2] Department of Metallurgical and Materials Engineering, Federal University of Technology Akure, Ondo State, PMB 704, Nigeria

[3] Department of Materials Science & Engineering, Obafemi Awolowo University Ile-Ife, Osun State, Nigeria

Abstract. Conventional duplex stainless steel degrade in wear and mechanical properties at high temperature. Attempts have been made by researchers to solve this problems leading to the dispersion of second phase particles into duplex matrix. Powder metallurgy methods have been used to fabricate dispersion strengthened steels with a challenge of obtaining fully dense composite and grain growth. This could be resolved by appropriate selection of sintering parameters especially temperature. In this research, spark plasma sintering was utilized to fabricate nanostructured duplex stainless steel grade SAF 2205 with 5 wt.% nano TiN addition at different temperatures ranging from 1000 °C to 1200 °C. The effect of sintering temperature on the microstructure, density, hardness and wear of the samples was investigated. The results showed that the densities and grain sizes of the sintered nanocomposites increased with increasing the sintering temperature. The microstructures reveal ferrite and austenite grains with fine precipitates within the ferrite grains. The study of the hardness and wear behaviors, of the samples indicated that the optimum properties were obtained for the sintering temperature of 1150 °C.

1. Introduction

Duplex stainless steels are widely used in aerospace, chemical, power generation and biomedical and nuclear industries because of high corrosion resistance, good mechanical strength and ductility, abrasion resistance, erosion resistance and a very good weldability [1-3]. However, these alloys may suffer considerable loss in wear and mechanical strength at elevated temperatures. One promising approach to improve the elevated temperature properties as well as wear and oxidation properties of stainless steels is by reinforcing it with second phase particles [4].

One of the processing methods that can be explored to fabricate dispersion strengthened stainless steels is powder metallurgy (PM) [5]. Hot extrusion (HE), hot isostatic pressing (HIP) and mechanically alloying (MA) are the most frequently used PM processes for the fabrication of dispersion strengthened composites [6-7]. Relatively, spark plasma sintering (SPS) have been reported to be a better technique due to its low energy consumption and very short sintering time. SPS utilizes a pulsed direct electrical current to perform high speed consolidation of powders. Rapid high heating and cooling rates which characterizes SPS process allow maintaining and improvement of the inherent properties of powders in their fully dense products [8-10]. SPS has been used to fabricate nanostructured composites, thermoelectric materials, ferroelectric materials, magnetic materials and nanograined materials [11].

Sintering temperature is an important process parameter in the consolidation of powders during SPS. Its appropriate selection could be used to control density, grain/crystallite size and porosity of PM parts which may improve mechanical properties [12]. Javanbakht et al., [13] investigated the effect of sintering temperature on structure and mechanical properties of AISI 316L stainless steel. The authors reported that residual porosity reduced with increasing sintering temperature while crystallite/grain sizes were enhanced. The optimum mechanical properties were obtained for the sintering temperature of 1150 °C. In a related study by Li et al. [14], with the increasing of sintering temperature from 950 °C to 1025 °C, the density and average grain sizes of 14Cr-ODS ferritic steels increased from 290 nm to 2420 nm. Sorour et al. [15] investigated the microstructure and tribology of Fe-Cr-B alloy sintered by SPS. They sinter at 1150 °C at a heating rate of 225 °C/min, holding time of 10 min and were able to achieve a hardness value of 991 HV with this alloy.

However, the effects of sintering temperature on nanostructured stainless steels have received attention in recent time. In this study, duplex stainless steels grade SAF 2205 and nano TiN was consolidated via SPS. This report documents a study of the effects of spark plasma sintering temperature on the micro-structure, hardness and wear of TiN nano structured SAF 2205 steel.

2. Experimental procedure

DSS 2205 (particle size about 22μm, supplied by Sandvik Osprey Ltd, UK) and TiN (particle size of 20nm, 97% purity, supplied by Nanostructured & Amorphous Materials, Inc.USA) powders were utilized as the starting materials for this research. The powders were mixed in a dry environment using the Turbula Shaker Mixer T2F at a mixing speed of 72rpm for 8h to ensure homogeneous mixing of powders.

The appropriate amount of mixed powders required to produce 20mm diameter and 5mm thick DSS-TiN nano structured composites were poured into a graphite die and then sintered using an automated spark plasma sintering machine (model HHPD-25, FCT GmbH Germany).

The phases present in the sintered specimen were characterized by X-ray diffraction (XRD) using a PANalytical Empyrean model with Cu Kα radiation and analyzed using Highscore plus software. The samples were etched using carpenters reagent for 20s and the microstructure was analyzed using a scanning electron microscope (FESEM, JSM-7600F, JEOL, Japan) incorporated with an EDX detector (Oxford X-Max) with INCA X-Stream2 pulse analyzer software, and Back Scattered Electron (BSE) detectors.

The Vickers microhardness (Hv) at room temperature was measured by a microhardness tester (Future-tech) at a load (P) 100 gf (1.0 N) and a dwell time of 10 s. Three repeat tests were carried out on the samples to ensure reliability of the data generated. The wear tests were conducted on the universal micro-tribotester (UMT-2, Center for Tribology, Inc.). The samples were fixed onto the precision linear stage. A reciprocating motion based on a crank-slider mechanism was exerted on the fixed samples by a ball under dry condition. The frictional force was measured by the load sensor integrated to the side of the ball holder of the tribotester. A load (P) of 5 N was selected; meanwhile, the sliding length (Ls) and the time of linear reciprocating sliding were fixed to 2 mm and 1000s, respectively.

3. Results and discussion

Figure 1a shows that the nano particles of the TiN were evenly distributed in the 2205 matrix. The XRD result (Figure 1b) of the mixed powders shows that only peaks corresponding to TiN and phases containing iron, chromium, and nickel which are the main constituent of DSS 2205 phases are detected in the patterns. The XRD result of the sintered composites with sintering temperature is presented in figure 2. It is observed from the XRD result that there phase change, and a new phase of

IOP Conf. Series: Materials Science and Engineering **272** (2017) 012030 doi:10.1088/1757-899X/272/1/012030

iron nitride with chemical formula $FeN_{0.068}$ was formed. There was slight change in relative intensity of peaks which is obviously due to the difference in sintering parameters of temperature.

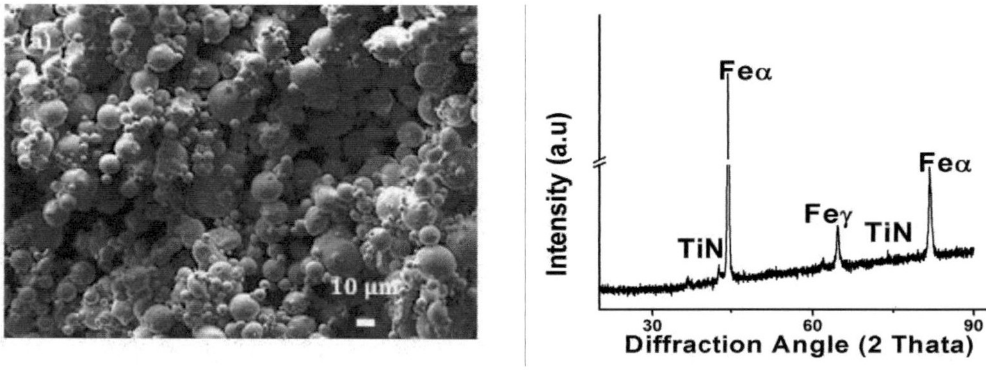

Figure 1. SEM (a) and XRD (b) results of the mixed 2205 duplex stainless steel and TiN powders.

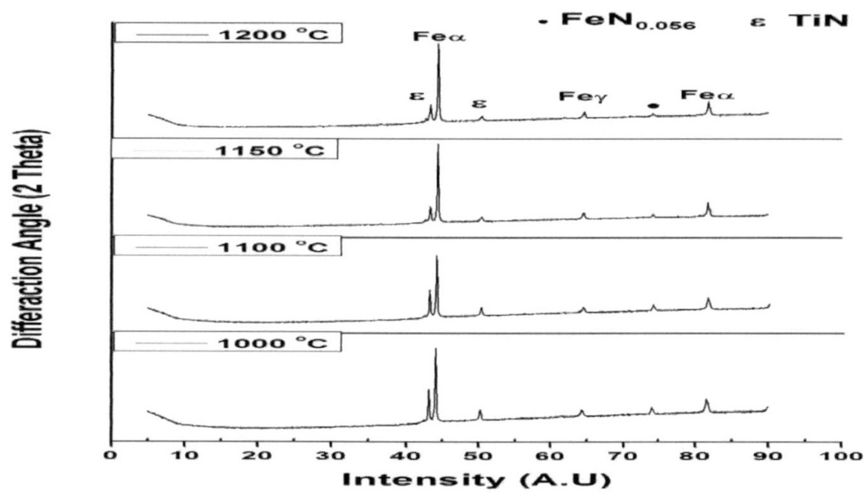

Figure 2. XRD result of sintered DSS 2205-5TiN composite at different temperatures.

The effect of sintering temperatures on the microstructure of the TiN nanostructure SAF 2205 composites is shown in Figure 3. The microstructure consists of austenite and ferrites. The reinforcing TiN phase was uniformly distributed in the duplex matrix. The grains were also observed to grow with increasing sintering temperature. At 1150 °C (Figure 3c), fine nitride precipitates were detected within the grains and grain boundaries with a higher packing density and an increase in the number of necks which indicates improved sintering. The presence of these precipitates accumulated within the grain and grain boundaries could significantly retard the grain boundary mobility and dislocation movement [5].

Figure 4 presents the density and micro-hardness results with increase in sintering temperature. The figure shows that the density and micro-hardness increased with sintering temperature up to 1150°C. The increase in densification could be attributed to improved powder contact and decrease in porosity level obtained at 1150°C. A reduction in density and hardness was however noted with an increase in sintering temperature to 1200 °C. The decrease in densification and hardness at 1200°C may be as a result of localized melting of powders. Toor et al., [16] optimized process parameters for Fe-18Cr-2Si, the reported a decrease in density and hardness at sintering temperature 1200 °C. The authors suggested that partial melting of stainless steel powders could occur at 1200 °C.

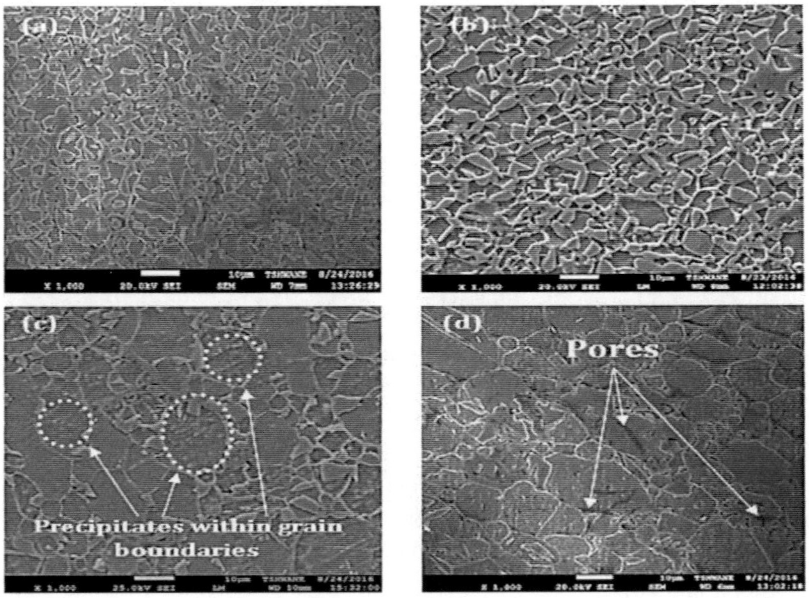

Figure 3. Microstructures of composites with 5 % TiN sintered at (a) 1000 (b) 1100 (c) 1150 and (d) 1200 °C.

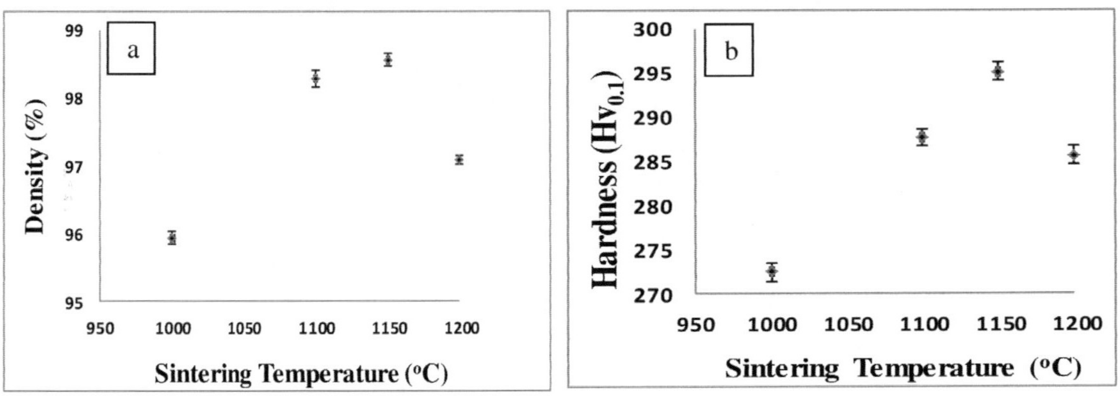

Figure 4. Effect of sintering temperature variation on the (a) densification and (b) hardness of SAF 2205-5TiN.

Figure 5 presents the influence of sintering temperature on coefficients of friction and wear depths of the TiN reinforced SAF 2205 composites over a period of 1000 s. Related frictional behaviors were noticed irrespective of temperature except for the grade sintered at 1000 °C which shows a high degree of instability. The coefficients of friction (Figure 5a) slightly decreased with increased sintering temperature. However, the coefficients of friction of the grades sintered at 1100 °C and 1150 °C were much lower than those of other frictional pairs and were maintained almost constant throughout the testing time. Upon increasing the sintering temperature (1200 °C), the mean coefficients of friction slightly increased from 0.02 to 0.03. Wear depths as shown in figure 5b linearly increased with increasing time. The composite grade sintered at 1150 °C exhibited the highest wear resistance. The low coefficient of friction observed for the composite sintered at 1150 °C could be as a result of its high hardness and densification reported in figure 4.

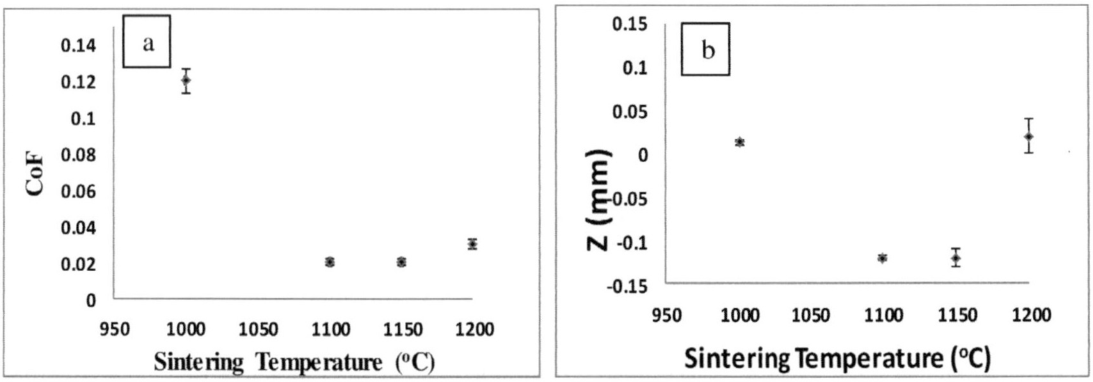

Figure 5. Variation of (a) COF and (b) wear depths of spark plasma sintered SAF 2205-5%TiN powders under different sintering temperature.

The worm surface as observed under the SEM for the grade sintered at 1150 °C is presented in Fig. 6. A series of parallel smooth grooves (Figure 6a) extending longitudinally toward the sliding direction is observed while the material is displaced to the edges, forming a topography that can be described as extruded ridges [17]. The smooth parallel grove observed is obviously due to plastic flow and daubed at the wear track by the sliding friction, which is in agreement with other studies [18].

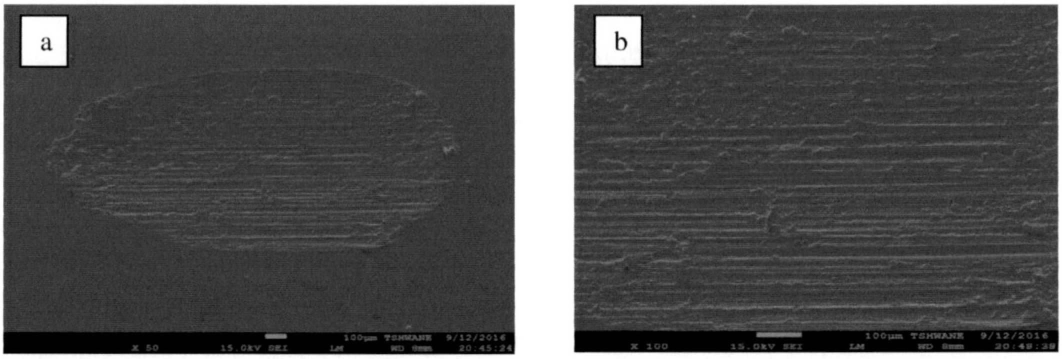

Figure 6. SEM images of the worn surface of the SAF 2205-5TiN composites at 1150 °C.

4. Conclusion

DSS 2205-5%TiN composite had been synthesized by SPS. SPS process parameters were optimized for the composites in terms of their effect on density, hardness and Wear. The results can be concluded as follows;

1. The density and hardness increased with increase in sintering temperature and holding time with maximum densification and hardness achieved at 1150°C
2. The grains were noticed to increase with increasing sintering temperature with fine nitride precipitates within the ferrite grains.
3. For wear, grade sintered at 1150 °C has the highest wear resistance compared to other nano-composite grades.

5. Acknowledgement

The authors would like to extend appreciation to National Research Foundation (NRF) of South Africa and the Global Excellence and Stature of the University of Johannesburg, South Africa for their financial support.

IOP Conf. Series: Materials Science and Engineering **272** (2017) 012030 doi:10.1088/1757-899X/272/1/012030

6. References

[1] T. Karahan, H. E Emre, M Tümer, R. Kaçar, "Strengthening of AISI 2205 duplex stainless steel by strain ageing'', Mater. Des. Vol. 55, pp. 250–256, 2014.
[2] L.A. Dobrzanski, Z. Brytan, M. Actis Grande, M. Rosso, "Properties of duplex stainless steels made by powder metallurgy'', Arch. Mater. Sci. Eng. Vol. 28, pp. 217–223, 2007.
[3] G. Chail, P. Kangas, "Super and hyper duplex stainless steels: structures, properties and applications'', Procedia Structural Integrity, vol. 2, pp. 1755-1762, 2016.
[4] A .Fedrizzi, M. Pellizzari, M. Zadra, E. Marin, "Microstructural study and densification analysis of hot work tool steel matrix composites reinforced with TiB_2 particles'', Mater. Charact. Vol. 86, pp. 69– 79, 2013.
[5] M. Javanbakht, M. Hadianfard, E. Salahinejad, "Microstructure and mechanical properties of a new group of nanocrystalline medical-grade stainless steels prepared by powder metallurgy'', J. Alloys Compd. Vol. 624, pp. 17–21, 2015.
[6] S.R. Oke, O.O. Ige, O.E. Falodun, B.A. Obadele, M.B. Shongwe, P.A. Olubambi. "Optimization of process parameters for spark plasma sintering of nano structured SAF 2205 composite'', J. of Materials Research and Technol, http://dx.doi.org/10.1016/j.jmrt.2017.03.004, 2017
[7] S. Cheng, E. Ma, Y. Wang, L. Kecskes, K. Youssef, C. Koch, "Tensile properties of in situ consolidated nanocrystalline Cu'', Acta Mater. Vol. 53 1521–1533, 2005.
[8] Z..H. Zhang, Z.F. Liu, J.F. Lu, X.B. Shen, F.C. Wang, Y.D. Wang, "The sintering mechanism in spark plasma sintering – Proof of the occurrence of spark discharge'' Scripta Materialia, vol. 81, pp. 56–59, 2014.
[9] S. Diouf, A. Molinari, "Densification mechanisms in spark plasma sintering: Effect of particle size and pressure'', Powder Technology, vol. 221, pp. 220 –227, 2012.
[10] R. Marder, R. Chaim, G. Chevallier, C. Estournes, "Densification and polymorphic transition of multiphase Y2O3 nanoparticles during spark plasma sintering'', Materials Science and Engineering A, vol. 528, pp. 7200-7206, 2011.
[11] I. Sulima, P. Putyra, P. Hyjek, T. Tokarski, "Effect of SPS parameters on densification and properties of steel matrix Composites'', Advanced Powder Technology, vol. 26, pp. 1152–1161, 2015.
[12] G. Ji, F. Bernard, S. Launois, T. Grosdidier, Processing conditions, microstructure and mechanical properties of hetero-nanostructured ODS FeAl alloys pro-duced by spark plasma sintering, Mater. Sci. Eng. A vol. 559, pp. 566 – 573, 2013.
[13] M. Javanbakht, E.Salahinejad, M.J. Hadianfard, "The effect of sintering temperature on the structure and mechanical properties of medical-grade powder metallurgy stainless steels'' Powder Technology, vol. 289, pp. 37–43, 2016.
[14] Z. Li, Z. Lu, R. Xie, C. Lu, C. Liu, "Effect of spark plasma sintering temperature on microstructure and mechanical properties of 14Cr-ODS ferritic steels'', Materials Science & Engineering A, vol. 660, pp. 52– 60, 2016.
[15] A.A. Sorour, H.W. Strauss, R.R. Chromik, M. Brochu, Microstructure and tribology of spark plasma sintered Fe–Cr –B metamorphic alloy powder, Tribol. Lett. Vol. 44, pp.269 –278, 2011.
[16] I. Toor, J. Ahmed, M.A. Hussein, N. Al-Aqeeli, "Optimization of process parameters for spark plasma sintering of nano-structured ferritic Fe-18Cr-2Si alloy'', Powder Technology, vol. 299, pp. 62–70, 2016.
[17] A. Jourani, S. Bouvier S, "Friction and Wear Mechanisms of 316L Stainless Steel in Dry Sliding Contact: Effect of Abrasive Particle Size", Tribology Transactions, vol. 58, no. 1, pp. 131-139, 2015.
[18] P. Tian, N.W. Khun, S.B. Tor, E. Liu, Y. Tian, "Tribological behavior of Zr-based bulk metallic glass sliding against polymer, ceramic, and metal materials'', Intermetallics, vol. 61, pp. 1-8, 2015.

IOP Conf. Series: Materials Science and Engineering **272** (2017) 012031 doi:10.1088/1757-899X/272/1/012031

Effect of SO_2 concentration as an impurity on carbon steel corrosion under subcritical CO_2 environment

MGR Mahlobo[1], K Premlall[1] and PA Olubambi[2]

[1]Chemical, Metallurgical and Materials Engineering Department, Tshwane University of Technology, Pretoria, 0001, South Africa

[2]Chemical Engineering Department, University of Johannesburg, Johannesburg, 2000, South Africa

Abstract. Carbon dioxide (CO_2) is considered to be easier to transport over moderate distances when turned into supercritical state (dense phase) than at any other state. Because of this reason, the transportation of CO_2 during carbon capture and storage requires CO_2 to be at its supercritical state. CO_2 temperature profile from different regions causes CO_2 to deviate between supercritical and subcritical state (gas/liquid phase). In this study the influence of sulphur dioxide (SO_2) on the corrosion of carbon steel was evaluated under different SO_2 concentrations (0.5, 1.5 and 5%) in combination with subcritical CO_2. Scanning Electron Microscopy (SEM), X-Ray Diffraction (XRD) and Energy-Dispersive X-ray Spectroscopy (EDS) were used to characterize the CO_2 corrosion product layer formed on the carbon steel surface. The weight loss results showed that corrosion rate increased with SO_2 concentration with corrosion rate up to 7.45 mm/year while at 0% SO_2 the corrosion rate was 0.067 mm/year.

1. Introduction

Carbon capture and storage (CCS) is a recent technology that has been developed to mitigate gradual increase of greenhouse gases by capturing CO_2 from the emission point and storing it into geological media [1-9]. During the process of CCS the captured CO_2 gas is compressed to the dense phase or supercritical state where temperature and pressure are over 31.1°C and 74.14 bar, respectively, in order to make it easier and less costly to transport [10-11]. According to the CO_2 phase diagram calculated by [11], the two-phase or subcritical CO_2 phase exists between pressures 7 to 73 bar and temperatures -50°C to 30.978°C. The variation of underground properties increases the likelihood of the subcritical CO_2 phase in the pipeline and in the presence of sulfur dioxide (SO_2) as one of the emitted CO_2 constituents, a corrosive environment can prevail. While most reported work have focused on the supercritical CO_2 environment during pipeline transportation, the necessity to understand the subcritical CO_2 environment has gained large interest from other researchers. [4] & [12] reported that the present free H_2O in the pipeline will react with SO_2 to form a corrosive sulfuric acid which will lead to the passivation of the steel pipeline surface.
Subcritical CO_2 thus presents a new challenge in the CCS operation and the current study was aimed at addressing this challenge by identifying carbon steel corrosion potential under subcritical CO_2 in the presence of various SO_2 concentrations as an impurity. Weight loss method was used to carry out corrosion experiments through a high pressure reactor. Scanning electron microscope (SEM) and X-Ray Diffraction (XRD) were used to study the morphology of the corroded carbon steel surface.

2. Experimental procedure

2.1. Materials

Carbon steel specimen with the chemical elemental composition (weight %) of 0.044 C, 0.001 Mo, 0.237 Mn, 0.035 Al, 0.025 P, 0.0007 V, 0.009 S, 0.0001 Nb, 0.01 Si, 0.001 Ti, 0.01 Cu, 0.0002 B, 0.005 Ni, 0.0038 N, 0.01 Cr and Fe balance was cut to 16 × 10 × 5 mm and only 16 × 10 mm exposed to the corrosion media (subcritical CO_2/SO_2 environment) as the tested surface inside the high pressure reactor (shown in figure 1). The cut samples were grinded to 120 - 1200 grits and polished with the 1 μm using silicone discs. Prior to the experiments, the high pressure reactor was first deoxygenated by purging with N_2 for 24 hours and denitrogenated by purging pure CO_2 into the reactor for 4 hours. Deionized water was added to adjust the H_2O concentration of 1000 ppmv (0.1%) which was well above the H_2O concentration (50 ppmv) needed to enhance the corrosion process inside the steel pipeline [13]. The polished carbon steel samples were rinsed, dried, weighed (precision of 0.0001 g) and stored in a desiccator. CO_2 was charged into the reactor at constant 60 bar and 25°C (subcritical region). Different SO_2 concentrations were added into the subcritical CO_2 (as shown in Table 1), carbon steel samples were then exposed to this environment for a period of 336 hours. The chemical composition of the carbon steel used in this study is shown in table 1 with iron (Fe) being the balance.

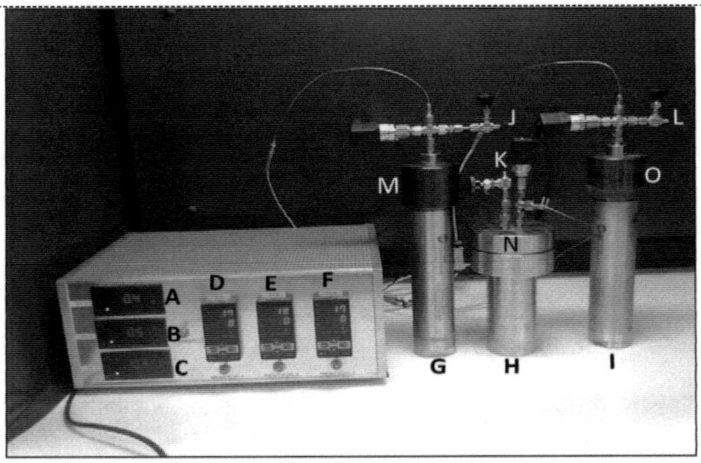

Figure 1. High pressure reactors (bolted and OC reactors).

Table 1. Experimental conditions

Sample #	Pressure (bar)	Temperature (°C)	Exposure Time (hours)	CO_2 (%)	SO_2 (%)
1	60	25	336	100	0
2	60	25	336	99.5	0.5
3	60	25	336	98.5	1.5
4	60	25	336	95	5

2.2. Steel Surface Characterization

A 5 mega pixel (auto focus) Lumia 520 digital camera was used to take photographic images of the carbon steel specimen at low magnification. After the experiments, the samples were removed from the reactor and weighed again for the weight loss calculations using the weight loss given in equation (1). The steel surface morphology analysis was carried out using JEOL JSM-7600F Field Emission Scanning Electron Microscopy (SEM) and the phase analysis was carried out using EMPYREAN PANalytical X-Ray Diffraction (XRD). After characterization the samples were washed with a concentrated hydrochloric acid and acetone to remove the corrosion products on the steel surface in order to identify the type of corrosion that occurred.

$$CR = 8.76\times 10^{4}\frac{\Delta m}{\rho \cdot A \cdot t} \qquad (1)$$

where CR is the corrosion rate, mm/year; Δm is the weight loss of the specimen, g; ρ is the density of the specimen, g/cm^3; *A* is the area of the specimen, cm^2; t is the immersion time, hour.

3. Results

3.1. Weight loss Results

Figure 1 shows the relationship between corrosion rate of carbon steel and SO_2 concentration at the specified operating conditions. Figure 2 shows that the corrosion rate increased significantly when 0.5% SO_2 was introduced into the system which saw the corrosion rate to be 1.84 mm/year. At this point it is already observed that SO_2 has a huge impact on the corrosion behaviour in comparison to the pure CO_2 scenario. This becomes more evident when the SO_2 concentration is increased to 1.5% further to 5% which gave corrosion rate values of 3.64 mm/year and 7.45 mm/year. A linear trend is observed on the corrosion rate versus SO_2 concentration. The mechanism behind SO_2 increasing corrosion rate so drastically can be explained by the formation of $FeSO_3$ when SO_2 is added in the supercritical CO_2 as shown in equations (2) & (3):

$$SO_2 + H_2O \rightarrow SO_3^{2-} + 2H^{+} \qquad (2)$$

$$Fe^{2+} + SO_3^{2-} \rightarrow FeSO_3 \qquad (3)$$

$FeSO_3$ is one of the corrosion products that form on the steel surface when SO_2 is added therefore increasing the corrosion rate drastically [12]. On the experiments conducted by [12] at 80 bar, 50°C and 24 hours exposure using a carbon steel specimen, the corrosion rate increased dramatically from 0.38 mm/year to 5.6 mm/year when 1% SO_2 (0.8 bar SO_2 partial pressure) was added. The corrosion behaviour upon SO_2 addition is similar to the one obtained in the current study, the highest corrosion rate obtained by [12] was 5.6 mm/year while in this study is 7.45 mm/year indicating similar trends on SO_2 effect towards corrosion rate.

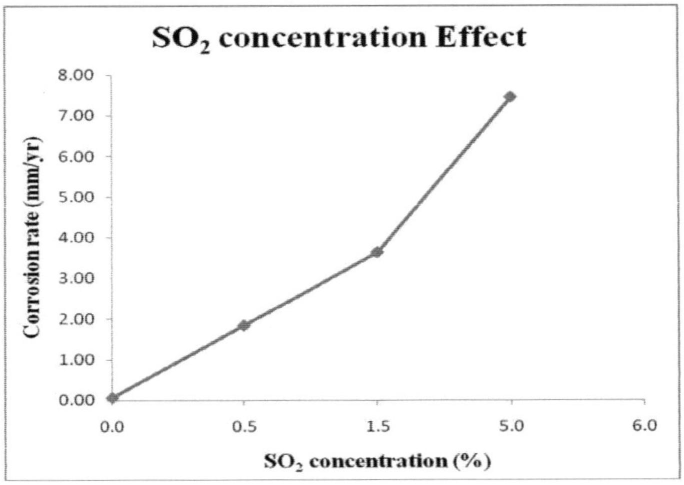

Figure 2. Relationship between corrosion rate and SO_2 concentration under subcritical CO2 environment.

3.2. Photographic Images on Steel Surface

The type of corrosion attack that took place on the carbon steel surface was identified after corrosion product removal. The green compact scale observed on the samples exposed to subcritical CO_2/SO_2, i.e. images a) to d) from figure 3, is believed to be iron sulphite hydrate which is in line with the observation made by [8]. Uniform corrosion was observed on the carbon steel sample exposed to pure subcritical CO_2 (0% SO_2), i.e. image e) from figure 3. This is a similar observation made by [14] after exposing carbon steel to the environment that is almost similar to the one used in this study, i.e. supercritical CO_2 (100 bar and 25°C) with 1220 ppmv H_2O, 0 ppmv SO_2 for 336 hours.

The introduction of SO_2 into the subcritical CO_2 environment appeared to have caused pitting corrosion as pit areas started to be visible when the SO_2 concentration in subcritical CO_2 was increased from 0 to 0.5% (image f) from figure 3). [14] also observed pit spots on the carbon steel surface when he increased the SO_2 concentration from 0 to 344 ppmv in the supercritical CO_2 with 1220 ppmv H_2O and 336 hours of exposure period. The introduction of SO_2 into the subcritical CO_2 environment provided the sulphite ions, which are very aggressive to passive metal surfaces generally consisting of chromium and nickel such as, the carbon steel used in this study. [4] also observed pitting corrosion in the presence of sulphate, sulphite and thiosulphate or sulphide.

3.3. Surface Morphology

Figure 4 shows the SEM micrograph of the carbon steel sample exposed to subcritical CO_2 with different SO_2 concentrations. The SEM image in figure 4 A) shows carbon steel microstructure exposed to pure subcritical CO_2 with neither SO_2 (0% SO_2) nor any impurity present. A slight metallic luster was observed on the steel surface exposed to pure subcritical CO2 (figure 4 a)) which suggested that the corrosion scale did not form on the entire carbon steel surface. This agrees well with the low corrosion rate (0.07 mm/year) obtained from the weight loss results under pure subcritical CO_2 with 0% SO_2. Furthermore, the introduction of SO_2 as an impurity in the subcritical CO_2 environment appears to be even more destructive than pure CO_2 as far as corrosion is concerned. The SEM images shown in figure 4 b), c) and d) when SO_2 concentration was increased to 0.5, 1.5 and 5% respectively also revealed that the increase in SO_2 concentration leads to the formation of more corrosion products ultimately destructing carbon steel surface. It can also be observed that corrosion scale on the carbon steel surface becomes more compact with the increase in SO_2 concentration with pit area starting to be visible on the surface exposed to 5% SO_2. [8] presented SEM images that showed a similar corrosion

IOP Conf. Series: Materials Science and Engineering **272** (2017) 012031 doi:10.1088/1757-899X/272/1/012031

behavior but much stronger steel surface deformation as he operated at supercritical (100 bar) rather than subcritical (60 bar) CO_2 with SO_2 concentrations of 0.2%, 0.7%, 1.4% and 2.0%.

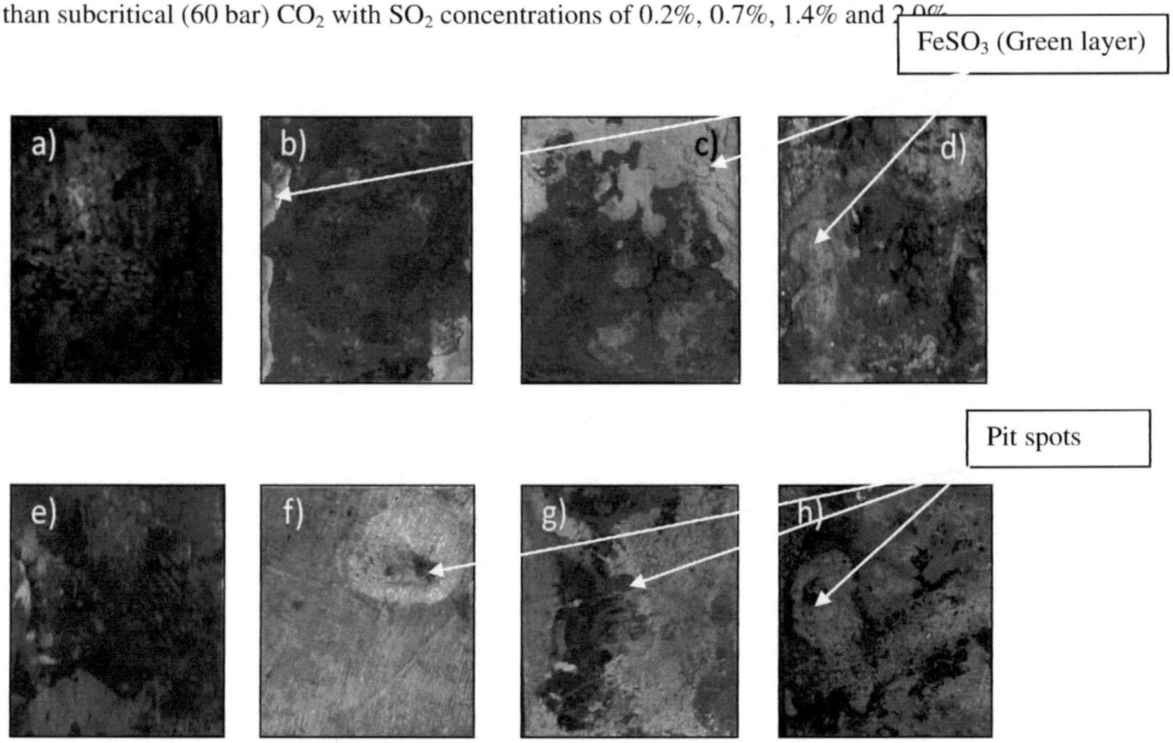

Figure 3. Photographic images (20 cm) of carbon steel surface exposed to subcritical CO_2 before corrosion products removal at; a) 0% SO_2, b) 0.5% SO_2, c) 1.5% SO_2 & d) 5% SO_2. After corrosion products removal at; e) 0% SO_2, f) 0.5% SO_2, g) 1.5% SO_2 & h) 5% SO_2

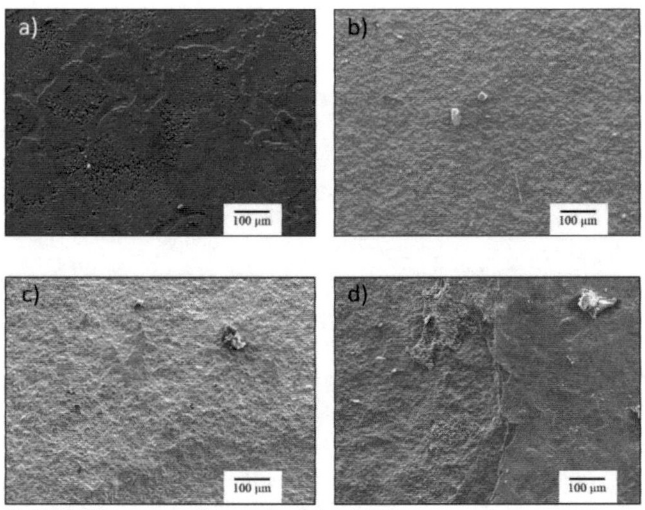

Figure 4. SEM images after subcritical CO_2 corrosion experiments with; a) 0% SO_2, b) 0.5% SO_2, c) 1.5% SO_2 & d) 5% SO_2.

3.4. Phase Analysis

Figure 5 shows the XRD spectra of the carbon steel samples exposed to subcritical CO_2 environment with different SO_2 concentrations. Iron sulphite hydrate ($FeSO_3 \cdot 3H_2O$) was detected in all tested SO_2 concentrations which confirms the observation made on the photographic images of the carbon steel samples at low magnification before corrosion product removal. The formation of $FeSO_3 \cdot 3H_2O$ on the carbon steel surface was due to the presence of SO_2 in the subcritical CO_2 environment and can be explained by equation (4) [8]:

$$Fe + SO_2 + (x+1)H_2O \rightarrow FeSO_3 \cdot xH_2O + H_2 \quad (4)$$

The XRD analysis conducted by [4], [8] & [10] under supercritical CO_2 with different SO_2 concentrations reported $FeSO_3 \cdot 3H_2O$ as the major component on the tested steel surfaces. The dominance of $FeSO_3 \cdot 3H_2O$ can be explained by the high SO_2 solubility at high pressures (supercritical state) leading to greater production of sulphite ions [8].

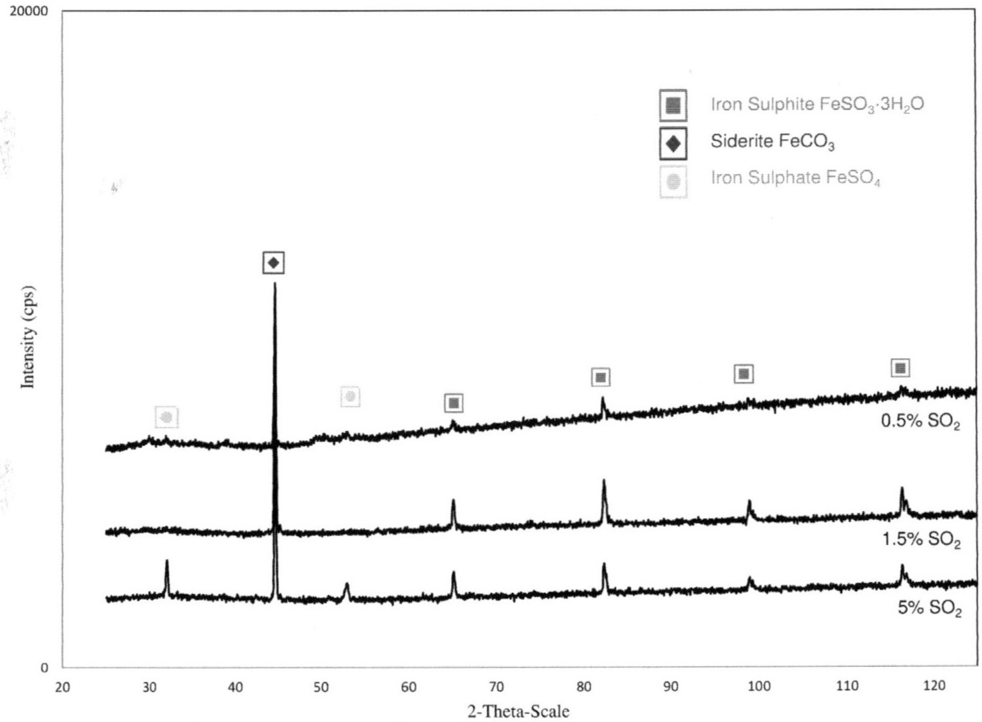

Figure 5. Superimposed diffractions of 0.5%, 1.5% and 5% SO_2 under subcritical CO_2 environment.

4. Conclusion

- An increase in SO_2 concentration in subcritical CO_2 environment rapidly increases the corrosion rate of carbon steel.
- In the presence of SO_2 in the CO_2 environment, $FeSO_3 \cdot 3H_2O$ was observed to be another significant component as one of the corrosion products under subcritical CO_2 environment.
- In the absence of SO_2 in the subcritical CO_2 environment, $FeCO_3$ appears to be the dominant component on the carbon steel surface.
- The introduction of SO_2 in the subcritical CO_2 environment gave rise to the pitting corrosion while only uniform corrosion was observed on the carbon steel surface under pure subcritical CO_2.

5. Acknowledgements
This work was supported financially by National Research Foundation (NRF) - South Africa. Many thanks to Ms. Tiisetso Seema for assisting with conducting the sample preparation and Prof. Nicola Wagner - University of Johannesburg for providing high pressure equipment used in this work.

6. References
[1] Ruhl AS, Goebel A and Kranzmann A. 2012. Corrosion behavior of various steels for compression, transport and injection for carbon capture and storage. *Energy Procedia*: 216-225.

[2] Ruhl AS and Kranzmann A. 2012a. Corrosion behavior of various steels in a continuous flow of carbon dioxide containing impurities. *International Journal of Greenhouse Gas Control*: 85-90.

[3] Ruhl AS and Kranzmann A. 2012b. Corrosion in supercritical CO_2 by diffusion of flue gas acids and water. *The Journal of Supercritical Fluids*, **68**: 81-86.

[4] Ruhl AS and Kranzmann A. 2013. Investigation of corrosive effects of sulphur dioxide, oxygen and water vapour on pipeline steels. *International Journal of Greenhouse Gas Control* **13**: 9-16.

[5] Pfennig A, Zastrow P and Kranzmann A. 2013. Influence of heat treatment on the corrosion behaviour of stainless steels during CO_2-sequestration into saline aquifer. *International Journal of Greenhouse Gas Control*: 213-224.

[6] Pfennig A, Linke B and Kranzmann A. 2011. Corrosion behaviour of pipe steels exposed for 2 years to CO_2-saturated saline aquifer environment similar to the CCS-site Ketzin, Germany. *Energy Procedia*: 5122-5129.

[7] Capelle J, Furtado J, Azari Z, Jallais S and Pluvinage G. 2013. Design based on ductile–brittle transition temperature for API 5L X65 steel used for dense CO_2 transport. *Engineering Fracture Mechanics*. 110: 270-280.

[8] Xiang Y, Wang Z, Xu C, Zhou C, Li Z and Ni W. 2011. Impact of SO_2 concentration on the corrosion rate of X70 steel and iron in water-saturated supercritical CO_2 mixed with SO_2. *The Journal of Supercritical Fluids*: 286-294.

[9] Xiang Y, Wang Z, Xu M, Li Z and Ni W. 2013. A mechanistic model for pipeline steel corrosion in supercritical CO_2–SO_2–O_2–H_2O environments. *The Journal of Supercritical Fluids*: 1-60.

[10] Choi YS and Nesic S. 2011. Determining the corrosive potential of CO_2 transport pipeline in high pCO_2–water environments. *International Journal of Greenhouse Gas Control*: 788-797.

[11] Goos E, Riedel U, Zhao L and Blum L. 2011. Phase diagram of CO_2 and CO_2-N_2 gas mixtures and their applications in compression processes. *Energy Procedia* **4**: 3778-3785.

[12] Choi YS, Nesic S and Young D. 2010. Effect of impurities on the corrosion behavior of CO_2 transmission pipeline steel in supercritical CO_2- water environments. *Environmental Science and Technology*: 9233-9923.

[13] Paschke B and Kather A. 2012. Corrosion of pipeline and compressor materials due to impurities in separated CO_2 from fossil-fuelled power plants. *Energy Procedia*: 207-215.

[14] Dugstad A, Halseid M and Morland B. 2013. Effect of SO_2 and NO_2 on corrosion and solid formation in dense phase CO_2 pipelines. *Energy Procedia* **37**: 2877-2887.

IOP Conf. Series: Materials Science and Engineering **272** (2017) 012032 doi:10.1088/1757-899X/272/1/012032

Effect of CO_2 partial pressure and different CO_2 phases on carbon steel corrosion

MGR Mahlobo[1,2], K Premlall[1] and PA Olubambi[2]

[1]Chemical, Metallurgical and Materials Engineering Department, Tshwane University of Technology, Pretoria, 0001, South Africa

[2]Chemical Engineering Department, University of Johannesburg, Johannesburg, 2000, South Africa

Abstract. Carbon capture and storage (CCS) is the recent promising technology aimed at reducing greenhouse gas emission. Like many other developed technologies, CCS is faced with great challenges such as pipeline transportation failure due to corrosion. There are many factors contributing to steel corrosion during the pipeline transportation of carbon dioxide (CO_2). This study focuses on CO_2 partial pressure and different phases of CO_2 as some of the factors contributing to steel corrosion. Carbon steel was used as a testing specimen. High pressure reactor was used in this study to compress CO_2 from low to high pressures ultimately changing the CO_2 from gaseous phase to gas/liquid phase (subcritical) and to dense phase (supercritical). Weight loss method was employed to determine the corrosion rate while scanning electron microscopy (SEM) and X-Ray diffraction (XRD) were used to study the carbon steel morphology and phase analysis. Using low magnification digital camera, the type of corrosion that took place on the carbon steel surface was identified.

1. Introduction

Carbon dioxide (CO_2) corrosion has recently been identified as one of the disastrous concern facing the successful operation of carbon capture and storage (CCS) [1] – [2]. During the process of CCS, CO_2 is transported from the capturing point to the permanent storage sites through a steel pipeline system [3]. CO_2 is first compressed to its supercritical state (74.14 bar and 31.14°C) which makes it denser thus easier to transport over significantly long distances [4]. Many researchers have explained that dry gaseous CO_2 is not corrosive but the presence of moisture in the CO_2 stream leads to significantly severe corrosion threats on the pipeline steel [5]. This is because increasing the CO_2 pressure increases the CO_2 solubility which, in the presence of water, leads to the formation of the corrosive carbonic acid (H_2CO_3) [6]. This behavior has attracted a lot of attention to the CO_2 partial pressure not only as it influences the CO_2 solubility but potentially changes the CO_2 phase from gaseous to supercritical phase. Very little information is known on supercritical CO_2 corrosion as far as steel pipeline transportation is concern.

Many researchers have already published work discussing some of the factors influencing the CO_2 corrosion on steel pipeline. These factors include CO_2 partial pressure, temperature, pH, steel exposure time, impurities, etc [6] – [7]. One of the advantages of supercritical fluids is that they are easy to transport as they exist at higher densities thus lower viscosity. In reference to the CO_2 transportation during the CCS operation, supercritical CO_2 rather than gaseous CO_2 would be the best to optimize the transportation of CO_2 over long distances.

The aim of this study was to investigate the influence of CO_2 partial pressure on the steel pipeline corrosion. A high pressure reactor was used to compress CO_2 from low pressure to high pressure which covers 3 (three) phases of CO_2 thereby shedding a light also on corrosion of steel pipeline under different CO_2 phases. Carbon steel was selected as the testing specimen and was exposed to the varied CO_2 partial pressures. Weight loss method was used to calculate the corrosion rate while the steel surface morphology and phase analysis were carried out by scanning electron microscopy (SEM) and X-Ray diffraction (XRD) respectively.

2. Experimental Procedure

2.1. Materials

Carbon steel specimen with the elemental composition (weight %) of 0.044 C, 0.001 Mo, 0.237 Mn, 0.035 Al, 0.025 P, 0.0007 V, 0.009 S, 0.0001 Nb, 0.01 Si, 0.001 Ti, 0.01 Cu, 0.0002 B, 0.005 Ni, 0.0038 N, 0.01 Cr and Fe balance was cut to 16 × 10 × 5 mm. The surface area measuring 16 mm × 10 mm was exposed to the corrosion media (subcritical CO_2/SO_2 environment) as the tested surface inside the high pressure reactor (shown in Figure 1). The cut samples were grinded to 120 - 1200 grits and polished with the 1 μm using silicone discs. Prior to the experiments, the high pressure reactor was first deoxygenated by purging with N_2 for 24 hours and denitrogenated by purging pure CO_2 into the reactor for 4 hours. Deionized water was added to adjust the H_2O concentration of 1000 ppmv (0.1%) which was well above the H_2O concentration (50 ppmv) needed to enhance the corrosion process inside the steel pipeline [8]. The polished carbon steel samples were rinsed, dried, weighed (precision of 0.0001 g) and stored in a desiccator. CO_2 was charged into the reactor at constant 60 bar and 25°C (subcritical region). Different CO_2 concentrations were added into the subcritical CO_2 (as shown in Table 1), carbon steel samples were then exposed to this environment for a period of 336 hours. The chemical composition of the carbon steel used in this study is shown in table 1 with iron (Fe) being the balance.

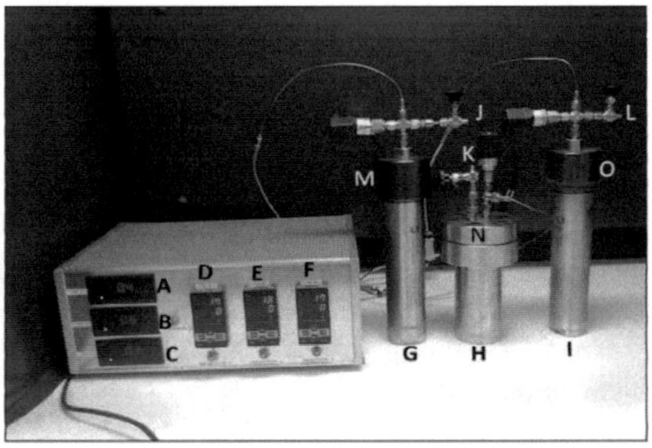

Figure 1. High pressure reactors (bolted and OC reactors).

Table 1. Experimental conditions

Sample #	Pressure (bar)	Temp. (°C)	Time (days)	H_2O (ppm)	CO_2 (Phase)
1	1	25	336	1000	Gas
2	50	25	336	1000	Subcritical
3	60	25	336	1000	Subcritical
4	70	25	336	1000	Subcritical
5	95	25	336	1000	Supercritical

2.2. Steel Surface Characterization

A 5 mega pixel (auto focus) Lumia 520 digital camera was used to take photographic images of the carbon steel specimen at low magnification. After the experiments, the samples were removed from the reactor and weighed again for the weight loss calculations using the weight loss given in equation (1). The steel surface morphology analysis was carried out using JEOL JSM-7600F Field Emission Scanning Electron Microscopy (SEM) and the phase analysis was carried out using EMPYREAN PANalytical X-Ray Diffraction (XRD). After characterization the samples were washed with a concentrated hydrochloric acid and acetone to remove the corrosion products on the steel surface in order to identify the type of corrosion that occurred.

$$CR = 8.76 \times 10^4 \frac{\Delta m}{\rho \cdot A \cdot t} \quad (1)$$

where CR is the corrosion rate, mm/year; Δm is the weight loss of the specimen, g; ρ is the density of the specimen, g/cm^3; A is the area of the specimen, cm^2; t is the immersion time, hour.

3. Results

3.1. Weight loss Results

The first data point, i.e. corrosion rate; pressure = 0.000 mm/year; 1 bar, is in good agreement with the fact that dry gaseous CO_2 is not corrosive. The prepared carbon steel in this instance was pressured with CO_2 at 1 bar for a period of 336 hours at 25°C and no change of mass was observed implying no corrosion took place on the carbon steel surface. Figure 2 again shows a gradual increase in corrosion rate as CO_2 partial pressure increased to 50 bar, a region where CO_2 is in a subcritical phase (gas/liquid) with corrosion rate increasing from 0-0.0533 mm/year. Reference [4] also published a study showing a corrosion rate of approximately 0.3 mm/year at a CO_2 pressure of 60 bar exposed for 24 hours. This study shows a rather similar finding as the corrosion rate was found to be 0.1419 mm/year under similar experimental conditions (CO_2 partial pressure of 60 bar). The corrosion rate increased further to 0.5604 mm/year at the CO_2 partial pressure of 70 bar. The highest corrosion rate of 0.7806 mm/year was obtained at supercritical CO_2, i.e. 95 bar, concluding that an increase in CO_2 pressure increases the corrosion rate of carbon steel. According to [4], [7] and [9], the possible explanation for this behavior is that liquid or dense CO_2 interacts differently from gaseous CO_2. An increase in CO_2 pressure increases the concentration of H_2CO_3 which accelerates the cathodic reaction and ultimately enhancing the corrosion rate. The corrosiveness in aquous phases coexisting with sub and supercritical CO_2 is a major driving force behind the trend observed when increasing the CO_2

partial pressure [7]. Furthermore, [7] explained that density of CO_2 at supercritical state can increase up to 0.4 to 0.5 g/cm^3 which makes the CO_2 solubility to be much higher than that at lower pressures.

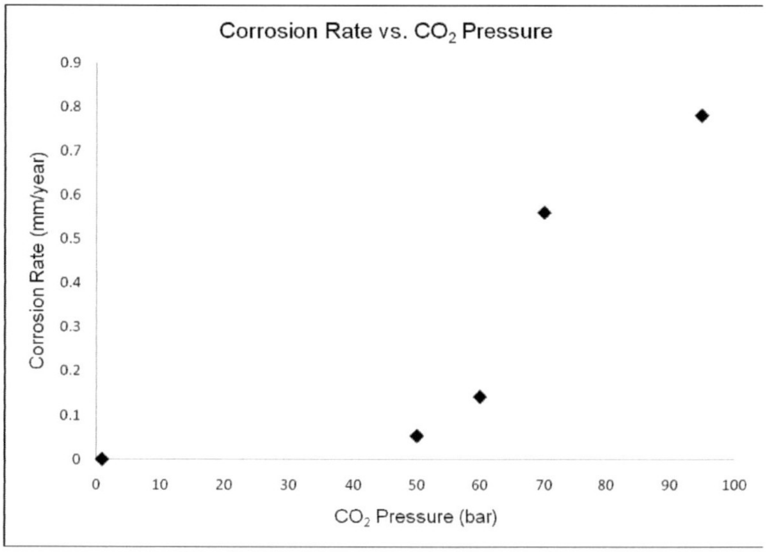

Figure 2. Relationship between corrosion rate and CO_2 partial pressure.

3.2. CO_2 Phase Effect

To investigate the effect of CO_2 phase on corrosion behaviour, pure CO_2 was charged at three (3) phases, i.e. gas, gas/liquid (subcritical) and liquid (supercritical) phase. In order to achieve the above phases, the relationship between pressure and temperature in reference to the CO_2 physical properties were considered. The literature done in this study states that CO_2 at the pressure of 73.82 bar (and above) and temperature of 31.1°C is no longer in the gas phase but dense phase and behaves as a liquid meaning it possesses the properties of liquids. Below 73.82 bar and 31.1°C CO_2 falls within the region called subcritical where it possesses both gas and liquid properties referred to in this study as gas/liquid phase. At pressures between 1 to 50 bar and temperatures below 25°C CO_2 is in a gas phase. Taking note of these CO_2 properties, it was then pressurized to 1 (dry gas), 60 (subcritical) and 95 bar (supercritical) into the reactor to investigate the effect of CO_2 phase. Figure 3 shows the data obtained at all three CO_2 phases.

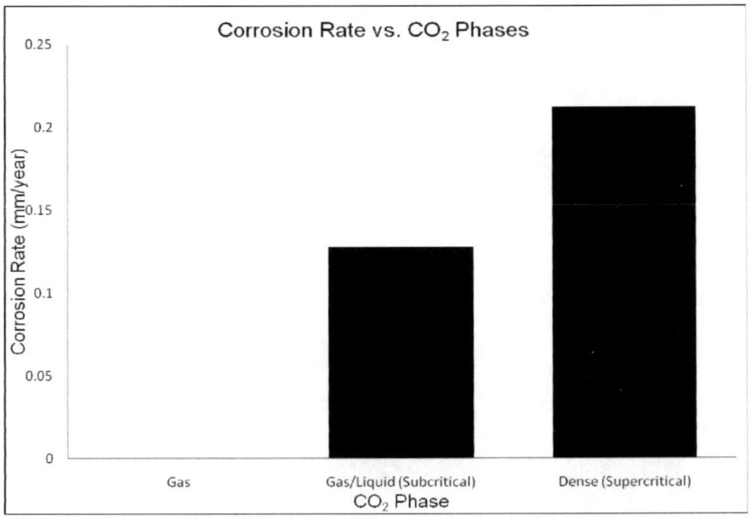

Figure 3. Relationship between corrosion rate and pure CO_2 phases.

Dry gaseous CO_2 resulted into almost no corrosion (0.0002 mm/year) which is in good agreement with literature. This indicates that dry pure CO_2 has very little negative influence towards corrosion of the steel pipe. Figure 3 shows a much higher corrosion rate of 0.1276 mm/year when CO_2 is at subcritical region indicating that as CO_2 changes from gaseous to liquid phase corrosion rate increases. Furthermore, corrosion rate increases even much higher when CO_2 is completely in the dense phase (supercritical state) with a corrosion rate of 0.2138 mm/year. This behavior is attributed to the CO_2 solubility changes at different conditions which then either accelerates or decreases the reaction to form corrosion products (mainly $FeCO_3$).

3.3. Steel Surface Morphology

The SEM analysis carried out on all carbon steel corroded surfaces before and after corrosion experiments are shown in figure 4. In an attempt to clearly identify the corrosion impact on the carbon steel surface, a blank carbon steel specimen (prior to corrosion experiment) was first analyzed then used as a reference basis. The blank carbon steel specimen is shown in figure 4 a) and no corrosion scale is observed. Scratches from sample preparation are however observed on the blank carbon steel surface. SEM image b) from figure 4 shows a carbon steel surface that is clear off any corrosion scale and also almost similar to the blank carbon steel SEM image. This makes a good agreement with the results obtained from weight loss (0 mm/year at 1 bar CO_2) in this study, i.e. gaseous CO_2 is not corrosive. As the CO_2 partial pressure is increased (from 1 to 50, 60, 70 and 95 bar) massive corrosion scale is observed on the carbon steel surface. The corrosion scale on SEM images c), d), e) and f) also shows the impact of increasing CO_2 partial pressure as it (corrosion scale) appears to be more compact and greater with an increase in CO_2 partial pressure. The observations from the CO_2 partial pressure behavior also concludes that CO_2 partial pressure increase change the CO_2 solubility which results in an accelerated formation of $FeCO_3$ on the carbon steel surface.

IOP Conf. Series: Materials Science and Engineering **272** (2017) 012032 doi:10.1088/1757-899X/272/1/012032

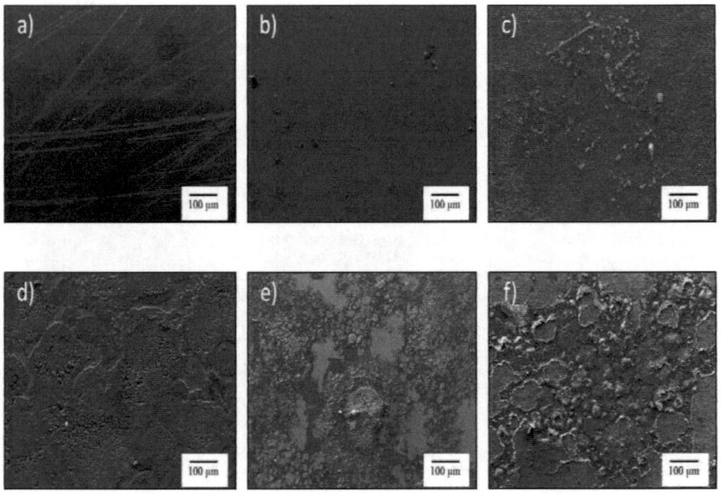

Figure 4. SEM images after corrosion experiments with CO_2 partial pressures: a) Blank sample, b) 1 bar, c) 50 bar, d) 60 bar, e) 70 bar & f) 95 bar.

Figure 5 shows photographic images of carbon steel specimen before and after corrosion experiments under various CO_2 environments. In order to identify the type of corrosion attack on the carbon steel surface, corrosion scale was removed from all carbon steel specimen. Localized corrosion attack is observed on the carbon steel surface when exposed to pure CO_2 under various partial pressures while pitting corrosion attack occurs in the presence of impurities under various CO_2 partial pressures [2].

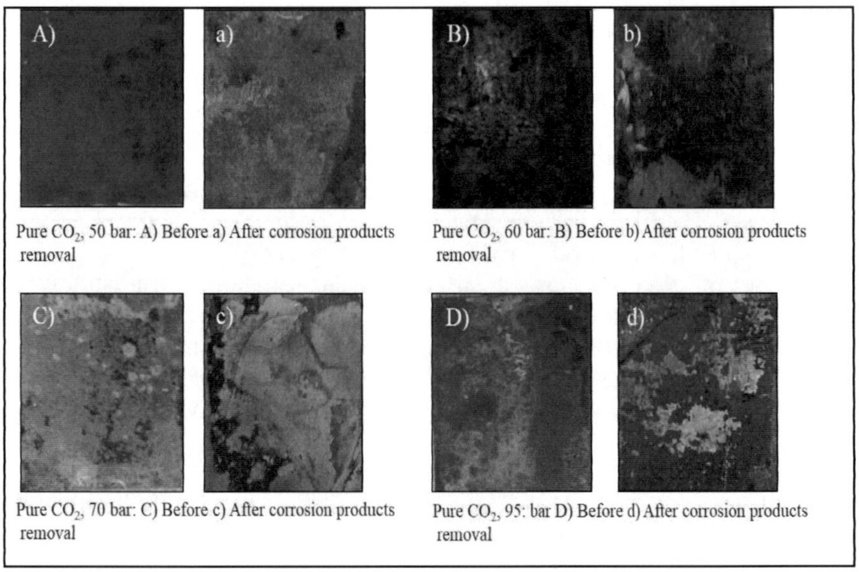

Figure 5. Photographic images of carbon steel specimen before and after corrosion.

3.4. Phase Analysis

Figure 6 shows the overlaid XRD spectra of the carbon steel specimen under various CO_2 partial pressures. Only iron (Fe) was detected on the carbon steel surface exposed to 1 bar (gaseous CO_2) indicating no formation of corrosion scale. This, again, shows a good agreement with the weight loss and SEM results discussed earlier in this study, i.e. gaseous CO_2 is not corrosive. A solid compact iron

IOP Conf. Series: Materials Science and Engineering **272** (2017) 012032 doi:10.1088/1757-899X/272/1/012032

carbonate ($FeCO_3$) was detected as a major component on the carbon steel surface as CO_2 partial pressure was further increased. Wet iron oxide ($Fe_2O_3{\cdot}H_2O$) also made an appearance on the carbon steel surface which normally occurs in the presence of oxygen (O_2) within the CO_2 environment. Recalling that in this study the high pressure reactors were deoxygenated prior to the experiments, the oxidation of iron can only be explained by exposure of the carbon steel specimen to air during sample preparation prior to SEM and/or XRD analysis. A very important observation made from the XRD spectra in figure 6 is the fact that Fe was not detected almost at all under 95 bar CO_2 (supercritical phase). This implies that the entire surface of carbon steel was filled with corrosion scale implying that the further increase of CO_2 partial pressure to 95 bar increased the CO_2 solubility which resulted to more formation of mainly $FeCO_3$ according to the overall equation (2) shown below:

$$Fe + H_2CO_3 \rightarrow FeCO_3 + H_2O \quad (2)$$

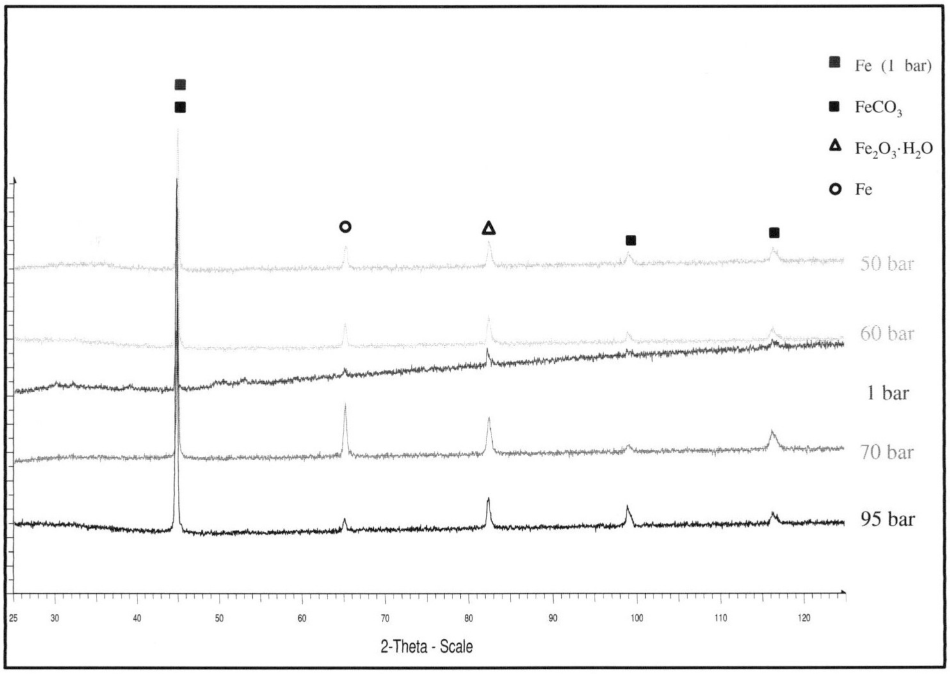

Figure 6. XRD spectra of carbon steel specimen corroded with different CO_2 partial pressures: a) 1 bar, b) 50 bar, c) 60 bar, d) 70 bar & e) 95 bar.

4. Conclusion

Based on the results obtained and discussed on this study, the following conclusions were drawn:

- An increase in CO_2 partial pressure increases the carbon steel corrosion rate.
- Gaseous CO_2 showed no corrosion on the carbon steel while significant corrosion was observed under subcritical CO_2 with the highest corrosion being occurring under supercritical CO_2.
- Localized corrosion attack was observed on the carbon steel surface when exposed to pure CO_2 under various partial pressures.
- Iron carbonate ($FeCO_3$) was the dominant component on the corrosion scale formation on the carbon steel surface.

Acknowledgment
This work was supported financially by National Research Foundation (NRF) - South Africa. Many thanks to Prof. Nicola Wagner - University of Johannesburg for providing high pressure reactor.

References

[1] Pfennig A and Kranzmann A. 2009. Effects of saline aquifere water on the corrosion behaviour of injection pipe steels 1.4034 and 1.7225 during exposure to CO_2 environment. *Energy Procedia*: 3023-3029.
[2] Pfennig A, Zastrow P and Kranzmann A. 2013. Influence of heat treatment on the corrosion behaviour of stainless steels during CO_2-sequestration into saline aquifer. *International Journal of Greenhouse Gas Control*: 213-224.
[3] Pfennig A and Kranzmann A. 2011. Reliability of pipe steels with different amounts of C and Cr during onshore carbon dioxide injection. *International Journal of Greenhouse Gas Control*: 757-769.
[4] Choi YS and Nesic S. 2011. Determining the corrosive potential of CO_2 transport pipeline in high pCO_2–water environments. *International Journal of Greenhouse Gas Control*: 788-797.
[5] Yin ZF, Zhao WZ, Bai ZQ, Feng YR and Zhou WJ. 2008. Corrosion behavior of SM 80SS tube steel in stimulant solution containing H_2S and CO_2. *Electrochimica Acta*: 3690-3700.
[6] Pfennig A and Kranzmann A. 2012a. Effect of CO_2 and pressure on the stability of steels with different amounts of chromium in saline water. Corrosion Science **65**: 441-452.
[7] Zhang Y, Pang X, Qu S, Li Z and Gao K. 2012. Discussion on the CO_2 corrosion mechanism between low partial pressure, supercritical condition. *Corrosion Science*: 1-53.
[8] Paschke B and Kather A. 2012. Corrosion of pipeline and compressor materials due to impurities in separated CO_2 from fossil-fuelled power plants. *Energy Procedia*: 207-215.
[9] Gao K, Yu F, Pang X, Zhang G, Qiao L and Chu W. 2008. Mechanical properties of CO_2 corrosion product scales and their relationship to corrosion rates. *Corrosion Science*: 2796-2803.

IOP Conf. Series: Materials Science and Engineering **272** (2017) 012033 doi:10.1088/1757-899X/272/1/012033

Using Small Punch tests in environment under static load for fracture toughness estimation in hydrogen embrittlement

B Arroyo[1], J A Álvarez[1], R Lacalle[1,2], P González[1] and F Gutiérrez-Solana[1]

[1] LADCIM, University of Cantabria. Avda. Los Castros 44, 39005 Santander, Spain.
[2] INESCO INGENIEROS. CDTUC Module 009A phase B, 39005 Santander, Spain.

Abstract. In this paper, the response of three medium and high-strength steels to hydrogen embrittlement is analyzed by means of the quasi-non-destructive test known as the Small Punch Test (SPT). SPT tests on notched specimens under static load are carried out, applying Lacaclle's methodology to estimate the fracture toughness for crack initiation, comparing the results to K_{IEAC} fracture toughness obtained from C(T) precracked specimens tested in the same environment; SPT showed good correlation to standard tests. A novel expression was proposed to define the parameter $K_{IEAC-SP}$ as the suitable one to estimate the fracture toughness for crack initiation in hydrogen embrittlement conditions by Small Punch means, obtaining good accuracy in its estimations. Finally, Slow Rate Small Punch Tests (SRSPT) are proposed as a more efficient alternative, introducing an order of magnitude for the adequate rate to be employed.

1. Introduction

A critical aspect concerning mid and high strength steels is their resistance to Stress Corrosion Cracking (SCC) and Hydrogen Embrittlement (HE). The recommendations presented by various research groups over the last few decades have been collected in the standard ISO 7539 [1], but there are particular situations where standards cannot be followed characterize in-service components, due to the impossibility of machining specimens fitting the dimensions. The Small Punch Test (SPT) is one of the most notables solutions, which estimates parameters such as the yield stress, ultimate tensile strength and fracture toughness of metallic materials with high reliability. Recently the validity of the SPT when used in HE and SCC has been proved, being faster and easier and cheaper to perform than standard tests [2,3,4,5]; the ultimate research, which is applied to this work, consists on SPT tests performed in environment under static load [6]. The aim of this work is to estimate the crack initiation toughness of the material in embrittlement conditions, $K_{IEAC-SP}$, caused by a liquid aggressive environment.

2. Materials and environments employed

2.1. Three materials employed

- A quenched and tempered Cr-Ni-Mn high-strength, which has a tempered martensite microstructure. It is employed in the manufacture of large anchor chains for off-shore platforms.
- A rolled X80 medium-strength steel, which has a ferritic-pearlitic micrPosctructure. This steel is commonly employed for the manufacture of gas and petroleum pipes at low temperatures.

- A weldable thermo-mechanically treated S420 medium-strength steel, which has a ferritic-pearlitic micrososctructure. It is mainly used in the construction of pressure vessels.

2.2. Environment for Hydrogen Embrittlement (HE) simulation

An environmental condition known as cathodic charge, or anodic polarization, was employed, it causes substantial embrittlement on the steel by the action of the hydrogen going through and getting trapped in it. It consists of the interconnection, via an acid electrolyte, of a noble material (platinum in this case) and the steel, which will protected due to the fixed current interposed. In this study an environmental condition, previously used [6,7] to reproduce cathodic charge situations, was proposed. It is consisting of an 1N H_2SO_4 solution in distilled water containing 10 drops of CS_2 and 10mg of As_2O_3 dissolved per liter of dissolution. The solution of As_2O_3 was prepared using Pessouyre's method [7]. A platinum grid was used as an anode. The PH was controlled in the range 0,65 - 0,80 during the tests and at room temperature 20°C - 25°C. An embrittlement level of 5mA/cm^2 was employed.

3. The Small Punch Test

The quasi-non-destructive SPT technique allows to test in-service structures since the extraction of a sample does not compromise the component's integrity. It has a European Code of Practise, CWA 15627, edited by CEN in 2007 [8]. It has been successfully employed in the evaluation of tensile [9], fracture [9,10] and creep properties, and it was recently applied to HE scenarios [5,6]. The SPT consists of punching a plane specimen of small dimensions deforming it until fracture; during the test the force as well as the displacement made by the punch are registered continuously.

When characterizing materials in HE situations the testing rate is an important parameter to take in account in order to reproduce the subcritical processes taking place [11]. The ultimate research in the SPT field [6, 11] advises to perform static load tests in order to allow hydrogen to cause all of its embrittling power. In this type of testing three zones can be distinguished in the punch displacement vs time register (d-t); zone I: indentation and settlement, zone II: quasi-constant punch rate caused by the flexibility variation of the system produced by an increasing cracking in both radial and thickness directions of the specimen, and zone III: final and rupture.

4. Experimental Methodology

4.1. Standard validation fracture mechanics K_{IEAC} determination tests

Fracture mechanics tests were carried out in order to determine K_{IEAC} for each of the materials in the aforementioned environmental condition (cathodic polarization at a level of 5mA/cm^2). B=25mm C(T) specimens according to [12] were employed, being precracked up to a/w=0.53. The specimens were subjected to hydrogen absorption during 48 hours, and the test was performed subsequently by applying 6E-9 m/s of constant solicitation rate; recommendations of Standard ISO-7539 [1] were followed submerging the samples up to the crack tip. The methodology proposed by ASTM E-1820 [12] was employed for the K_{IEAC} value calculation.

4.2. Static load SPT tests for crack initiation toughness in environment determination, $K_{Jth\text{-}EAC}$

SPT notched samples were used to determine the material toughness necessary for a crack initiation from the edge of the notch, $K_{Ith\text{-}EAC}$. The environment employed was cathodic polarization at a level of 5mA/cm^2. Prior to the test, the specimens were subjected to hydrogen charging during 2 hours [6, 11]; subsequently the load was softly applied by an endless screw system on the specimen subjected to the environment. For each material several SPT notched samples were tested using decreasing loads, which produced decreasing punch rates in the zone II of the curve [6] (figure 1-right), up to that load that was not enough to produce any cracking departing from the edge of the notch. The sample geometry employed, according to [8,10,11], presented on figure 1-left, was consisting on a plate 10mmx10mm of section and 0.5±0.01mm of thickness including a lateral notch machined by wire electro-erosion of 0,15 mm of radius. The orientation of the notches in SPT and C(T) samples for validation tests was the same.

From each sample the material toughness for crack initiation from the edge of the notch, K_{Jth}, was calculated using Lacalle's methodology [10], which is based on the CTOD parameter for crack initiation on a SPT notched sample, δ_{SPi}, and allows to calculate using the expression (1) the initiation toughness J_{Ic}. Then, using the expression (2) the toughness values in terms of J_{Ic} can be transformed into terms of K_{Jth}. In (1), d_n is a parameter tabulated in [13] (figure 2-top) which is function of the material properties; δ_{SPi} can be obtained in function of the punch displacement that causes the crack initiation form the edge of the notch, d_i, using the graph proposed in [9] (figure 2-right).

$$J_{Ic} = \frac{\delta_{SPi} \cdot S_y}{d_n} \quad + \quad K_{Jth} = \sqrt{J_{Ic} \cdot E} \qquad (1) + (2)$$

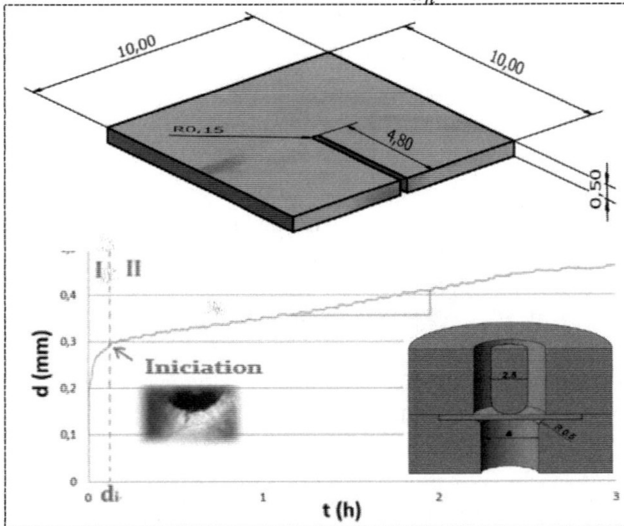

Figure 1. Top: SPT notched samples employed; **Bottom:** Determination of the punch displacement, di, for crack initiation between zones I and II.

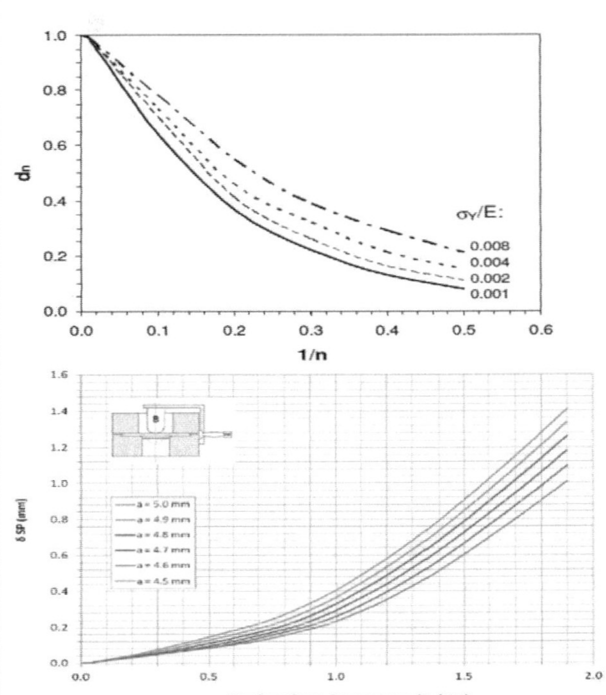

Figure 2. Top: Determination of d_n [13]; **Bottom:** δsp in function of punch displacement, d [9].

5. Experimental results

5.1. Determination of K_{IEAC} by standard fracture mechanics tests

Figure 3 shows the results of the fracture mechanics tests performed on C(T) specimens in environment.

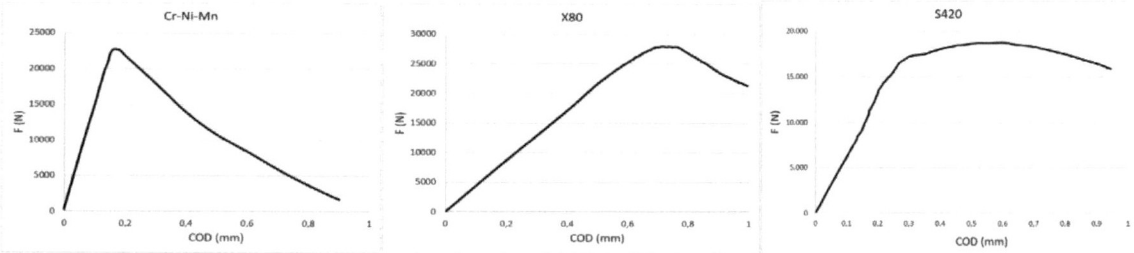

Figure 3. P-COD registers of the standard test performed on the three steels to obtain K_{IEAC}.

5.2. Determination of $K_{Jth-EAC}$ by SPT static tests in environment

In figure 4 the d-t registers from the SPT tests performed on the three materials studied are presented; accompanying them the punch rates developed by the system in the zone II and pictures of the crack initiation area of the samples are shown. In table 1 the values of $K_{Jth-EAC}$ for a specific material and environment obtained by SPT means, defined as a lower bound (sample where propagation didn't take place), are presented and compared to the results of K_{IEAC} obtained from the standard tests in environment.

It can be observed that in all cases the SPT values are higher (around the double) of their homologous K_{IEAC} obtained from standard tests. The relationship shown in equation (3) is proposed to obtain an initiation fracture toughness parameter by Small Punch means, $K_{IEAC-SP}$, as an optimization to Lacalle's methodology previously exposed [9,10] for embrittlement scenarios. Looking for the best fit for (3), showed in figure 5, a value of 0.498 was found for α, obtaining the values of $K_{IEAC-SP}$ showed in the penultimate column of table 1.

$$K_{IEAC-SP} = \alpha \cdot K_{Jth-EAC} \tag{3}$$

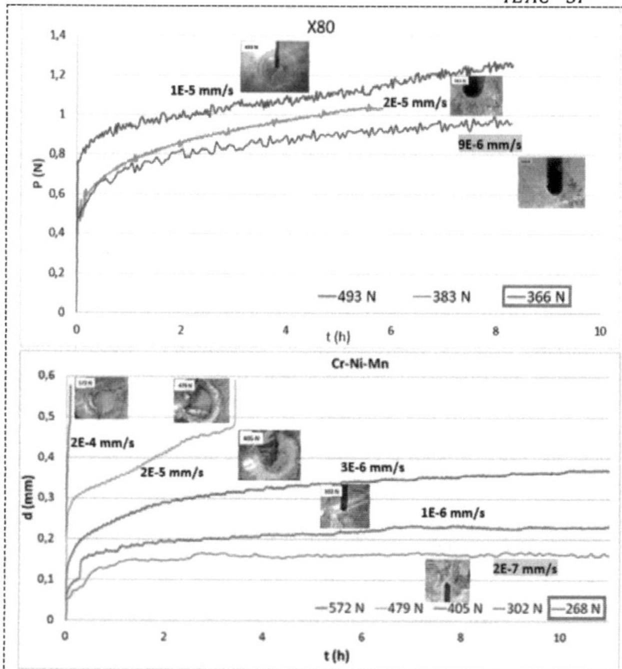

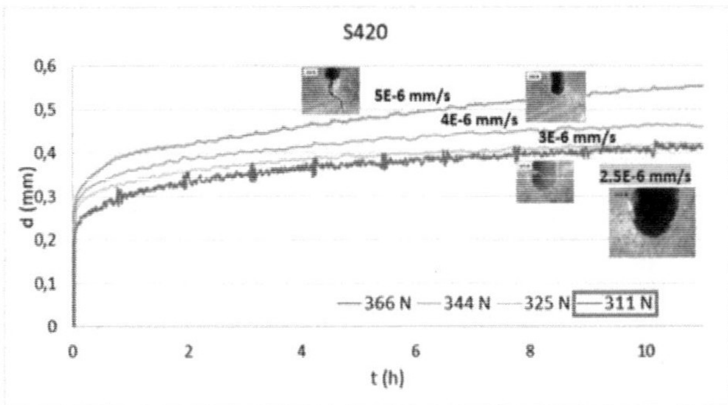

Figure 4. Punch displacement vs time (d-t) registers of the SPT static load tests in environment performed on the three steels; **Top:** Cr-Ni-Mn steel; **Mid:** X80 steel; **Bottom:** S420 steel;

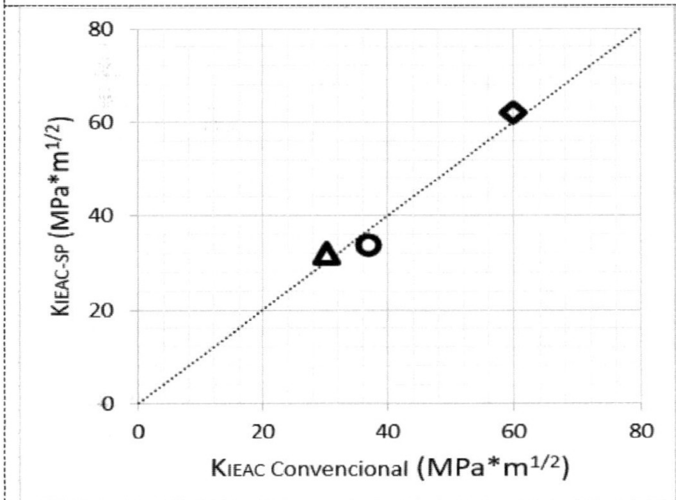

Figure 5. K_{IEAC} results from C(T) vs $K_{EAC\text{-}SP}$ SPT technique proposed results for the materials studied.

Table 1. $K_{Ini\text{-}EAC}$ results obtained from SPT static load tests vs K_{IEAC} results from C(T) standard tests.

Material	Load (N)	di (mm)	δ_{spi} (mm) (fig. 2)	J_{Ic} (kN/m) (expr.(1))	K_{Jth} (MPa$\sqrt{m}$) (expr.(2))	$K_{IEAC\text{-}SP}$ (MPa$\sqrt{m}$) (expr.(3))	K_{IEAC} (MPa$\sqrt{m}$)
Cr-Ni-Mn	268	0.110	0.021	20.13	**64.23**	**31.96**	**30.08**
X80	366	0.479	0.111	73.378	**124.13**	**61.76**	**59.89**
S420	311	0.231	0.039	21.76	**67.60**	**33.63**	**36.92**

5.3. Discussion about toughness estimation and punch rates

The tests where propagation did not take place were just the ones employed in each material to define $K_{EAC\text{-}SP}$; in this cases the punch rates developed were around 1E-6 or 1E-7 mm/s (4 orders of magnitude lower than usually employed in air). Based on this a more efficient way of testing may be to perform Slow Rate Small Punch Tests (SRSPT) of this order of magnitude. An upper bound for SPT testing rates should be defined in order to reproduce the subcritical micromecanisms where the interaction between a given material and environment governs the process; this will be a faster and easier testing method.

6. Conclusions and future work

6.1. Conclusions

A methodology to estimate the fracture toughness for crack initiation using notched Small Punch samples was proposed. It is based on Lacalle's research, using the CTOD parameter for crack initiation, and includes a lineal expression (to correlate its results to the validation tests on C(T)'s) to obtain the parameter $K_{IEAC-SP}$ that showed good accuracy.

Finally, the performance of Slow Rate Small Punch Tests (SRSPT) was proposed as a more efficient way of testing to determine the aforementioned parameter, but the first step in this research will be finding the appropriate rates, that should be in the order or magnitude of 1E-6 or 1E-7 mm/s.

6.2. Future work

On the one hand, the methodology to estimate fracture toughness by SPT tests in environment proposed should be extended and validated in a wider range of materials and embrittlement conditions. On the other hand, a suitable upper bound for SRSPT punch rates should be found. It could be the one that produces in the edge of the notch the same CTOD growing rate taking place, for instance, in a C(T) pre-cracked specimen during a fracture mechanics test in environment at a slow rate, reproducing like this in both cases the same plasticity conditions in the crack tip.

7. References

[1] ISO 7539:2011 Parts 1 to 9 "Corrosion of metals and alloys"
[2] Tao B and Kaishu G 2013 Material and Design **561** 849-860.
[3] García TE, Rodríguez C, Belzunce FJ, Peñuelas I and Arroyo B 2015 Mater Sci Eng A **626** 342-351.
[4] Arroyo B, Álvarez JA and Lacalle R 2016 Theorethical and Applied Fracture Mechanics **86** 61-68.
[5] García TE, Arroyo B, Rodríguez C, Belzunce FJ and Álvarez JA 2016 Theorethical and Applied Fracture Mechanics 86 89-100
[6] Arroyo B, Álvarez JA, Lacalle R, Uribe C, García TE and Rodríguez C 2017 Materials Science and Engieneering A **691** 180-194.
[7] Pressouyre GM 1997 PhD Thesis Carnegie Mellon University.
[8] CWA 15627:2008 "Small Punch Test for Metallic Materials" European Comittee for Standarization
[9] Lacalle R 2012 PhD Thesis University of Cantabria.
[10] Lacalle R, Álvarez JA and Gutiérrez-Solana F 2008 ASME PVP **66** 1363-1369.
[11] Arroyo B 2017 PhD Thesis, University of Cantabria.
[12] ASTM E-1820-15a 2015
[13] Shih CF 1981 Journal of the Mechanics and Phisics of Solids **29** 305-326.

IOP Conf. Series: Materials Science and Engineering **272** (2017) 012034 doi:10.1088/1757-899X/272/1/012034

Fiber Reinforced Concrete (FRC) for High Rise Construction: Case Studies

KooroshGharehbaghi[1] and Rhea Chenery[2]

[1]RMIT University, Australia. MIEAust, MITE, MIIE, MCIAust, MATSE
[2]RMIT University, Australia

E-mail: Koorosh.gharehbaghi@rmit.edu.au, Rhea.chenery@rmit.edu.au

Abstract. Due to its material element, Fiber Reinforced Concrete (FRC) could be stronger than traditional Concrete. This is due to FRC internal material compounds and elements. Furthermore, FRC can also significantly improve flexural strength when compared to traditional Concrete. This improvement in flexural strength can be varied depending on the actual fibers used. Although not new, FRC is gradually gaining popularity in the construction industry, in particular for high rise structures. This is due to its flexural strength, especially for high seismic zones, as it will provide a better solution then reinforced Concrete. The main aim of this paper is to investigate the structural importance of FRC for the high rise construction. Although there has been numerous studies and literature in justifying the FRC for general construction; this paper will consider its use specifically for high rise construction. Moreover, this paper will closely investigate eight case studies from Australian and United States as a part of the FRC validation for high rise construction. In doing so, this paper will examine their Structural Health Monitoring (SHM) to determine their overall structural performance.

1. Introduction

Generally, composite materials are made up of individual constituents, including matrix and reinforcement [1]. Conversely, the matrix material surrounds and supports the reinforcement materials by maintaining their relative positions; and the reinforcements pass on their special mechanical and physical properties to enhance the matrix properties [2]. While the composite could be formed into various shapes, the matrix can be introduced to the reinforcement, before or after the reinforcement material has commenced being manufactured. [3].

Many commercially produced composites use polymer matrix material, often called a resin solution, as their matrix. There are many different polymers available, depending on the starting raw ingredients. The reinforcement materials are often fibers, but can also be commonly ground minerals. Figure 1 below shows fiber elements and the finished product.

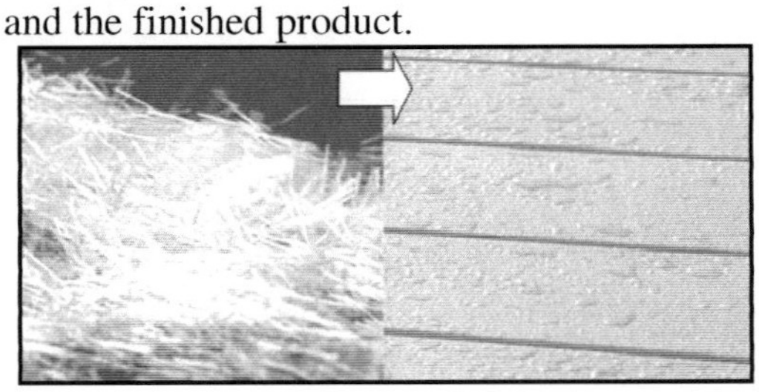

Figure 1. From raw fiber to finished material [3].

In construction and building structures, fiber reinforced modules are utilized as composite materials [4]. Fiber reinforced materials are unconventional materials, which utilize substances such as Fiber as bonding agents [1]. With every increase in material waste, the utilization of such bonding agents can also further reduce the impact of recycling requirements [5]. More importantly, such innovative materials, can withstand the required loads (stress and strain) as per more conventional methods of material engineering [6].

FRC is an example of such innovative material engineering. Although this material has been used in general construction for some time, its use for high rise construction is gradually increasing. In the past, the reason for the reluctant use of FRC for high rise construction was due to industry's lack of trust with its functionality [7]. As per norm, material functionality is the most important factor to consider when a selection is being carried out [8]. One of the most significant facets of material functionality is its ability to fit the purpose which it is required to do. As per traditional Reinforced Concrete (RC), such materials are utilized as load bearing components of the structures. With the development of the FRC, it is believed that this material cannot match the functionality of the RC. However, many studies have been conducted to support the opposite to this view [8]. Furthermore, where low load bearing materials are required, FRC can easily be employed [9].

The intriguing aspect of FRC is the actual type of fiber which is being used as there are many fibers to select from, such as Polymer, Steel, Glass and even Natural fibers (vegetable origin) [10]. Figures 2a and 2b below illustrate Steel and Glass fibers.

Figure 2A. Steel Fiber [9].

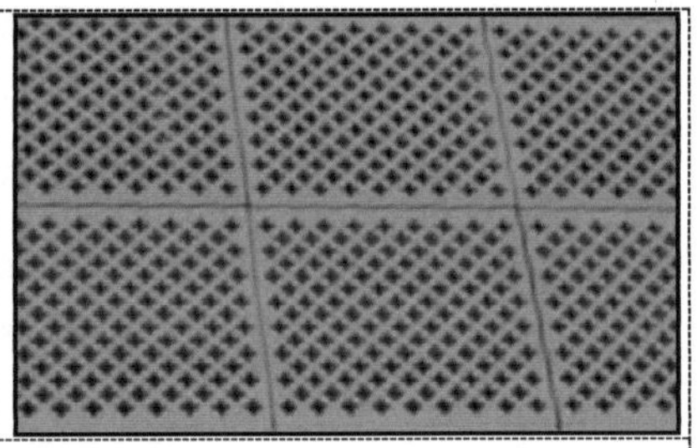

Figure 2B. Glass Fiber [4].

Moreover, another important aspect of FRC is its composition progression, including the interface and bonding processes [9], both of which are discussed further in the following section.

2. FRC Composition Progression

As already discussed, Fiber Reinforced materials are unconventional materials, which utilize substances such as Fiber as bonding agents. The properties of the Fiber-Concrete interface have a critical effect on the properties of FRC [[4]. Also as previously conversed, a key aspect of FRC is its composition progression, including its interface and bonding processes. While the bonding process includes the actual bond strength, the interface comprises of the actual cross point between quantities of fiber and concrete[11]. Table 1 (below) further depicts these two key processes.

TABLE 1. Interface and Bonding process of FRC

Processes	Actions	Effect
Interface	Fibers such as steel create specific zones and boundaries where the actual interface is being created	Gradual mixing occurs: for example for glass Fiber it takes up to 28 days
Bonding	The joining action will take place and its curing time depends on the Fiber type and the environment	Three types of bonding process can occur: elastic, frictional and mechanical

As it can be observed, both processes are complex and subjective to many factors. More importantly, the bonding process can be achieved via three different processes [8]:

- Elastic, that is the Fiber sticking to the matrix.
- Frictional, which is creating friction and also providing resistance to pull out force. Frictional bonding controls the significant element of FRC stress-strain behavior, and highlights the cracking value and its subsequent performance [9].
- Mechanical, which ensures the fibers and their composition correctly interlock with the matrix. Of the three boding process, mechanical is the most complex and complicated procedure. As Gharehbaghi (2015) correctly argued, crimping the fibers to provide the mechanical bonding can effectively increase the average bond stress value by an approximate "factor of four".

Although there are many literatures discussing which method of bonding is more effective, as a general rule, the performance of the bonding is rather dependent on the actual fiber type, for example, steel, glass and so on [11]. Although the bonding process is a considerable issue to contemplate, ultimately it is the fiber type which could determine the structural performance of the FRC [12]. Determination of the structural performance is a vital issue to contemplate when considering the use of FRC for high rise construction.

3. FRC For High Rise

As outlined earlier, there is much literature regarding the use of the FRC for general construction. However, this paper highlights the importance of FRC specifically for high rise Construction. Figures 3a and 3b below exemplify the assembly and erection of FRC for high rise construction.

Figure 3A. Retaining walls (for high rise) made from FRC.

Figure 3B. Assembly of FRC panels.

While figure 3a illustrates retaining walls made from FRC together with connection bars as additional support; figure 3b demonstrates the assembly of FRC panels for high rise construction. Structural Health Monitoring (SHM), which refers to the process of implementing a damage detection and characterization strategy for buildings, is a key tool in determining the structural performance of the FRC for high rise construction.

4. Structural Health Monitoring (SHM)

In determining damage detection and monitoring of structures important techniques such as Structural Health Monitoring (SHM) are utilized [13]. The SHM technique ensures that effective monitoring of structures is carried out due to long-term movement and degradation of their materials [14]. Commonly, the main causes of deterioration in structures such as buildings include wear and tear together with additional movements [7]. For buildings in-particular the main deterioration includes cracking due to significant structural loads consisting of Imposed and Seismic loads, together with the Environmental loads (including Temperature changes, and so-on). Moreover, a period of dramatic environmental changes, such as increased Seismic activities, could further complicate the SHM process and analyses; and also reduce the Building's structural performance [11].

As Gharehbaghi (2017) correctly claimed, the traditional damage detection strategies consist of visual inspection and localized non-destructive evaluation techniques, such as vibration and impact tests. Moreover, the utilization of Young's modulus can further assist with determining any structural

IOP Conf. Series: Materials Science and Engineering **272** (2017) 012034 doi:10.1088/1757-899X/272/1/012034

deterioration and damage. An important aspect for FRC for high rise construction is to be sure that it maintains adequate structural integrity. Correspondingly, using SHM to be sure that FRC adequately meets the structural integrity is an important damage detection and monitoring consideration. In addition, SHM improves the functionality of building together with timely warning of impending failures, and thus will provide appropriate and cost-effective maintenance regimes [14]. To further examine the importance of the FRC for high rise construction, a case study of eight buildings from Australian and United States was conducted.

5. Case Studies

To systematically examine FRC for high rise construction, a case study consisting of a variety of high rise buildings was carried out. The status of each structure and its fiber information was captured and is shown in table 2 below.

TABLE 2. Comparison of Fiber for high rise

Structure	Fiber Type	Fiber Bonding Process	Fiber Insertion
Building I (USA)	Steel	Mechanical	Internal Walls
Building II (AUST)	Steel	Combination of Frictional and Mechanical	Internal Walls
Building III (USA)	Steel	Combination of Frictional and Mechanical	Internal Walls
Building IV (USA)	Polymer	Mechanical	External Walls
Building V (AUST)	Steel	Mechanical	Internal Walls
Building VI (USA)	Polymer	Mechanical	External Walls
Building VII (USA)	Steel	Combination of Frictional and Mechanical	Internal Walls
Building VIII (AUST)	Polymer	Mechanical	External Walls

As it can be noticed, most of the buildings utilized Steel as the basis of their FRC. In addition, their fiber bonding process generally consists of a combination of Frictional and Mechanical bonding. This is perhaps due to these two processes being the most commonly used in the material engineering fields. As a part of the Structural Health Monitoring and Modeling, SensorConnect provided all the necessary and up-to-date data for the analysis purposes. Figure 4 provides a SensorConnect output for one of the Buildings studied.

IOP Conf. Series: Materials Science and Engineering **272** (2017) 012034 doi:10.1088/1757-899X/272/1/012034

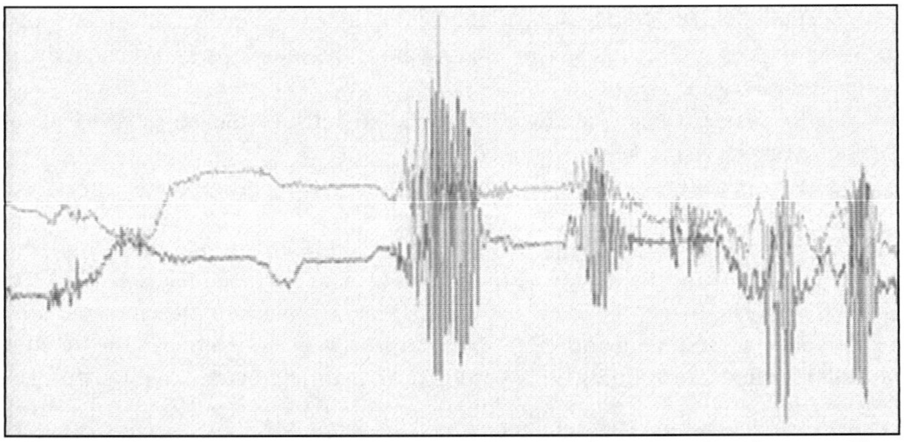

Figure 4. *SensorConnect* output for one of the buildings (acceleration vs roll) .

In addition, to further examine these structures, each building's structural performance was analyzed using SensorConnect software, and their summary is shown in table 3 below.

TABLE 3. Comparison of SHM performance

Structure	Structure Age (years) [approximately]	Imposed Loads	FRC Component	SHM Performance (Overview)
Building I (USA)	8	Intermediate Seismic	Non-load bearing	Outstanding and exemplary
Building II (AUST)	15	Standard	Non-load bearing	Due for a some minor upgrade and improvements
Building III (USA)	21	Minor Seismic	Non-load bearing	Due for a minor upgrade and improvements (some cracking due to movements)
Building IV (USA)	9	Standard	Limited load bearing	Outstanding and exemplary
Building V (AUST)	11	Standard	Non-load bearing	Satisfactory and on-going
Building VI (USA)	7	Standard	Non-load bearing	Minor structural failure (displacement)
Building VII (USA)	18	Minor Seismic	Non-load bearing	Due for a some minor upgrade and improvements
Building VIII (AUST)	2	Standard	Limited load bearing	The structure is to new to notice any wear and tear

While the Structure age is only an approximation, it is important to note that all the buildings were Monocoque structures. This includes the involvement of additional structural reinforcement especially for demanding areas such as Seismic areas. Generally, FRC was utilized for non-load bearing structural components, especially for Seismic areas. As it can be noticed from table 3, the majority of buildings involved the use of Steel as the basis of FRC. The SHM process involved accessing the available and allowable database.

As already discussed the SHM technique involved:

- Using SHM logs to collect the sensor data from all the buildings and then analyzing the data on a high end platform.
- Accessing the maintenance and inspection logs to retrieve the periodic (depending on the inspection regimes) visual inspection records.
- Accessing the dedicated data-acquisition SensorConnect Software for sensor configuration and data collection.

As already discussed, as a part of the Structural Health Monitoring and Modeling process, SensorConnect provided all the necessary and up-to-date data for analysis purposes. This analysis further provided the information to be used for the SHM Performance column in the table 3 above. The SHM performance includes information about the overall structural health together with the performance of the buildings, including the observation of any maintenance and rehabilitation regimes and schematics.

6. Summary

The general principle of composite materials is to engineer materials which are stronger, lighter, and possibly less expensive when compared to traditional materials. In high rise construction, the composite materials usually include, concrete, and FRC among other materials. Much research has been conducted to demonstrate that due to its material elements, FRC is rather stronger than traditional Concrete. Although not novel, FRC is steadily gaining recognition in the construction industry in-particular for high rise structures such as buildings.

Fittingly, the main aim of this paper was to investigate the structural significance and functioning of FRC for high rise construction. Although there has been abundant investigation into justifying the use of FRC for broad-spectrum construction, this paper provided a number of case studies (from Australia and United States) to further encourage its usage, explicitly for high rise structures.

Finally, this paper examined the case studies using SHM, to further emphasize their overall structural integrity. Close examination of the case studies found that all the buildings were Monocoque structures. This included the involvement of additional structural reinforcement particularly for demanding areas such as Seismic areas. Moreover, FRC was thus utilized for non-load bearing structural component, predominantly for Seismic areas.

7. References

[1] K. Gharehbaghi, Material Advancements for High-rise Construction (Proceedings of the 5th International conference on the constructed environment, University of Pennsylvania, Philadelphia, 16-17 October 2014).

[2] R. C. Hibbeler and K. S. V. Sekar, Mechanics of materials (9th edition, worldwide adaptation edition, SI edition, Singapore: Pearson Education South Asia Pty Ltd, 2014).

[3] M. S. Mamlouk and J.P. Zaniewski, Materials for Civil and Construction Engineers (3rd edition, Prentice Hall, London, 2011).

[4] J. Gustavo, High Performance Fiber Reinforced Cement Composites 6 (International RILEM conference on high performance fiber reinforced cement composition, 2012)

[5] F. Ching, Building structures Illustrated Patterns, Systems, and Design (2nd edition, Hoboken: Wiley, 2013).

[6] S. Goodhew, Sustainable construction processes: a resource text (Hoboken: John Wiley and Sons, 2016).

[7] K. Gharehbaghi, Advancements in Concrete Technology in Australia: Geo-Polymer Concrete (International Journal of the Constructed Environment, Vol 7, issue 1, pp 19-29, 2015).

[8] J. Gere, and B. Goodno, Mechanics of materials (8th edition, Stamford, Conn.: Cengage Learning, 2013).

[9] E. Cuerca, On Shear Bahavior of Structural Elements made of Steel Fiber Reinforced Concrete (Springer: USA, 2014).

[10] P. Domone and J. Illston, Construction Materials: Their nature and Behaviour (4th edition, Spon Press, London, 2010).
[11] P. Xincheng, Super-high-strength high Performance Concrete (Hoboken: Taylor and Francis, 2012).
[12] A. Ramdasi and N. Tande, Lateral load analysis of High Rise Structures (International Journal of Engineering Science and Technology, Vol 6, Iss 6, pp 268-275, 2014).
[13] H. Hugo, Performance Based Building Design 1 From Below Grade construction to Cavity Walls (Hoboken: Wiley, 2012).
[14] K. Gharehbaghi, Modular high-rise construction as an alternative building system, (international Journal of the Constructed Environment, Vol 8, Iss 3, pp 15-25, 2017).

Comparative performance of conventional OPC concrete and HPC designed by densified mixture design algorithm

Trong-Phuoc Huynh[1], Chao-Lung Hwang[2], and Shu-Ti Yang[2]

[1] Department of Rural Technology, College of Rural Development, Can Tho University, Campus II, 3/2 St., Ninh Kieu Dist., Can Tho City 900000, Vietnam
[2] Department of Civil and Construction Engineering, National Taiwan University of Science and Technology, No. 43, Keelung Rd., Da'an Dist., Taipei City 10607, Taiwan

E-mail: htphuoc@ctu.edu.vn

Abstract. This experimental study evaluated the performance of normal ordinary Portland cement (OPC) concrete and high-performance concrete (HPC) that were designed by the conventional method (ACI) and densified mixture design algorithm (DMDA) method, respectively. Engineering properties and durability performance of both the OPC and HPC samples were studied using the tests of workability, compressive strength, water absorption, ultrasonic pulse velocity, and electrical surface resistivity. Test results show that the HPC performed good fresh property and further showed better performance in terms of strength and durability as compared to the OPC.

Keywords: Conventional OPC concrete, High-performance concrete (HPC), Densified mixture design algorithm (DMDA), Compressive strength development, Durability.

1. Introduction

Concrete is one of the major materials that used with a large quantity of many construction activities. Recently, the design consideration for concrete proportion has not only focused on short-term strength but also emphasized on long-term performance (strength and durability). It is reported that the amount of cement and mixing water used in the concrete mixture is closely associated with many problems occurred during its life-cycle. Therefore, the cement and water content in the mixture should be limited in order to maintain the durability of the concrete [1–3].

High-performance concrete (HPC) is the concrete with good workability and mechanical properties and high resistance ability to chemical attack as compared to conventional concrete [4]. The incorporation of various supplementary cementitious materials, such as fly ash (FA), ground granulated blast-furnace slag (GGBS), rice husk ash (RHA), etc. in the HPC mixture has been studied by many researchers with a general result showing an improvement of concrete performance in both fresh and hardened stages [4–6]. The application of the DMDA, which was developed by the Hwang's research group at National Taiwan University of Science and Technology in the HPC mix design is promoted in this study. In DMDA, the maximum density of the concrete can be obtained by well packing all solid particles with different sizes to a dense structure. Thus, the DMDA-concrete is expected to have high strength and good durability. The DMDA has been successfully applied to design the concrete that used for many projects in Taiwan [4, 7, 8]. An evaluation on engineering properties and durability of the HPC in comparison with those of the conventional OPC concrete will be performed and discussed in this study.

2. Materials and experimental works

2.1. Materials

The physical properties and chemical compositions of binder materials used in this study are given in Table 1. For the preparation of concrete samples, natural crushed sand (fineness modulus 3.0, density 2640 kg/m^3, and water absorption 1.0%) and natural crushed stone (12 mm maximum size, density 2670 kg/m^3, and water absorption 0.7%) were used as fine and coarse aggregates, respectively. The workability of fresh concrete was controlled using type-G superplasticizer (SP) with a specific gravity of 1.34. Local tap water was used as mixing water.

Table 1. Characteristics of source materials.

	Items	Cement	GGBFS	FA
Physical properties	Specific gravity	3.15	2.92	2.26
	Mean particle size (μm)	19.1	8.8	21.8
	Specific surface area (m^2/g)	0.78	1.68	0.66
Chemical compositions (wt.%)	SiO_2	20.0	39.1	64.1
	Al_2O_3	4.2	13.0	22.1
	Fe_2O_3	3.1	0.2	5.6
	CaO	62.4	37.5	2.7
	MgO	4.1	7.1	0.9
	Na_2O	0.3	-	0.8
	K_2O	0.4	0.2	1.3
	Others	3.6	2.0	1.9

Table 2. Mixture proportions for the preparation of concrete samples.

Mixture	w/b	Concrete ingredients proportion (kg/m^3)						
		OPC	GGBFS	FA	Sand	Stone	SP	Water
OPC	0.35	509.5	-	-	751.0	994.0	1.7	180.2
HPC		265.6	103.3	93.3	854.5	942.0	2.8	160.4

2.2. Mix design concept and ingredient proportions

In this study, the mix proportions of conventional OPC concrete and HPC were designed in accordance with the traditional ACI method and the DMDA technology, respectively. The Hwang's research group at the National Taiwan University of Science and Technology developed the DMDA from the hypothesis that physical properties of a concrete will be optimum when its physical density is high. Thus, the DMDA uses FA as a filler to fill the voids between aggregate particles and hence increases the density of the aggregate system. Additionally, the DMDA also uses FA and GGBFS as pozzolanic materials that may consider as a cement substitution. Hence, the cement paste of DMDA mixture can be reduced as compared to that of ACI mixture. It is important to note that the following criteria should be controlled in the DMDA mix design in order to guarantee the strength and long-term durability of the concrete: (i) water-to-cement ratio is greater than 0.42 (to prevent autogenous shrinkage or chemical contraction due to the cement hydration and/ or pozzolanic reaction); (ii) mixing water is less than 160 kg/m^3 (to ensure the volume stability and prevent the large drying shrinkage); and (iii) using as less cement as possible (to reduce the harmful expansion or swelling due to the alkali-aggregate reaction or sulfate attack). The detailed mix design procedures, as well as the design consideration, for both the ACI and DMDA methods are described in references [9] and [3], respectively. The ingredient proportions for these concrete mixtures are shown in Table 2.

2.3. Samples preparation and test methods

The measurement of fresh concrete workability was performed in accordance with ASTM C143. The cylindrical concrete samples with a diameter of 100 mm and a height of 200 mm were prepared for the tests of compressive strength, ultrasonic pulse velocity (UPV), and electrical surface resistivity (ESR). The compressive strength test was conducted at 7, 14, 28, and 56 days in accordance with ASTM C39. Whereas, the UPV and ESR values were measured at 28 and 56 days by following the ASTM C597 guidelines and using a four-point Wenner array [10], respectively. The water absorption capacity (WAC) of the concrete samples was also determined at 28 and 56 days in accordance with ASTM C1585 using samples with a diameter of 100 mm and a height of 50 mm.

3. Results and Discussion

3.1. Workability of fresh concrete
Workability of fresh concrete including slump, slump flow spread, and flow time is presented in Table 3. Different SP dosages (as shown in Tables 2 and 3) were used to achieving the desired slump value of greater than 200 mm. Table 2 shows that the OPC concrete uses more cement, therefore, it normally requires more water in order to produce a certain workability because the flocs of cement particles trap a part of mixing water. Whereas, the HPC uses more SP to limit the water content, one of the design considerations of the DMDA as aforementioned, but still remaining a similar workability as that of the OPC. Additionally, the HPC incorporating FA with spherical particles may significantly reduce the friction of the particles system, resulting in a greater workability. As the results, both the OPC and HPC mixtures exhibited good flowability without bleeding or segregation.

Table 3. Workability of fresh concrete mixtures.

Mixture	w/b	SP (%)	Slump (mm)	Slump flow (mm)	Flow time (sec.)
OPC	0.35	0.33	225	510	9
HPC		0.61	245	580	17

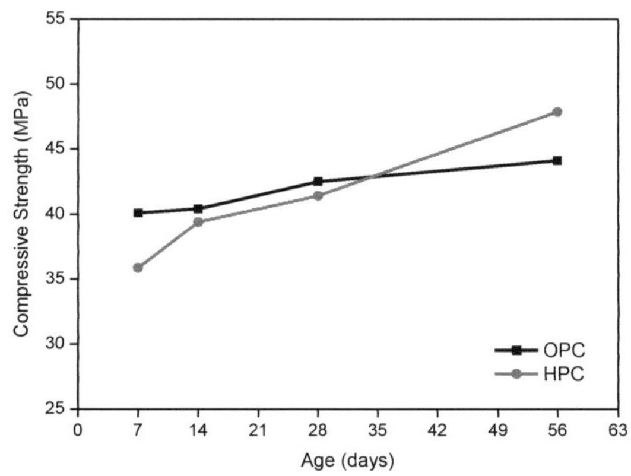

Figure 1. Compressive strength development of the concrete samples.

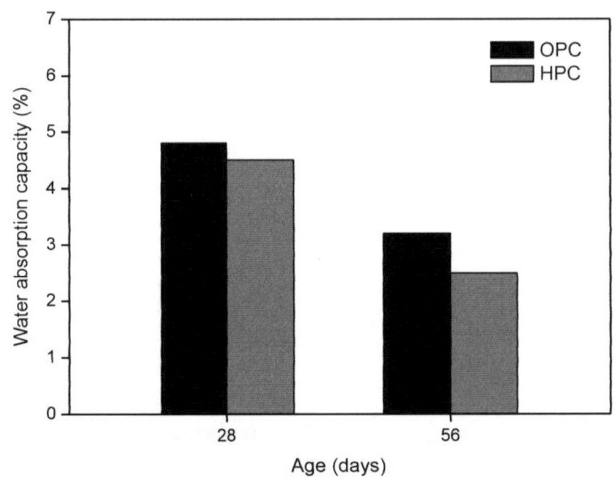

Figure 2. Water absorption capacity of the concrete samples.

3.2. Compressive strength development
Compressive strength is one of the important indicators of concrete quality. In this study, the compressive strength development of the concrete samples is presented in Figure 1. As expected, the compressive strength increases with curing age of the concrete samples. At 28-day age, the compressive strength values of the OPC and HPC were 42.5 and 41.4 MPa, respectively. After 56 days of curing, the compressive strength of the OPC and HPC reached the values of 44.1 and 47.9 MPa, respectively. In addition, Figure 3 clearly shows that the OPC obtained higher strength than the HPC at the early age. However, the reversed trend was observed at the later age of the concretes. The compressive strength of the 56-day-old HPC was approximate 1.1 times higher than that of the OPC at the same age. Moreover, the rate of strength increment (from 28 days to 56 days) of the HPC was 15.7%, which was much greater than that of

the OPC (only 3.8%). The higher long-term strength attributed to the positive effects of both aggregates packing and hydration of both cement and pozzolanic materials (FA and GGBFS) at later ages [11].

3.3. Water absorption capacity

The water absorption capacity (WAC) of all concrete samples is presented in Figure 2. The WAC of the OPC at 28 days and 56 days were 4.8 and 3.2%, while that of the HPC at the respective ages was 4.5 and 2.5%. Thus, the WAC of the 28-day-old and 56-day-old OPC samples was 6.7 and 28.0% higher than that of the HPC at the respective ages. The test result of WAC supported the development in compressive strength of concrete as the WAC decreased with the increased compressive strength along with its curing age. Such reduction in WAC is primarily due to the decrease in the porosity of the concrete samples.

3.4. Ultrasonic pulse velocity

The value of ultrasonic pulse velocity (UPV) is considered as an indicator of concrete durabilities such as porosity and permeability. A higher UPV value is generally associated with a denser structure of concrete, indicating a good concrete quality. The UPV values of all tested samples are presented in Figure 3. As can be seen from the figure, the UPV value of both the OPC and HPC increased with the age of the sample. Moreover, the recorded UPV values for the HPC at 28-day and 56-day ages were approximate 1 and 3% higher than those of the OPC, respectively. The OPC had a higher porosity, which is indicated by the high WAC as mentioned above, than the HPC. Therefore, the UPV value of the OPC is lower than that of the HPC because of the increased travel time of the pulse through the samples [11]. Overall, the UPV values of all concretes were significantly higher than 3500 m/s after 28 days. Thus, these samples exhibited good quality and durability [12], in which the HPC exhibited a better durability performance, indicated by higher UPV value than the OPC.

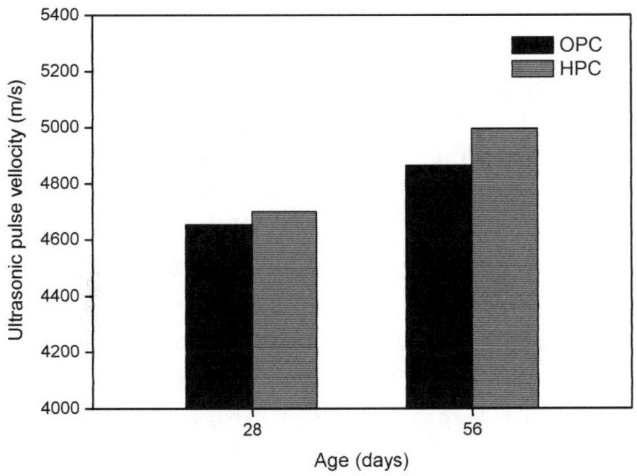

Figure 3. UPV values of the hardened concrete.

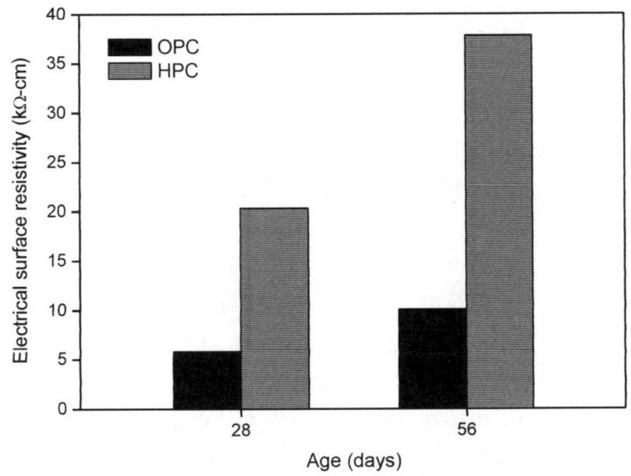

Figure 4. ESR values of the hardened concrete.

3.5. Electrical surface resistivity

Electrical surface resistivity (ESR) is another indicator of concrete durability [13] as higher ESR associated with the lower potential of chloride penetration [3]. Figure 4 shows the test results of electrical surface resistivity (ESR) of all concrete samples used in this investigation. Similar to compressive strength development and UPV, the ESR was found to be increased with concrete ages. As the results, the OPC and HPC had a respective ESR value of 5.8 and 20.3 kΩ.cm at 28 days and reached to 10.1 and 37.8 kΩ.cm after 56 days of curing. It is suggested by some researchers that concrete with ESR value of greater than 20 kΩ.cm exhibited an excellent anti-erosion ability [13–15]. Thus, the corrosion may not occur in the HPC. Moreover, the ESR values of the HPC at 28-day and 56-day ages were 3.5 and 3.7 times higher than those of the OPC, respectively. Therefore, the HPC had better corrosion endurance than the OPC.

4. Conclusions

The present study evaluates the performance of the conventional OPC concrete designed by ACI method and the HPC designed by DMDA. Based on the test results, the following conclusions may be drawn:
1. The fresh HPC mixture with the use of SP and the incorporation of FA provided a better workability than the OPC one.
2. The HPC designed with less cement and water contents obtained higher long-term compressive strength and lower water absorption rate than the OPC.
3. The ESR of the OPC was under the threshold limit value of 20 kΩ.cm, whereas that of the HPC was above this limit. On the other hand, the UPV values of these concrete samples significantly reached above 3500 m/s. Thus, the HPC can be considered to be more durable and good corrosion endurance than the OPC.
4. The results of this study promote the application of the DMDA in the mix proportion design of the HPC in order to achieve better performance of the concrete.

5. Acknowledgements
This research was financially supported by the Ministry of Transportation and Communications of Taiwan. The experimental works were conducted at the Construction Material Research Laboratory of the Taiwan Tech (NTUST) with the valuable assistance of the Hwang's research group.

6. References
[1] Mather B Use less cement 2000 *Concr. Int.* **22** 55–56
[2] Neville A Water and concrete: a love–hate relationship 2000 *Concr. Int.* 22 34–38
[3] Hwang C L and Hung M F Durability design and performance of self-consolidating lightweight concrete 2005 *Constr. Build. Mater.* **19** 619–26
[4] Tu T Y, Chen Y Y, and Hwang C L Properties of HPC recycled aggregates 2006 *Cem. Concr. Res.* **36** 943–50
[5] Isaia G C, Gastaldini A L G, and Moraes R Physical and pozzolanic action of mineral additions on the mechanical strength of high-performance concrete 2003 *Cem. Concr. Compos.* **25** 69–76
[6] Chang P K An approach to optimizing mix design for properties of high-performance concrete 2004 *Cem. Concr. Res.* **34** 623–29
[7] Chang P K, Hwang C L, and Peng Y N Application of high performance concrete to high-rise building in Taiwan 2001 *Adv. Struct. Eng.* **4** 65–73
[8] Tsai C T, Li L S, Chang C C, and Hwang C L Durability design and application of steel fiber reinforced concrete in Taiwan 2009 *Arab. J. Sci. Eng.* **34** 57–59
[9] ACI Committee 211 1996 *Standard practice for selecting proportions for normal, heavyweight, and mass concrete* ACI Manual of Concrete Practice
[10] Kessler R J, Powers R G, Vivas E, Paredes M A, and Virmani Y P 2008 Surface resistivity as an indicator of concrete chloride penetration resistance, *2008 Concrete Bridge Conference* (St. Louis, MO)
[11] Bui L A T, Hwang C L, Chen C T, Lin K L, and Hsieh M Y Manufacture and performance of cold bonded lightweight aggregate using alkaline activators for high performance concrete 2012 *Constr. Build. Mater.* **35** 1056–62
[12] Malhotra V M 1976 *Testing hardened concrete: Nondestructive methods* American Concrete Institute
[13] Buenfeld N R, Newman J B, and Page C L The resistivity of mortars immersed in sea-water 1986 *Cem. Concr. Res.* **16** 511–24
[14] Wang H Y Durability of self-consolidating lightweight aggregate concrete using dredged silt 2009 *Constr. Build. Mater.* **23** 2332–37
[15] Broomfield J P 2007 *Corrosion of steel in concrete: Understanding, investigation and repair* Taylor & Francis Group (New York)

Institute of Physics
Dirac House, Temple Back
Bristol BS1 6BE UK

ISSN: 1757-8981
ISBN 978-1-5108-5404-8

CURRAN ASSOCIATES INC.
proceedings.com